THE EVOLUTION AND EXTINCTION OF THE DINOSAURS

THE EVOLUTION AND EXTINCTION OF THE DINOSAURS

David E. Fastovsky
University of Rhode Island

David B. Weishampel
Johns Hopkins University

With original illustrations by
Brian Regal, Tarbosaurus Studio

CAMBRIDGE
UNIVERSITY PRESS

Published by the Press Syndicate of the University of Cambridge
The Pitt Building, Trumpington Street, Cambridge CB2 1RP
40 West 20th Street, New York, NY 10011-4211, USA
10 Stamford Road, Oakleigh, Melbourne 3166, Australia

© Cambridge University Press 1996

First published 1996

Printed in the United States of America

Library of Congress Cataloging-in-Publication Data
Fastovsky, David E.,   1954–
The evolution and extinction of the dinosaurs / David E.
Fastovsky, David B. Weishampel.
p.   cm.
ISBN 0-521-44496-9 (hc)
1. Dinosaurs–Evolution. 2. Extinction (Biology)   I. Weishampel,
David B.,   1952– . II. Title.
QE862.D5F38   1996
567.9'7–dc20   95-6002
                                                              CIP

A catalog record for this book is available from the British Library

ISBN 0-521-44496-9 Hardback

All line drawings and color illustrations, unless otherwise noted
© 1995 Renaissance Studios, Inc.

Design and production by Bookworks

To Judy, Sarah, Amy,
My Sweet Honey, and
Poor Robert, because...

CONTENTS

PREFACE

THE IDEA BEHIND THIS TEXT is simple: to use dinosaurs as an attractive vehicle to understand aspects of natural history. Although we are neither terrible nor lizards, we are nevertheless passionate that dinosaurs – when properly understood – can illuminate not only the prehistoric but also the present. A proper treatment (this text?) in the proper atmosphere (a university course?) may thus transform dinosaurs from arcane litanies of extinct life to real insight into the ebb and flow of evolution.

We have attempted to present dinosaurs as professionals understand them. Dinosaurs are *not* a list of unpronounceable Latin names. The study of dinosaurs has much to do with the history of life and of the earth, with the nature of nature, and with who we are. Considerable insight is required, therefore, into much that is not exactly dinosaurs. For us, the addition to this book of "much that is not exactly dinosaurs" enriches our understanding of dinosaurs, and underlines our belief that all human knowledge is fundamentally connected. An awareness of those connections distinguishes an educated person from one who is not, just as it distinguishes paleontologists from enthusiasts who confuse science with the recitation of endless Linnaean binomials like *Tyrannosaurus rex*. Because our experience as teachers suggests to us that students tend to compartmentalize knowledge and do not explicitly make the connections that would make their educations richer, we rejoice in those connections throughout this book.

A WORD TO THE READER. *The Evolution and Extinction of the Dinosaurs* is designed to be readable, but also a resource. The chapters build sequentially, reflecting the nature of our science. Still, risking the very compartmentalization that we seek to avoid, we have written each chapter as a kind of essay. Particular chapters treat particular subjects.

Dinosaurs are presented here in a phylogenetic context. To investigate dinosaurs as professional paleontologists do, we must apply all the intellectual tools of the trade; to do less shortchanges our readers. The prose of phylogenetic systematics, however, can be deadly. For this reason, the chapters in which the great groups of dinosaurs are discussed individually – in particular, Chapters 6–12 – are organized in a consistent fashion. The enthusiast wishing to skim the descriptions and systematic paleontology may go directly to the section in each of these chapters entitled "Paleobiology and Paleoecology."

ACKNOWLEDGMENTS. In our work, we have been aided by many generous, yet critical, friends and colleagues. We wish to extend our thanks and appreciation to colleagues who provided us with information and to the reviewers of these pages, among them A. K. Behrensmeyer, M. J. Benton, J. A. Cain, K. Carpenter, P. Dodson, C. A. Forster, P. M. Galton, R. E. Heinrich, C. Janis, J. S. McIntosh, M. B. Meers, J. H. Ostrom, H.-D. Sues, and L. M. Witmer. P. Dodson and K. Padian waded through the entire thing in rough draft retaining, against incalculable odds, the ability to critique constructively. L. A. Di Panni and K. Andrews helped with a variety of indispensable odds and ends, and J. Murray did the lion's share of the index as well as a formidable proofreading job. Ben Creisler's marvelous lexicon of dinosaur names was an indispensable companion throughout the writing of this book. Naturally, all these individuals are held blameless for any errors in this book. Those errors that have crept in are clearly Dave's fault.

Several colleagues, museums, and other institutions provided photographs for use in this book. We greatly appreciate the courtesy of reproducing these photos.

We are also grateful to Robert Sugar and Dorothy Duncan of Bookworks for their thoughtful readings and careful formatting, as well as to Cambridge University Press, particularly Dr. Robin Smith, for help and encouragement throughout this project.

Each of us has individuals to whom special thanks are due. For Fastovsky, this is an opportunity to acknowledge S. P. Welles for his generous support and encouragement in those early days when enthusiasm was inversely proportional to experience. Thanks are also due to M. A. Morales, for "invent-ing" the dinosaur class from which these pages fundamentally stem, and for the invitation to partake as a teaching assistant in those early days. Finally, special thanks are due to his parents, Ashley and Jean, for whom Fastovsky's potential career in paleontology was an option worthy of encouragement.

From Weishampel come thanks to his parents, Wilbur and Mildred, who first took him to see the dinosaurs at the Cleveland Museum of Natural History when he was five, and six, and seven . . . (couldn't really have stopped him). Thanks also go to Jim Collinson, Aurele LaRocque, Jerry Downhower, Toby Gaunt, Gord Edmund, Chris McGowan, Peter Dodson, and . . . well, everyone.

THE
EVOLUTION
AND EXTINCTION
OF THE DINOSAURS

Sir Richard Owen, the brilliant 19th-century English
anatomist and father of the term "Dinosauria." *Photograph courtesy of the American Museum of Natural History.*

PART I

SETTING THE STAGE

CHAPTER 1

INTRODUCTION

DO WE REALLY NEED ANOTHER BOOK ABOUT DINOSAURS? We live in dinosaur-crazy times. We can watch a sitcom predicated on dinosaurs' having a suburban home life. Or, we can endure "Barney," a purple, neotenic biped with a talent for lulling children to tranquillity. For the more active (among the passive), there are dinosaur documentaries that deal with dinosaur extinctions, dinosaur descendants, dinosaur habits, and all else dinosaurian. We have feasted on *Jurassic Park*, a technical masterpiece with dinosaurs good, bad, and ugly. Godzilla – not quite a dinosaur, but then not obviously anything else, either – has been returning since 1954 a bit like a bad meal. You can enjoy dino toys and you can savor dino candy. Popular music, never missing a beat, has supplanted older bands like "T. rex" and "Birdsongs of the Mesozoic" with "Dinosaur Jr." and kiddie disks like *Dinosaur Rock*. There are coloring books, erasers, stick-ons, refrigerator magnets, wooden and plastic models, pen covers, clothes . . . ! Tabloids fill us with the latest dinosaur "research," and books abound. Even a few paleontologists – practitioners of a profession previously stereotyped by mild-mannered bookishness – have become minor media personalities. Do we really need another book about dinosaurs?

The long answer and short answer are both yes. Misinformation, disinformation, and noninformation abound in this gorging of things dinosaurian. And in it all, dinosaurs, that marvelous group of extinct archosaurs, have been lost in the hype.

THIS BOOK. Our goal in writing this book is to attempt to present dinosaurs as professional paleontologists view the group. Because dinosaurs have been known since 1818, a good deal is understood; by the same token, a 20-year-old revolution in methods of studying them has only in the last 10 years (or less) really begun to overturn long-held ideas about them and their 160-million-year stay on earth. Much of the recent media excitement is a reflection of that revolution, and books older than 20 years must be approached with caution. We take pains to present the most up-to-date ideas, and we present divergent viewpoints (as is expected and desirable in science). The give-and-take of scientific dialogue are well recorded in these pages, for this is the most accurate reflection of how paleontologists see dinosaurs.

The ideas here are not the final answer about all things dinosaurian; science is a process and not a static solution. Our field would lose its vibrancy if time and further research didn't modify what we present here. These pages contain only our best call on what is known about dinosaurs. The fossil record may be written in stone, but its interpretation is not.

So what follows is on one level a tale of dinosaurs; who they were, what they did, and how they did it. But on a more significant level, it is a tale of natural history. In writing of dinosaurs, we are really developing a much fuller concept of the biosphere; that is, the sum total of the earth's organisms. Commonly, we think of the biosphere as occurring only in the present; a kind of three-dimensional insulation wrapping the globe. But in fact the biosphere has a 3.5-billion-year history, and we and all the organisms around us are products of that fourth dimension: its history. The history of life is a pageant, and to be unaware of it is to be unaware of who we are. Part of our goal in this book, therefore, is to explore the relationships of organisms to each other and to the biosphere. Historically, humanity has maintained a distorted sense of its position in the biosphere; with wilderness ever diminishing, we would do well to refine (or even redefine) our understanding of our relationship to the earth and its inhabitants. Dinosaurs have some significant information to impart in this regard; as we learn who dinosaurs really are, we can better understand who *we* really are.

Finally, ours is a tale of science: what it is, how it functions, and how scientific data are considered. We live in an increasingly technical world, and we can no longer afford to ignore science and the way it impinges on our lives. But science is often poorly understood. Indeed, one of us remembers a little ditty from a 1959 record called *Space Songs* that was supposed to describe science:

> It's a Scientific Fact – A Scientific Fact!
> It has to be Correct! It has to be Exact!
> Because it is, because it is a Sci-en-ti-fic Fact!

Yikes! If science were as it is portrayed in *Space Songs*, neither of us would have become scientists, either by disposition or by inclination. Science is a creative enterprise. In the following pages, therefore, we hope to build a sense of the beauty of science and a feel for the meaning of scientific data.

Like any other discipline, paleontology can only be mastered in a stepwise fashion. It has its own language and its own concepts, quite apart from even the related disciplines of geology and biology. So, although our book is written as a series of individual essays on selected topics relating to dinosaurs, it should be recognized that the development of concepts and ideas is sequential and that each chapter builds upon the previous ones.

THE WORD "DINOSAUR" IN THIS BOOK. The term "dinosaur" (*deinos* – terrible; *sauros* – lizard) was established in 1842 by the English anatomist Sir Richard Owen to describe a few fossil bones of large, extinct "reptiles." With modifications (for example, "large" no longer applies to all members of the group), the name proved resilient. It has become clear in the past 10 years, however, that not all dinosaurs are extinct; most vertebrate paleontologists now agree that *birds are living dinosaurs.*[1] This leaves us with a problem, because much of what we discuss concerns the **non-avian dinosaurs**, that is, all dinosaurs *except* birds. We could use the cumbersome, technically correct "non-avian dinosaurs," but it is far easier to use the term "dinosaurs" as a kind of shorthand for "non-avian dinosaurs." The distinction between non-avian dinosaurs and all dinosaurs is most relevant when we discuss the origin of birds in Chapter 13, where we will avoid confusing terminology. Throughout this book, therefore, the word "dinosaur" really refers to non-avian dinosaurs.

FOSSILS

That we even know that there ever were such creatures as dinosaurs is due to blind luck: Some members of the group happened to be preserved. Exactly what gets preserved and how that occurs provide insight into the kinds of information we can expect to learn about these extinct beasts. Dinosaurs last romped on this earth 65 million years ago. This means that their **soft tissues** – muscles, blood vessels, organs, skin, fatty layers, and so forth – are long gone. If any vestige remains at all, it is usually **hard parts** such as bones, teeth, or claws. Hard parts are not as easily degraded as the soft tissues that constitute most of the body. Obviously, the kinds of changes that organic remains undergo are of great interest. The study of those changes, and in fact the study of all of what happens to organisms after death, is called **taphonomy**.

Taphonomy is extremely important because by looking at a group of fossils and the rocks in which they are encased, we can learn something about the deaths (and lives) of the animals represented. Because all dinosaurs were land-dwelling vertebrates, our primary interest will be in the taphonomic changes that occur in terrestrial settings. Such settings generally consist of deserts, wetlands and lakes, river systems (channels as well as floodplains), and to a lesser extent, deltas – that is, where the rivers contact, and interface with, marine environments. Each of these settings is an environment in which bones can be preserved (see Chapter 3).

Taphonomy

BEFORE BURIAL. Consider what happens to a dinosaur – or any land-dwelling vertebrate – when it dies. If it is killed, it can be **disarticulated** (dismembered), first by the animal that killed it, and then by scavengers. In modern environments, these scavengers could be creatures like vultures or hyenas, but they could also be smaller animals, such as carrion-eating beetles (not to mention bacteria that pungently feast on rotting flesh!). The bones are commonly stripped clean of meat by scavenging and left to bleach in the sun. Some bones might get carried off and

---

[1]Ample justification for this statement can be found in Chapters 4 and 13.

gnawed somewhere else. Sometimes the disarticulated remains are trampled by herds of animals, breaking and separating them further. The more delicate the bones, the more likely they are to be destroyed. A result of these kinds of activities is that it is not uncommon to walk across a meadow and to find a few disarticulated, bleached bones lying in the grass, which are all that remains of the animal that died there.

If the animal dies a nonviolent death (old age, drowning, and disease all qualify) and is left intact and untouched, it is not uncommon for the carcass to simply swell up (bacterial decomposition produces gasses that inflate it), eventually release the gas (this can be more or less grotesque), and then (if not in water) dry out, leaving bones, tissues, ligaments, tendons, and skin hard and inflexible. The tissues shrink as they dry, bending limbs and pulling back heads and lips into hideous grimaces (giving the appearance that the animal died in agony, when it may not have). Under such conditions, the creature has become something like beef jerky, and the carcass can be exposed for a very long time without further decomposition.

Many vertebrates travel in herds and, occasionally, catastrophic things happen to gregarious animals. The most common catastrophes are floods, which can catch the herd as it crosses swollen rivers and drown many individuals within a very short period of time. The carcasses may bloat and float, and eventually be piled up along the edge of a channel, up on the floodplain, or most commonly, on the surface of a point bar at bends in the waterway. Left alone, the bones may be stripped, partially disarticulated, and bleached in piles of skeletons of one type of animal (Figure 1.1).

BURIAL. Sooner or later the bones get destroyed or buried. Destruction commonly comes from weathering; eventually, the minerals in the bones break down and the bones disintegrate. In certain cases, however, the bones get buried. At this point, they become **fossils**. The word "fossil" comes from the Latin word *fodere* (to bury) and has come to refer to anything once living that is now buried. Notice that there is no implication of how much time the remain has been buried; your dog burying his bone is technically producing a fossil! A **body fossil** is said to be preserved when a part of an organism is buried. We distinguish these from **trace fossils**, which are impressions in the sediment left by an organism.

Burial can take several forms. The simplest is when a bone or accumulation of bones simply gets covered by sediment. This can happen wherever there is **deposition**, that is, net addition of sediment to a land surface. Deposition takes place in all environments. For example, in a desert, it might occur when one sand dune migrates over the top of another, covering anything that was there before. Deposition can also occur during floods; for example, when the basements of houses have been filled with silt as flood waters recede. A more subtle type of deposition can occur, however, when sediments are **reworked**, which means that they are eroded from wherever they were originally deposited, and are redeposited somewhere else.

Rivers provide an important instance of reworking. When flooding occurs, rivers breach their channels, in the process eroding the edges of the channel. If any buried bones are within the eroded part of the channel margin, they too will be eroded with the channel margin material and will be swept along by the flood

Figure 1.1 **Bones.** A wildebeest carcass, partly submerged in mud and water and on its way to becoming permanently buried and fossilized. If the bones are not protected from scavengers, air, and sunlight, they decompose rapidly and are gone in 10–15 years. *Photograph courtesy of A. K. Behrensmeyer.*

waters. Thus, fossils that we find in a channel are not necessarily the remnants of a community that actually lived (and died) together in the channel. Instead, it is simply a reworked **assemblage** (collection) of bones that includes fossils from the floodplain, as well as from the channel, jumbled together.

Part of our job as paleontologists is to interpret how the bones got the way that we find them, because that may tell us something of how dinosaurs lived. Different types of concentrations of bones can come about through different death and post-mortem scenarios. When the bones are **articulated** (connected together), this suggests that they have not been transported far from where the animal died. None of the destructive forces we described earlier – scavenging and reworking – have had much effect on the fossils. Articulated bones are said to be largely **in-place**, or not reworked. On the other hand, a collection of disarticulated bones of several types of vertebrates in an ancient stream channel deposit suggests that the deposit is reworked and that the bones got there through sedimentary processes sometime after death and burial. Then there are the **bonebeds**: accumulations of bones of many individuals of a very few *kinds* of organisms, sometimes articulated and sometimes not. A bonebed with one or two species may represent a herd subjected to a catastrophic event; alternatively, if the bonebed accumulated over a long period of time, then it might represent a location where many animals of the same type chose to live (and die). Finally, isolated

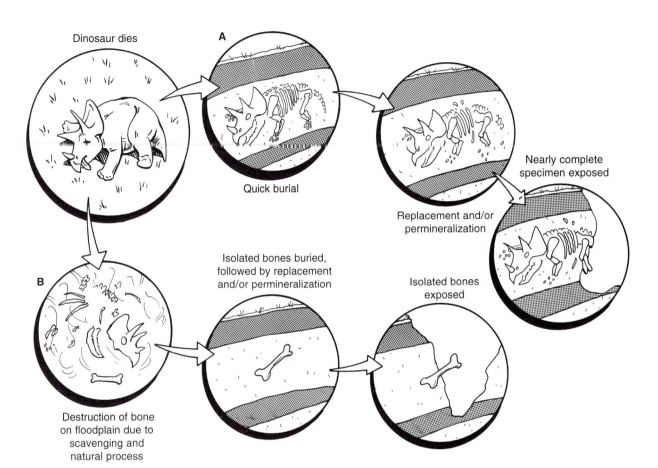

Figure 1.2 **Two endpoint processes of fossilization.** In both cases, the first step is the death of the animal. Some decomposition occurs at the surface. In the upper sequence (A), the animal dies, the carcass undergoes quick burial, followed by bacterial decomposition underground, and permineralization and/or replacement; finally, perhaps millions of years later, exposure. This kind of preservation yields bones in the best condition. In the lower sequence (B), the carcass is dismembered on the surface by scavengers and perhaps trampled and distributed over the region by these organisms. It may then be carried or washed into a river channel and buried, replaced and/or permineralized, eventually to be exposed perhaps millions of years later. Under these conditions, when the fossil is exhumed, it is disarticulated, fragmented, and the fossil bones may show water wear and/or the gnaw marks of ancient scavengers. Different conditions of fossil preservation tell us something about what happened to the animals after death.

finds – a thighbone here or a vertebra there – could represent almost any of the possibilities that we described previously. In Figure 1.2, two taphonomic sequences, leading to two different results, are shown.

AFTER BURIAL. Bone is made out of a complex mineral, *calcium (sodium) hydroxy apatite*, which reacts at temperatures and pressures at, or near, the surface of the earth. This means that the mineral changes with time, which in turn means that most fossil bones no longer have original bone matter present. This is especially likely if the bone is bathed in the variety of fluids that are associated with burial in the earth. Since bones are porous, it is common for the spaces in fossil bones to be filled up with some mineral. This process is called **permineralization**. Different from permineralization is **recrystallization**, where the original mineral (in the case of bone, hydroxy apatite) is dissolved and reprecipitated, retaining the exact original form of the fossil. Recrystallization can be very obvious – for example, when small crystals are replaced by large ones – but it can also be very subtle and occur on a microscopic scale. Bones can also undergo **replacement**, in which the original bone mineral is replaced by another mineral. If no fluids are present throughout the history of burial, the bone could remain **unaltered**, which is to say that original bone mineralogy remains. This situation is not that common and is progressively rarer in the case of older and older fossils.

In general, fossil bones undergo a combination of replacement, permineralization, and recrystallization. The resultant fossil, therefore, is chemically and texturally rather unlike the original bone matter, although the shape and features may remain. For example, fossil bones, because they are at least partly permineralized as well as replaced, tend to be much heavier than their living counterparts. Moreover, they are more brittle. Despite nearly perfectly mimicking the original, most fossil dinosaur bone is really more rock than bone in its texture and composition.

In general, the more quickly a bone is buried, the better the chance it has of being preserved. This is because quick burial generally inhibits the weathering processes that would normally break down the bone minerals.[2] In fact, it is not uncommon to find evidence of weathering *before* fossilization. For example, dinosaur bones that appear to have been transported in water show waterworn features that are exactly the same as those found in modern bones that have been transported by water: The bones are rounded and, commonly, their surfaces are partly or completely worn off.

So if the fossils are buried, how is it that we find them? The answer is, luck and skill: If fossil-bearing sedimentary rocks happen to be eroded, and a paleontologist happens to be looking for fossils at the moment that one is sticking out of an actively eroding sedimentary rock, the fossil *may* be observed and *may* be collected. Does this mean that there are great numbers of fossils that lie buried within sedimentary rocks that happen not to be eroding at the earth's surface? Undoubtedly. Have fossils been eroding out of rocks since the very first fossil was formed? No question. Paleontology is a relatively new human activity. This means that although fossils with hard parts have been produced for 570 million years, most of those few that happened to be fossilized in the first place and then happened to be exposed at the earth's surface were never collected; they simply weathered away with the rest of their host rock.

Dinosaurs first appeared about 228 million years ago, and all but birds went extinct 65 million years ago. We may be sure that throughout the 160-million-year existence of dinosaurs on earth, their fossils were constantly eroding out of sediments and were lost for eternity.

---

[2]An exception to this is when a bone ends up buried in an active soil; under certain conditions, the bone is then destroyed by biotic and abiotic soil processes.

You can see, then, that the odds are stacked against fossilization. And, although many fossils are found, it is clear that most creatures – even those with hard parts – that have lived on earth are not preserved. Some paleontologists estimate that somewhere between 8% and 25% of all the *genera* of dinosaurs that ever lived have been found as identifiable fossils. Estimates of the total number of genera of dinosaurs that ever lived range from a low of 900 to a high of greater than 3400. With these estimations as widely divergent as they are, who knows what percentage we now know of all the dinosaurs that ever lived?

OTHER KINDS OF FOSSILS. We commonly think of bones as the only remains of dinosaurs. This is far from the case. Occasionally the fossilized feces of dinosaurs and other vertebrates are found. These sometimes-impressive relics are called **coprolites** and can be informative as to the diets of their producers. Likewise, as we shall see later in this book, eggs and skin impressions have also been found. But the single most important type of dinosaur fossil, other than the bones, is trace fossils. Trace fossils (sometimes also called **ichnofossils**; *ichnos* – track or trace) are impressions, burrows, or other sedimentary structures left by organisms. In the case of vertebrates (including dinosaurs), the most common trace fossils are isolated **footprints** or complete **trackways** (groups of footprints) left as vertebrates walked across a substrate (Figure 1.3). The footprint must have been made in material that can hold an impression (mud is the most common,

Figure 1.3 Theropod dinosaur tracks from the Late Triassic, near Culpeper, Virginia.

although trackways are not uncommon in fine sand). Again, rapid burial is the most common way to ensure preservation. Two kinds of ichnofossils are possible: **molds**, which, in the case of tracks, represent the original impression itself, and **casts**, which are made up of material filling up the mold. Thus, a cast of a dinosaur footprint would be a three-dimensional object that formed inside the impression.

It is virtually impossible to link particular dinosaurs known from bones with their tracks. For this reason, footprints have their own names and are classified separately as **ichnotaxa**, or footprint types. This is an important way of keeping different kinds of data distinct.

Dinosaur trackway specialist M. G. Lockley of the University of Colorado (Denver) has recently calculated that over fifteen hundred dinosaur tracksites are known from around the globe. They have been used in a variety of studies, including demonstrating without doubt that dinosaurs walked erect (much as mammals and birds do today), indicating the position of the foot (up on the toes in some cases and flat-footed in others), and determining the speeds at which dinosaurs traveled (see Box 14.2). Trackways have had surprising applications in other areas of dinosaur paleontology. For example, trackways can give some indication of creatures that lived in a particular locality. Likewise, trackways have been used to suggest the stalking of a herd of herbivores by a carnivorous dinosaur. The idea of herds of dinosaurs has been strongly reinforced by trackways.

COLLECTION

Obtaining dinosaur fossils is a three-step process. The first step is **prospecting**; that is, hunting for them. The second step is **collecting** them, which means getting them out of whichever (commonly remote!) locale they are situated. The final step is making them available for study and/or display in a museum. This step involves **preparing** and **curating** them; that is, getting them ready for viewing and incorporating them into museum collections. These steps involve different kinds of skills and, frequently, different individuals.

Dinosaurs are expensive beasts. Suppose you would like to have a real fossil dinosaur skeleton displayed in your house. Let's say that you decided to bag a specimen of *Triceratops*, a moderately large, but common (as dinosaurs go), herbivorous dinosaur from the western part of the United States (so getting there and finding it would not be extraordinarily difficult). Presuming you do not have to buy the fossil (as one sometimes must if it is found on private land), you should budget 1.0–1.5 months for prospecting (if you and your crews know what you are doing), about 1–2 months for collecting (because it is so *big*), and somewhere between 12 and 24 months for preparation and mounting the specimen for display. More than one person will be involved in all of these steps, and it is likely that after salaries (for your field crews and preparation staff), equipment, transportation, and preparing the display, your dinosaur would cost you upward of $200,000–$250,000. Dinosaurs are neither for the faint at heart nor for the meager of purse!

Figure 1.4 **Paleontologists prospecting for fossils in Montana, U.S.A.**

Prospecting

A question that is commonly asked of paleontologists is, How do you know where the dinosaur fossils are? The simplest answer is, We don't. There is no secret, magic formula for finding dinosaurs except long, hard hours of persistent searching. On the other hand, you can make an educated guess about where you might search and can thus greatly increase your odds of finding dinosaurs. Even among the hard working, however, some collectors have fared better than others, and something of a "feel" seems to be involved, as well as an experienced eye and basic dumb luck (Figure 1.4).

Some basic criteria constrain the search. In general, the chances of finding dinosaur fossils are increased if

1. the rocks are sedimentary;
2. the rocks are of the right age; and
3. the rocks are terrestrial.

Of the three major kinds of rocks, **igneous**, **metamorphic**, and **sedimentary**, only sedimentary rocks have the potential to preserve fossils to any reasonable degree. This is because igneous rocks are derived from molten material from below the earth's crust – obviously not a suitable habitat for dinosaurs or for the preservation of their bones. Metamorphic rocks are formed by the intense folding and recrystallization of sedimentary rocks, and the process of creating new minerals – *metamorphism* – tends to destroy bones. But sedimentary rocks form in, and represent, sedimentary environments, many of them places where dinosaurs lived.

Obviously, one must look in rocks of the right age before one could hope to find dinosaurs. Dinosaurs first appeared 228 million years ago and became extinct 65 million years ago. This, then, is the window of opportunity for finding dinosaurs; older and younger rocks may yield all kinds of interesting and wondrous vertebrate fossils. But so far, no dinosaurs.

Dinosaurs were terrestrial beasts. This means that the number of environments in which they lived and are preserved, is restricted to river systems, deserts, and deltas. Dinosaur remains, however, are known from lake deposits and from nearshore marine deposits. Clearly they lived neither in lakes nor in the ocean. In nearshore cases, if the bones are articulated, it is thought that bloated carcasses may have floated out into the water and may have eventually sunk and been buried. If the bones are isolated, it is assumed that they simply washed out of the mouths of rivers into the lakes and oceans.

There are aspects of modern environments that also affect the likelihood of finding fossils. The more surface area of rock that is exposed, the better the chances of finding fossils. For this reason, many of the richest fossil localities are in areas with considerable rock exposure, such as badlands. Fossil localities are common in deserts; plant cover on the rocks is minimized, and the dry air slows down the rates of weathering so that once a fossil is exposed, it isn't immediately weathered away. Paleontologists seldom go to the jungle looking for fossils; the weathering rates are too high and the rocks are poorly exposed.

This is not to say that all dinosaur material has been found in badlands or in deserts; far from it. As long as the three criteria previously listed are met, there is a possibility of finding dinosaurs. Still, the number of places that one might search for fossils is relatively restricted, and looking in the right place definitely enhances the probability of finding fossils. Once you find well-exposed rocks that match the three criteria, you simply start searching for bone weathering out of the rock.[3] Despite a variety of high-tech options such as the ground-penetrating radar used in *Jurassic Park,* there has not yet been found any substitute for a well-trained eye. You walk, or crawl, head down, covering as much area as efficiently as possible, looking for exposed bone. If you're lucky and have a good eye, you'll spot something.

Collecting

Collecting is the arena in paleontology in which finesse meets brute force. Delicacy is required in preparing the fossils for transport; raw power is required for lifting blocks of bone and **matrix** (the rock that surrounds the bone) – commonly weighing many hundreds of pounds – out of the ground and into a truck or some other means of transportation.

---

[3]Notice that the term "dinosaur dig" is a misnomer. Nobody digs into sediment to find bones; bones are found because they are spotted weathering out of sedimentary rocks.

(a)

(b)

(c)

(d)

(e)

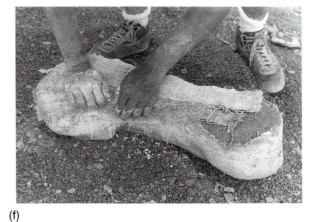

(f)

Figure 1.5 **Fossil collection. (a)** The lower part of the left thighbone (femur) of the Mongolian tyrannosaurid *Tarbosaurus*, ready to be collected; **(b)** cleaning it off and carving the pedestal; **(c)** applying layers of wet toilet paper; **(d)** applying the burlap and plaster jacket; **(e)** turning the specimen; and **(f)** jacketing the surface that was formerly the bottom, to ready it for transport. *Photographs courtesy of D. J. Nichols.*

The early steps are delicate. Once a bone is found sticking out of the ground, it needs to be cleaned off, so that its extent can be assessed (Figure 1.5a). If it is by itself, it needs to be exposed; if it is attached to other, buried, articulated bones, these, too, need to be exposed. Exposing bone can be done with a variety of tools, from small shovels, to dental picks, to fine brushes. As the bone is exposed, it is commonly "hardened," that is, impregnated with a glue that soaks into the fossil and then hardens.

When the bone is fully exposed (but not completely disinterred; some matrix must be left to support it), the rock around it is then scraped away. This is relatively easy to do if the rock is not too hard; but if the matrix is hard, this can be quite labor-intensive. For small fossils, a garden trowel might be adequate for removing matrix. For large fossils, however, this can mean taking off the face of a hill, in which case a backhoe is probably more appropriate. At any rate, this process continues until the bone or bones are literally sitting up on a **pedestal**, a pillar of matrix underneath the fossil (Figure1.5b).

The goal is to build a protective **jacket** around the fossil so that it can be transported. First, however, some padding must be placed between the fossil and jacket to cushion the fossil and to keep the jacket from sticking to it. The most cost-effective cushions are made from strips of toilet paper[4] that are wetted and then patted tightly against the fossil (Figure 1.5c). It takes a lot of toilet paper; for example, a 1-m thighbone (femur) could take upward of one roll. On the other hand, this is not a step where one cuts corners, because returning from the field with a shattered specimen, or one in which the plaster jacket is stuck firmly to the fossil bone, is counterproductive and embarrassing!

Next is the jacketing step. Jackets are made of strips of burlap soaked in plaster and then applied to the toilet paper–covered specimen, in the manner that casts are made for broken arms and legs. The plaster is mixed; then precut, rolled strips of burlap are soaked in it and unrolled onto the specimen and the pedestal (Figure 1.5d). The jacketed specimen is allowed to dry.

After the plaster jacket is hardened, the bottom of the pedestal is undercut, and the specimen is **turned** – that is, separated at the base of the pedestal from the surrounding rock (Figure 1.5e). A "cap" of plaster and burlap is then applied to the open bottom of the jacket. Now, the fossil is fully encased in the plaster-and-burlap jacket and is ready for transport from the field (Figure 1.5f). The plaster "casts," as the fully jacketed specimens are sometimes called, are quite rugged and can withstand a fair amount of jostling .

Transport out of the field can be difficult or not, depending upon the size (and weight) of the jackets. A small jacket (soccer-ball size) can be carried out easily enough. Many jackets, however, are considerably larger, and there are times when braces, hoists, winches, cranes, flatbed trucks, front-end loaders, and even helicopters are necessary. Fossils are not always conveniently located along roads, and more than one paleontologist has actually had to cut a road to get a truck to a remote specimen.

---

[4]Purchasing toilet paper for this purpose provokes raised eyebrows. Consider this often-repeated scene: A paleontologist and two or three crew members walk into a small town in a remote part of the world. They go to the local market and purchase a couple of *cases* of toilet paper rolls. "How long do you plan to be in the field?" the bewildered shopkeeper asks. "Oh, maybe a couple of weeks," the paleontologists reply. The shopkeeper is very impressed, but no less bewildered!

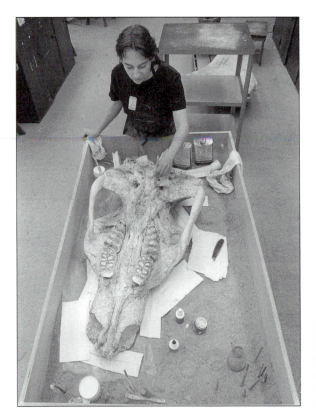

Figure 1.6 **Putting the finishing touches on a large fossil mammal skull in a preparation laboratory.** *Photograph courtesy of the American Museum of Natural History.*

Back at the Ranch

The fossil dinosaur bone is out of the field and back at the museum. The jacket must be cut open, and the fossil must be **prepared** (freed from the matrix). This is a complex process, for which there are a variety of techniques, from the simple to the highly sophisticated. Some fossils require little effort: The soft matrix is literally brushed away from the side of the well-preserved bone. In other cases, carbide-tipped needles must be used under a microscope as the matrix is painstakingly scraped from the fossil. In still other cases, air-powered vibrating scribes, called "zip scribes," are used to free the bone. Sometimes an "air dent," a kind of miniature sandblaster with a very fine spray of baking soda, is used to clear the matrix. A version of this tool is used in some dentists' offices to whiten teeth after cleaning. Finally, in certain circumstances, the matrix can be removed by "acid etching"; that is, dissolution in a bath of weak acid (Figure 1.6).

In many cases, the bone is softer than the matrix. This presents real problems, because the bone will be damaged before the matrix can be removed. A variety of nondestructive techniques are available for hardening the bone. Some involve various chemically sophisticated low-viscosity hardeners that soak into the bone and harden it. Sometimes carbowax is used; the specimen is heated, impregnated with carbowax, cooled, prepared, and then the carbowax is removed. There is almost no limit to the ingenuity required for preparation, and in the end, the success of the preparation is measured only by whether the fossil is freed of its matrix undamaged by the process.

Fossils commonly come fragmented, and part of the preparation is to put the pieces back together. Sometimes this is rather like a jigsaw puzzle, but sometimes the jacket holds adjacent pieces right next to each other and much of the jigsaw puzzle guesswork is removed. Pieces are stuck together by virtually any glue imaginable (depending upon the requirement): epoxies, superglue, and white glue. Interestingly enough, specimens are not always glued together when broken. This is because broken specimens can permit access to detailed parts of the skeleton – for example, the inside of the skull – that would be very hard to study if the specimen were intact.

People expect that the preparation process is completed only when a specimen is mounted as a free-standing display in a museum. Although this is once the way things were done, it is no longer the most modern way to approach dinosaur fossils. First, real fossil bone is delicate and needs to be supported. For this reason, a metal frame must commonly be welded to support the bones, a process that is time consuming and costly. Second, the process of mounting can destroy the fossil bone; the metal frame must be attached to the bone and, although a variety of techniques have been invented to minimize the destruction, it is clear that some destruction takes place. Third, if the specimens are mounted, they are not easily

Figure 1.7 **A dynamic mount of the sauropod *Barosaurus* and the theropod *Allosaurus*.** This mount is made of fiberglass and epoxy resin, cast from the bones of the original specimens. *Photograph courtesy of the American Museum of Natural History.*

studied by professionals, because the bones cannot be moved around and examined. Finally, mounted specimens commonly undergo damage over time; slight shiftings in the mounts because of the extraordinary weight, vibrations in the buildings in which the bones are housed, and/or museum patrons' lifting "insignificant" bits[5] have all diminished the quality of mounted specimens.

Most museums, therefore, cast bones in fiberglass and other resins, and mount and paint the casts. Such mounts are virtually indistinguishable from the originals if done by a skillful preparator. The casts have the advantage of being lighter than the original fossils. The use of casts allows internal frames to support the bones, which again contributes to more impressive mounts (Figure 1.7). Finally, mounted casts free the actual fossils so that the real bones can be optimally protected and studied under the most ideal conditions.

Do casts cheat the public of its right to see the originals? We think not. A cast is not a poor substitute for the real thing. Leaving the bones disarticulated, properly stored, and available for study maximizes returns on the very substantial investments that are involved in collecting dinosaurs. Paleontology is carried out in large part by public support, and mounted casts give the public the best value for its money.

Important Readings

Behrensmeyer, A. K., and A. P. Hill, eds. 1980. Fossils in the Making. University of Chicago Press, Chicago, 338 pp.

Cvancara, A. M. 1990. Sleuthing Fossils: The Art of Investigating Past Life. John Wiley and Sons, New York, 203 pp.

Gillette, D. D., and M. G. Lockley, eds. 1989. Dinosaur Tracks and Traces. Cambridge University Press, New York, 454 pp.

Kielan-Jaworowska, Z. 1969. Hunting for Dinosaurs. MIT Press, Cambridge, 177 pp.

Leiggi, P., and P. May, eds. 1944. Vertebrate Paleontological Techniques, Vol. 1. Cambridge University Press, New York, 344 pp.

Lessem, D. 1992. Kings of Creation: How a New Breed of Scientists Is Revolutionizing Our Understanding of Dinosaurs. Simon and Schuster, New York, 367 pp.

Moore, R. C., C. G. Lalicker, and A. G. Fischer. 1952. Invertebrate Fossils. McGraw-Hill, New York, 766 pp.

Preston, D. J. 1986. Dinosaurs in the Attic: An Excursion into the American Museum of Natural History. St. Martin's Press, New York, 244 pp.

Sternberg, C. H. 1985. Hunting Dinosaurs in the Bad Lands of the Red Deer River, Alberta, Canada. NeWest Press, Edmonton, Canada, 235 pp.

Walker, R. G., and N. P. James, eds. 1992. Facies Models: Response to Sea Level Change. Geological Association of Canada, Memorial University of Newfoundland, St. John's, Newfoundland, 409 pp.

[5]At the American Museum of Natural History in New York, tail vertebrae of a number of the mounted specimens have had to be replaced, because dinosaur enthusiasts availed themselves of the elements.

CHAPTER 2

THE MESOZOIC ERA:
Back to the Past

IN 1939, JUDY GARLAND as Dorothy Gale stepped out of her sepia farmhouse into the brilliance of the Land of Oz and opined to her dog and dazzled movie-goers, "Toto, I have a feeling that we're not in Kansas anymore!" A trip to the Mesozoic Era, the time in which the dinosaurs lived, would be no less astounding to us than Oz was to Dorothy. The animals were obviously different; tyrannosaurs and brontosaurs, instead of lions and elk, roamed the earth. But many other things differed as well. Until well into the age of dinosaurs, you couldn't awaken outdoors to the sound of birds singing in a meadow: There were no birds and there was no grass. You couldn't even receive roses as a token of true love; roses – in fact, *all* flowers – did not appear until the Mesozoic Era was over halfway over. It was a very different world.

Little on earth is permanent. Dorothy's home of Kansas was covered by an ocean throughout much of the Mesozoic. New mountain ranges came and went. Even the continents themselves were not always located where they are found today: They appear to have spent much of the 280 million years of the Mesozoic getting into positions more familiar to us today. Finally, all of these changes affected worldwide climates.

So the Mesozoic was a most strange and wondrous world, and we cannot really understand the dinosaurs until we learn something of it. In this chapter, therefore, we will explore a few aspects of the Mesozoic world. In general, our dis-cussion will be focused on the latter six-sevenths of the Mesozoic (the Late Triassic to the end of the Cretaceous), because this is when dinosaurs roamed the continents. And because dinosaurs roamed continents, and not oceans, we will primarily address ourselves to things terrestrial.

WHEN DID DINOSAURS LIVE AND HOW DO WE KNOW?

Dinosaur remains are found in layers – or *strata* – of sedimentary rocks. How old or young these layers (and the fossils they contain) are, is within the realm of **stratigraphy**, which is the study of the relationships of strata to each other and to the fossils they contain.

Sedimentary deposition – the deposition of sediments to produce strata – is an episodic process; periodically, earth materials are deposited (and, if not eroded, eventually become sedimentary rocks), but in any given place, nothing much happens most of the time. For example, the many years represented by the buildings, trees, grass, and houses where you live, are not recorded by sedimentary deposition. That is, where you live has no geological record that represents the present. Perhaps sedimentary deposition is occurring elsewhere (thereby recording this moment in time). Future geologists looking back on this time interval will assume, reasonably enough, that the time existed, but there is no rock record of it where you live.

This dichotomy, that on the one hand there is time, and on the other there are rocks (and/or fossils) that represent intervals of that time, is fundamental to stratigraphy (Figure 2.1). Geologists assume that time has passed, whether or not there are rocks (or fossils) preserved that represent that time. In this sense, geologists are not existentialists: Time exists independently, whether or not anything was there to record it!

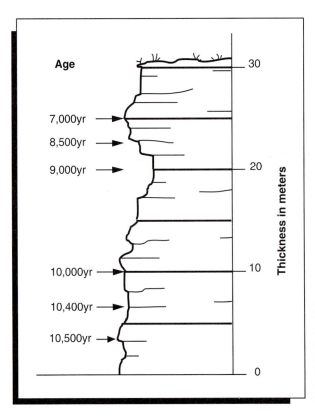

Figure 2.1 Rocks and time. An outcrop of rock is shown, and the ages of several of the layers are given. Note that the amount of time represented is not equivalent to the thickness of the rock.

So the sedimentary record and hence the fossil record (because, of course, fossils are found in rocks) are limited; the amount of time represented by rock deposition is orders of magnitude less than the amount of time that has actually elapsed. The challenge for stratigraphers is to correlate in time all the sedimentary rocks around the world, with the object of developing as complete and precise a representation of the history of the earth as is possible. Because, as we observed earlier, stratigraphers dichotomize rocks (or fossils) and time, stratigraphy is divided into **chronostratigraphy** (*chronos* – time), or time stratigraphy; **lithostratigraphy** (*lithos* – rock), or rock stratigraphy; and **biostratigraphy** (*bios* – organisms), or the relationships of organisms in time.

Chronostratigraphy

In the drama *Inherit the Wind*, Matthew Harrison Brady, defending religion against a supposed threat from evolution, proclaims, "I am more interested in the Rock of Ages than in the Age of Rocks!" Regardless of creed, however, *geologists* are interested in the age of rocks, the study of which they term chronostratigraphy. Geologists generally signify time in two ways: in numbers of years before present, and by reference to blocks of time with special names. For example, one might note that the earth was formed 4.6 billion *years before present*, meaning that it was formed 4.6 billion *years ago* and is thus 4.6 billion years *old*. Learning the age in years of a particular rock or fossil is not always possible. For this reason, time has been broken up into blocks with names, and appropriate rocks and fossils can be referred to these time intervals.

The vastness of geological time has been rhapsodized by people far more poetic than your authors. John McPhee, for example, termed it "deep time," a phrase that is redolent with the antiquity, mystery, richness, and unfathomable extent of earth history. Geological time occurs on a scale that is very literally not within our direct experience. How insignificant is the ephemeral flicker of human existence in the context of 3.5 billion years of life history!

Age in Years

Geologists are happiest when they can learn the age of a rock or fossil in *years before present*, a determination that is called its **absolute age**. Obtaining an absolute age, however, can be accomplished only in those instances in which rocks are found that contain minerals bearing certain radioactive elements. To understand how these are used to determine absolute ages, a little background in chemistry is needed.

CHEMISTRY. The earth is made up of elements. Many of these, such as hydrogen, oxygen, nitrogen, carbon, and iron, are familiar, whereas others, such as berkelium, iridium, and thorium, may be less familiar. All elements are made up of **atoms**, which can be considered the smallest particle of any element that still retains the properties of that element. Atoms, in turn, are made up of **protons**, **neutrons**, and yet smaller **electrons**, which are collectively termed **subatomic** ("smaller-than-atomic") particles. Protons and neutrons reside in the central core, or **nucleus,** of the atom. The electrons are located in a cloud surrounding the nucleus, and some electrons are more tightly bound around the nucleus than others. Those that are less tightly bound are, as one might expect, more easily removed than those that are more tightly bound. An atom is diagrammed in Figure 2.2.

Electrons and protons are electrically charged; electrons have a charge of -1 and protons have a charge of $+1$. Neutrons, as their name implies, are electrically neutral and have no charge. To keep a charge balance in the atom, the number of protons (positively charged) must equal the number of electrons (negatively charged). This number – which is the same for protons and electrons – is called the **atomic number** of the element and is displayed to the lower left of the elemental symbol. For example, the element carbon is identified by the letter C, and it has 6 protons and 6 electrons. Its atomic number is thus 6, and it is written $_6C$.

Along with having an electrical charge, some subatomic particles also have mass. Rather than work with the extremely small mass of protons and neutrons (one of them weighs about $1/6.02 \times 10^{23}$ grams!), we assign them masses of 1. Because the masses of electrons are negligible relative to protons and neutrons, the **mass number** of an element is composed of the total number of neutrons *plus* the total number of protons. In the case of the element carbon, for example, the mass number equals the total number of neutrons (6) plus the total number of protons (6), or 12. This is usually written ^{12}C and is called carbon-12. Note that ^{12}C has 6 protons and therefore must also have 6 neutrons, so its atomic number remains 6, and it is written $^{12}_6C$. Because the atomic number is always the same for a particular element, $^{12}_6C$ is usually abbreviated ^{12}C.

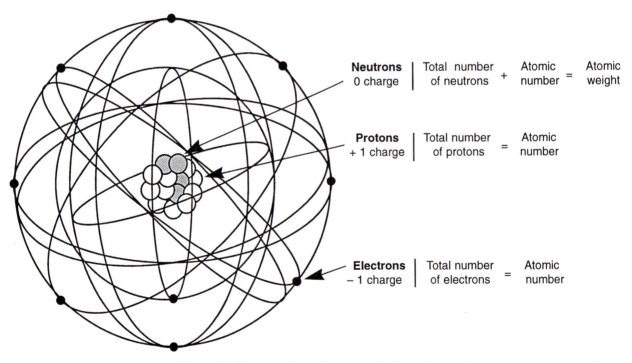

Neutrons 0 charge	Total number of neutrons	$+$	Atomic number	$=$	Atomic weight
Protons $+1$ charge	Total number of protons		Atomic number $=$		
Electrons -1 charge	Total number of electrons		Atomic number $=$		

Figure 2.2 **Diagram of a carbon atom.** In the nucleus are the protons and neutrons. In a cloud around the nucleus are the electrons, whose position relative to the nucleus is governed by their energy state.

Variations in elements exist in nature and those variations that have the same atomic number but different mass numbers are called **isotopes**. For example, a well-known isotope of carbon-12 (^{12}C) is carbon-14 (^{14}C). Since ^{14}C is an isotope of carbon, it has the same atomic number as ^{12}C (based upon 6 electrons and 6 protons). The change in *mass* number results from additional neutrons. Carbon-14 has 8 neutrons, which with the 6 protons increase its atomic mass to 14. Because it is carbon, of course, its atomic number remains 6.

ABSOLUTE AGE DATING.   Ages in years before present can sometimes be determined from the decay of **unstable isotopes**. An unstable isotope is an isotope that spontaneously decays from an energy configuration that is not stable (i.e., one that "wants" to change), to one that is more stable (i.e., one that will not change, but rather will remain in its present form).[1] The decay of an unstable isotope to a stable one can occur over a long or a short period of time, depending upon the particular element and isotope. Because decay does not occur instantaneously, isotopic decay can be used to tell time. The decay reaction is usually summarized as

$$\text{unstable "parent" isotope} \longrightarrow$$
$$\text{stable daughter isotope + nuclear products + heat}$$

Carbon again provides a good example. In the unstable isotope of carbon, ^{14}C, a neutron splits into a proton and an electron in the following reaction:

$$^{14}\text{C} \longrightarrow {}^{14}\text{N} + \text{heat}$$

Note that the atomic number in the decay reaction changes, and that with 7 protons and 7 electrons the stable daughter product is now nitrogen.

The *rate* of the decay reaction is critical to determining an absolute age. If we know (1) the original amount of parent isotope at the moment that the rock was formed or the animal died (before becoming a fossil), and we know (2) how much of the parent isotope is left, and we know (3) the rate of the decay of that isotope, we can estimate the amount of time that has elapsed. For example, suppose we know that 100% of an unstable isotope was present when a rock was new, but now only 50% remains. If we know the rate at which the element decays, we can estimate the amount of time that has elapsed since the rock was formed, that is, the *age of the rock*. This kind of relationship is shown in Figure 2.3. Because radioactive decay is the basis of the absolute age determination, unstable isotopic age estimations are sometimes called **radiometric** dating methods.

Since the rate of decay is constant (in any given stable isotope), it is convenient to summarize that rate by a single number. That number is called the **half-life** and is the amount of time that it takes for 50% of a volume of an unstable isotope to decay (leaving half as much parent as originally present). Two half-lives is the time that it takes for 75% of the volume of an unstable isotope to decay (leaving 25% of the original parent material), and so on. The half-life, then, is an indi-

---

[1] In some instances, a neutron splits into a proton and an electron, producing a different element. Because the daughter now has one more proton than did the parent, the atom now has a different atomic number. In other instances, different elements are produced as a result of the release of protons and neutrons.

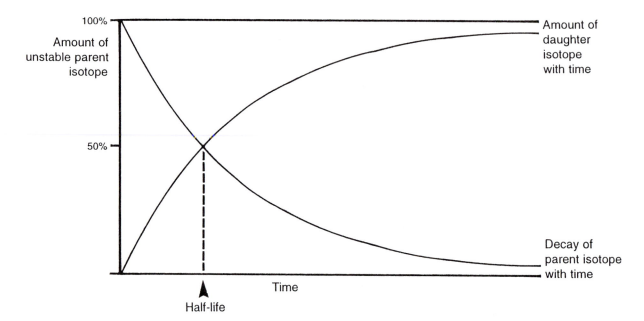

100%

Amount of
unstable parent
isotope

Amount of
daughter
isotope
with time

50%

Decay of
parent isotope
with time

Time

Half-life

Figure 2.3 **An isotopic decay curve.** Knowing the amount of unstable isotope that was originally present, as well as the amount of unstable isotope now present and the rate of decay of the unstable isotope, we can determine the age of a rock with that isotope in it. The diagram shows the half-life, that is, the amount of time that it takes for 50% of an unstable isotope to decay.

cator of decay rate. Dates obtained using radioisotopic methods can be accurate to an error of ±1%, astounding precision when one is attempting to learn the age of something hundreds of millions of years old!

Different unstable isotopes are selected depending upon the object being dated (and its presumed age). For example, if one dated human remains, not likely more than several thousand years old, the potassium/argon isotopic system ($^{40}K/^{40}Ar$), with its half-life of 1.3 billion years, would hardly be the ideal isotopic system. This would be a bit like timing a 100-m dash with a sundial. Likewise, if one dated dinosaur bones, such ages would be in the hundreds of millions of years, and ^{14}C, with a half-life of 5,730 years, would be unsuitable. Would it be meaningful to give your age in nanoseconds?

A problem with radioisotopes is not really their accuracy, but the fact that they cannot be used directly on dinosaur bone. Indeed, one must find a source of radioactive isotopes, which commonly occur in a variety of forms in igneous rocks, such as those blasted from a volcano. Then, the age of formation of the igneous rock can be obtained. No dinosaur lived within a volcano, and so the challenge is then to correlate in time the dinosaur bone with the volcanic event that produced the datable igneous rock. This – the relationships of one body of rock to another – is the province of lithostratigraphy. Lithostratigraphy involves the general study of all rock relationships; however, because we are primarily interested in sedimentary rocks, most of our discussion will apply to sedimentary rock units.

Lithostratigraphy

SEDIMENTATION AND SEDIMENTARY ROCKS. Sediments – sand, silt, mud, dust, and thousands of other, less familiar materials – are deposited in time and in space (geographically). Such deposition does not occur in strange or unpredictable shapes. Rather, it usually occurs in layers – strata – that can be broad and sheetlike, or narrow or ribbon-shaped, and in which each layer becomes progressively thinner until it is said to "pinch out." These shapes occur on scales of meters to hundreds of kilometers and are the direct result of some sedimentary process, such as flowing water, wind, or explosion from a volcano. One can envision a myriad of different *sedimentary environments* where such processes might take place. Indeed, virtually every geographic location we can think of – a river, a desert, a lake, an estuary, a mountain, the bottom of the ocean, the *pampas* – has sedimentary processes particular to it that will produce distinctive sediments.

Sediments harden through time by a variety of means to become **rock**. Thus, although we speak of sedimentary rocks, commonly we are actually talking about the *sediments* that they once were before they were **indurated**; that is, before they hardened.

The science that investigates the deposition of materials and their subsequent **induration** (the process of hardening into a rock) is called *sedimentology*. As sedimentologists, we *observe* rocks and the features they contain, from which we *infer* processes that were involved in the original depositional events. In turn, from these processes we attempt to *infer* the original environments in which the sediments were deposited. It is from this kind of study that we can learn much about the environments in which the dinosaurs lived and what kind of place the Mesozoic world was.

RELATIVE AGE DATING. The 17th-century Danish naturalist Nicholas Steno first recognized that

1. in any vertical, stacked sequence of sedimentary rocks, the oldest rocks are found at the bottom and successively younger-aged rocks are found above, with the youngest occurring at the top, an observation now termed the "law" of superposition; and
2. all sedimentary rock sequences were originally deposited horizontally (although subsequent geological events may have disrupted their original orientation in space).

Younger sediments are deposited *upon* older sediments, and thus, if a stratum lies above another (and the rocks have not been subsequently deformed by various geological processes), it is relatively younger than the one below it (Figure 2.4). We say "relatively younger" because without absolute ages, it is not obvious how much younger. And because we know the *relative ages* of the two strata (i.e., that one is older and the other is younger), we speak of **relative dating**: the type of dating that, although not providing age in years before present, provides age *relative* to another stratum. In historical terms, we might date antique furniture as post–Civil War era. This would tell us that the furniture was relatively younger than the Civil War, an event whose timing is known.

Figure 2.4. **Superposition of strata in the Grand Canyon, Arizona, U.S.A.** The rocks are oldest at the bottom and become successively younger as one gets closer to the top of the rim. The bluff in the foreground exposes more than 200 million years of earth history; a hike from the base of the Grand Canyon to its rim is a hike through more than 2 billion years of earth history.

Here, then, is part of the solution to dating dinosaur bone. Suppose that a stratum containing a dinosaur bone is sandwiched between two volcanic ash layers. Ideally, an absolute age date could be obtained from each of the ash layers. We would know that the bone was younger than the lower layer, but older than the upper layer. Depending upon how much time separates the two layers, the bone between them can be dated with greater or lesser precision.

Though all that looks fine in theory, datable ash layers rarely fall out of the sky because stratigraphers need them! More commonly, rocks for which absolute ages can be obtained are widely separated, not only in time, but also geographically. The challenge then becomes correlating the strata of known absolute age with those of unknown age.

Biostratigraphy

Biostratigraphy is a method of relative dating that uses fossil organisms. It is based upon the idea that a particular time interval can be characterized by specific organisms, because different creatures lived at different times. For example, if one knows that dinosaurs lived from 228 to 65 Ma,[2] then any rock containing a dinosaur bone fragment must fall within that age range. The question is, how precise a date can it really give?

Biostratigraphic correlation – the linking of geographically separated rocks based upon the fossils they contain – can be very precise. Although, like superposition, biostratigraphy cannot provide ages in years before present, the fact that many species of organisms have existed on earth for 1–2-million-year intervals enables them to be used as powerful dating tools. In our example, the group "dinosaurs" was used. Obviously, a particular dinosaur, say, *Tyrannosaurus rex*, is a more precise indicator of time than all dinosaurs considered as a group: *T. rex* itself lived for only about two million years (from 67 to 65 Ma). Thus, if we found a *Tyrannosaurus* fossil, we would know that the sediments in which it was found were deposited between 67 and 65 Ma.

The best organisms for biostratigraphy, therefore, are those that are geographically widespread, but which had relatively short species' durations – that is, the species existed on the earth for a relatively short period of time. A species that existed on earth for a long time affords little biostratigraphic precision. Likewise, a species that lived in a very restricted geographic area has limited utility for correlations across large areas.

Many organisms meet the criteria of good biostratigraphic indicators: wide geographic distributions and short species' durations. Dinosaurs and mammals are quite accurate indicators of time, especially in terrestrial sediments. Another important terrestrial biostratigraphic tool is fossil pollen, which, as anyone with hay fever knows, is ubiquitous! Bivalves, snails, corals, and a bewildering variety of other ancient invertebrates are important marine biostratigraphic indicators. Other organisms that are extremely important are single-celled creatures called **foraminifera** (see Figure 2.16); their wide distributions and short species' durations make them powerful biostratigraphic tools in the correlation of oceanic sediments.

Eras and Periods and Epochs, Oh My!

The oldest method of dating sediments is biostratigraphy. In the late 1400s, Leonardo da Vinci observed marine shells far inland where the ocean clearly was not; he correctly concluded that where he stood had once been inundated with ocean waters. By the early 1800s, the great French anatomist Georges Cuvier, studying strata around Paris, noted that higher strata had a greater proportion of fossil shells with living counterparts than did lower strata. The increasingly modern aspect of the fauna is due to the fact that the highest rocks are closest in time to the present, and thus the faunas that they contain are the most like those of today.

Within a generation of Cuvier, a revolution had been wrought in geological thinking. The **Phanerozoic** (*phaneros* – visible; *zoo* – life; 570–0 Ma) time interval, representing that interval of earth history during which there have been organisms with skeletons or hard shells present, was established. Using biostratigraphy as a time indicator, a variety of rock outcrops in northwestern Europe were designated as **type sections**, or original locations, where a particular interval of

---

[2]We use the expression "Ma," from the Latin *mill annos*, to mean million of years. Thus, 65 Ma is 65 million years ago.

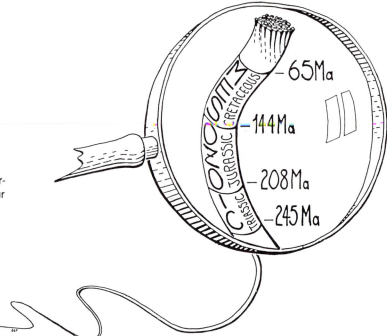

Figure 2.5 **The Mesozoic part of the geologic time scale.** The Mesozoic constitutes only a rather tiny fraction of the expanse of earth time. If you compacted earth time into a single year, from January 1 (the formation of the earth) to December 31 (the last 100,000 years, which, by this way of measuring earth history, would occur in less than a day!), then dinosaurs were on earth from about December 11 to December 25.

time is represented. The names of the **Eras**, the largest blocks of time within the Phanerozoic, came from a description of the life contained within each era. The Eras are, from oldest to youngest, the **Paleozoic** (*paleo* – ancient; *zoo* – life; 570–245 Ma), the **Mesozoic** (*meso* – middle; *zoo* – life; 245–65 Ma), and the **Cenozoic** (*cenos* – new; *zoo* – life; 65–0 Ma). Within each of these eras are smaller subdivisions (still consisting of tens millions of years, each) called **Periods**, and within the periods, in turn, are yet smaller subdivisions of time called **Epochs** (consisting of several millions of years, each). Figure 2.5 shows the currently understood distribution in time of the different periods within the Mesozoic Era, the time interval of special relevance in this book.

All intervals of time are hierarchically arranged; that is, large blocks of time are subdivided into smaller blocks of time, which are in turn subdivided into even smaller blocks of time. This is convenient, because the age of a rock or fossil can be designated with only as much precision as is known to the investigator. For example, we might know that a dinosaur fossil is 70 million years old, and thus we can say that it belongs to the interval called the "Maastrichtian."[3] Alternatively, we might not know its exact age and thus might be able to identify it only as Cretaceous, the period of time of which the Maastrichtian is a part. The Cretaceous includes more time than the Maastrichtian and is thus a less precise age designation.

Periods are also subdivided in other ways. "Early," "Middle," and "Late" mean oldest, middle, and youngest, respectively. Thus the Late Triassic, for example, represents the youngest years within the Triassic – that is, the major division

---

[3]Or alternatively, we can recognize it as Maastrichtian, thus restricting its age between 74 and 65 Ma.

of the Triassic that is closest in time to us. Remember that stratigraphers dichotomize rocks and time. Therefore, that same interval of time – or an interval approximating it – might also be referred to as the "Upper" Triassic. Here Steno's law applies, and the *uppermost* rocks of the Triassic would be those that are *youngest*. Hence, we commonly use "Lower," "Middle," and "Upper" to designate oldest, middle, and youngest, respectively. The Triassic and Jurassic are formally subdivided into three parts; however, the Cretaceous is formally subdivided into only two parts: Early and Late. Throughout this book, we rely upon these designations.

It is no surprise that in this book, our interest lies primarily in the Mesozoic Era. Dinosaurs (as defined in Chapter 1) are Mesozoic beasts, a fact that is reflected in the common, if inaccurate, designation of the Mesozoic as the "Age of Reptiles."[4] Dinosaurs lived from approximately 228 Ma to 65 Ma, which places their tenure on earth from the late Middle Triassic or early Late Triassic Period to the very end of the Cretaceous Period.

Growth of a Prehistoric Time Scale

One of the truly unsung great human achievements is the development of the time scale in the context of which major events in earth and organismal history can be understood. The job is hardly finished: The eminent stratigrapher W. B. N. Berry, writing a book on geological time, entitled it *Growth of a Prehistoric Time Scale*, emphasizing the idea that our concepts about time and the placement of events in earth history are constantly in a state of flux, as more (and more refined) techniques become available.

As things stand today, outcrops of sedimentary rock in virtually every part of the world are integrated into the time scale. Accurate global correlations, based upon biostratigraphy, lithostratigraphy, and/or chronostratigraphy have been established.

PLATE TECTONICS:
A driving force in evolution?

Throughout this chapter, we have made the claim that the earth is *dynamic* and that nothing is permanent except change itself. The movements of the continents are encompassed within the theory of **plate tectonics**. In effect, plate tectonics is the concept that the earth's surface is organized into large blocks, or **plates**, which migrate across the surface of the planet. The word "**tectonic**" refers to the behavior and positioning of the plates in geological time. By their movement, the plates constantly (if slowly) change the very shapes of the continents and oceans, as well as the positions of high and low regions on earth.

---

[4] The Paleozoic Era has been called the "Age of Fish" and the Cenozoic Era has been called the "Age of Mammals." These terms imply that initially there were "fish," then there were "reptiles," and then, finally, there were mammals. As we shall see in the chapters that follow, this kind of progression is a very inaccurate way to consider the history of life on earth. Moreover, the terms "Age of Fish," "Age of Reptiles," and "Age of Mammals" imply that vertebrates are somehow the defining creatures of a particular era, a view that inflates the significance of vertebrates in the ecosystem to a position that they have never occupied.

Plate tectonics is a relatively new concept; although it was first proposed by the German meteorologist Alfred Wegener as early as 1912, it did not come to be generally accepted until the late 1960s. As in the case of the theory of evolution by natural selection, acceptance of the theory was an intellectual revolution, not only in the field of earth sciences, but also, as we shall see, in evolutionary thinking.

WHY STUDY PLATE TECTONICS IN A BOOK ON DINOSAURS? Although much of the Western academic tradition tends to isolate areas of knowledge – classes are on a particular subject; we pursue careers in specific fields – the earth is a *far* more integrated system. Here we are interested in dinosaurs and their legacy, but it is clear that *where* dinosaurs lived and died and *when* they did so, was shaped by large-scale forces external to dinosaurs themselves. Indeed, evidence is accumulating that the positions of the continents, powered by forces deep within the earth, may be the single ultimate driving force in evolution. This may be no surprise, given that it has been known for many years that weather patterns are directly affected by the distributions of mountain ranges and low regions, as well as by the positions of the continents themselves. Weather aside, the isolation or connectedness of continental masses now appears to be a significant factor in the evolution of new species, as well as in the extinction of already existing populations.

Structure of the Earth

Although the earth is thought to have been originally homogenous in composition, we know that today the earth is stratified. Materials of different compositions and different densities are concentrated into concentric zones. The most dense material is found in the iron-rich, solid **inner core**, about 2,400 km in diameter. Around this is a liquid iron **outer core**, about 2,000 km thick. Beyond this is the **lower mantle** or **mesosphere**, which is about 2,700 km thick. This layer is composed of relatively dense materials such as magnesium and iron silicates. External to the lower mantle rests 250 km of the **upper mantle**, or **asthenosphere** (*asthen* – weak; feeble). The upper mantle is composed of materials less dense than those in the lower mantle, and because the materials in the upper mantle are more easily melted than those in the lower mantle, the upper mantle is in a partially molten state. This causes the asthenosphere to behave like a fluid, and therefore it is said to be *plastic*. The lightest material is concentrated in the outermost layer, the 100-km-thick **lithosphere** (*lithos* – rock). Relative to the asthenosphere, the lithosphere is rigid, a condition whose importance shall become evident shortly. The earth's **crust** is part of the lithosphere and consists of a comparatively thin, chemically distinct rind. There are two types of crust, **oceanic crust** and **continental crust**. Oceanic crust, which underlies the ocean basins, is relatively thin (~10 km in thickness) and is composed of dense minerals rich in magnesium and silica. Continental crust, the material from which the continents are made, is somewhat thicker (~40 km) and is composed of minerals rich in aluminum and silica. Continental crust is less dense than oceanic crust. These features of the earth are shown in Figure 2.6.

Motions of the Plates

The lithosphere is not a single, rigid sphere encircling the core and mesosphere, but rather is fragmented into 10 or so large *plates*. These plates ride atop the semi-

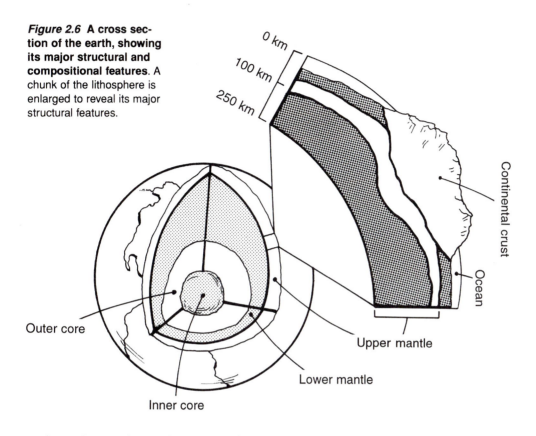

Figure 2.6 **A cross section of the earth, showing its major structural and compositional features**. A chunk of the lithosphere is enlarged to reveal its major structural features.

molten, plastic asthenosphere, in much the same way that scum rides at the surface of boiling soup. Remember how soup scum is continuously in motion and how new scum is constantly being added; so it is with lithospheric plates. Such additions take place at **spreading ridges**. These are elongate, raised linear features at which new oceanic crust is produced as new material emerges from the asthenosphere. This new material is added as older material is pulled away, and thus two stationary points located just across a spreading ridge *diverge* from each other gradually as the spreading continues. Obviously, if there is divergence in one place, there must be *convergence* somewhere else; where there is convergence, older lithosphere material must go somewhere. Indeed, it is very much like two sheets of plywood floating on water: If they collide and the force pressing them together proves irresistible, one must ride up over the other. When this occurs in the case of plates of lithosphere, the lower (or *down-going*) slab is said to be **subducted**. The subducted slab is driven downward, where its leading edge melts into the asthenosphere. Figure 2.7 illustrates this process.

Thus the earth can be thought of as a gigantic recycling machine: Lithosphere is produced at spreading ridges, conveyed for greater or lesser distances around the planet, and then recycled at *subduction zones*. As oceanic crust becomes older, it is more likely to encounter some kind of region of convergence (such as a continent), and for this reason, most oceanic crust that has not been thrust up and preserved on a continent is relatively young (post-Jurassic in age).

And what of the continents? Recall that continents are passengers riding on and within the lithosphere; they are gigantic granitic inclusions poking out of a

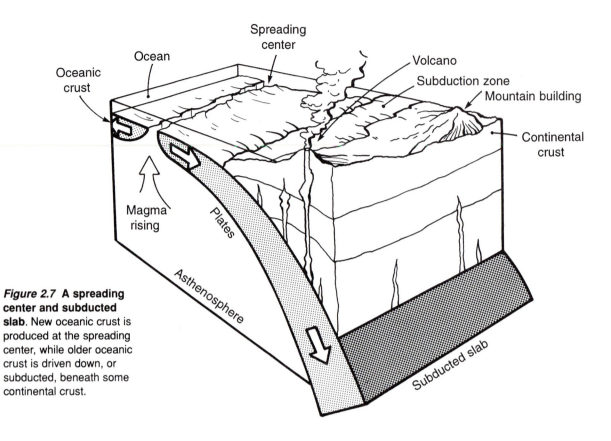

Figure 2.7 **A spreading center and subducted slab**. New oceanic crust is produced at the spreading center, while older oceanic crust is driven down, or subducted, beneath some continental crust.

rigid matrix of lithosphere. When lithosphere with oceanic crust collides with lithosphere bearing continental crust, the oceanic crust, being more dense, is commonly subducted beneath the continental mass and destroyed; however, there are those occasions in which the lithosphere with oceanic crust rides up over the continent and **accretes**, or bonds, to it. Either way, the tremendous forces involved in the impact and in the subduction or accretion cause deformation of the converging plates. Commonly, such deformation involves crumpling and/or shearing of the continental crust, and this is the means by which mountains are built. Thus, plate tectonics causes a redistribution of the oceanic and land masses, produces new mountain ranges, and alters the topography of the earth.

This is rather heady stuff: the movement on the asthenosphere of great slabs of lithosphere, the building of mountain ranges by the collision of the plates, and the melting and recycling of the down-going slabs. The energy sufficient to drive this drama is astounding: Temperatures in the asthenosphere are calculated to range from as high as 1,200°C at the base of the lithosphere to 5000°C at the interface between the mesophere and the core!

But what is driving the process of plate movement? Nobody fully understands all the mechanisms, but **convection** is thought to be at least one of the primary forces driving the plates. Again we turn to the kitchen for our analogy. When a pot of soup is heated, liquid nearest the burner heats sooner than the liquid at the top of the pot. As the lower liquid heats, it rises, displacing the liquid at the top, which is driven down. Now *that* liquid becomes warmer than the liquid above, and it rises, displacing the upper liquid. A cycle of convection is established. The motion of the convecting liquid moves the noodles around the pot. It is thought that, similarly, the plates are driven around the surface of the earth by a series of gigantic convection currents in the asthenosphere.

Plates through Time

BEFORE DINOSAURS. The oldest rocks known on earth are found in South Africa, Australia, India, and North America and are about 3.8 billion years old. These earliest rocks show distinctive features suggesting an earth in every conceivable feature very different from our own. Indeed, large continental masses such as we have today did not really begin to assemble until about 2.5 Ba (billion years ago). By 1 Ba, the earliest truly stable continental masses had formed. Such masses, termed **cratons**, constitute the stable interiors of the present-day continents.

Because here we are interested in dinosaurs, we will take the luxury of bypassing the mere 1.93 *billion* years of continental evolution that occurred from the formation of the earliest continents to the base of the Cambrian (570 Ma), when life with skeletons first appeared (still 360 million years before the first dinosaur!). By Early Cambrian time, the present-day African, Indian, Australian, and Antarctic cratons had coalesced into one large mass: the southern *supercontinent* of **Gondwana** (Figure 2.8). Gondwana was rather long lived; it finally completely dismembered only 20 million years ago.

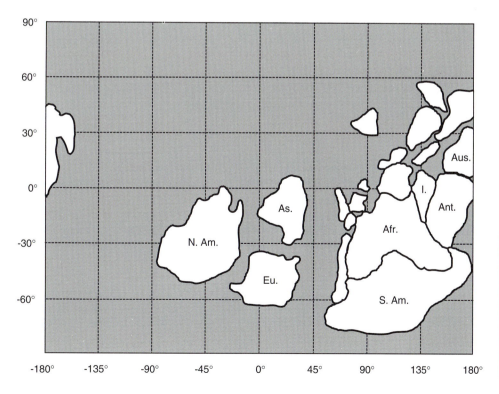

Figure 2.8 Rectilinear projection of plate positions in the beginning of the Paleozoic Era (Early Cambrian; ~560 Ma). Note formation of southern super continent of Gondwana. *Plate reconstruction courtesy of Paleogeographic Information System, M. I. Ross and C. R. Scotese.*

Map Abbreviations

Afr. – Africa
Ant. – Antarctica
As. – Asia
Aus. – Australia
Eu. – Europe
N. Am. – North America
I. – India
S. Am. – South America

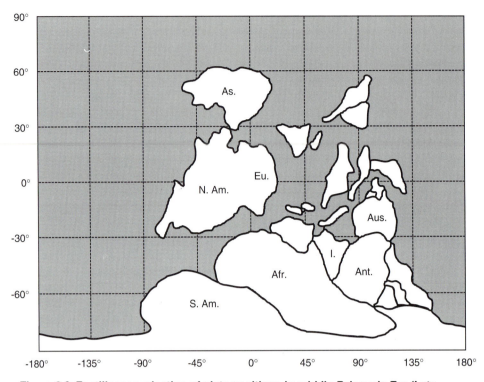

Figure 2.9 **Rectilinear projection of plate positions in middle Paleozoic Era (Late Devonian; ~367 Ma).** Note the appearance of the northern supercontinent of Laurasia. Map abbreviations as in Figure 2.8. *Plate reconstruction courtesy of Paleogeographic Information System, M. I. Ross and C. R. Scotese.*

By the time of the middle Paleozoic, the North American and European cratons were themselves coalescing into a supercontinent, sometimes called the Old Red Sandstone Continent, named for the middle-Paleozoic-aged Old Red Sandstone in England (a series of rocks whose presence records the collision between the two continents). The Late Paleozoic interval was characterized by a great amount of global tectonic activity, some of which brought the Siberian craton into contact with the Old Red Sandstone Continent to form a northern supercontinent called **Laurasia** (Figure 2.9).

A sort of tectonic climax was reached at the end of the Paleozoic Era, when Laurasia and Gondwana came together to make the mother of all supercontinents, **Pangaea** (Figure 2.10). It was at least theoretically possible to walk on land from any continent to any other. This, as you can imagine, had important ramifications for the possibility of migration by terrestrial organisms, as well as for the similarity of faunas all around the world. As we shall see, it also had an important influence on global weather patterns.

The dismemberment of Pangaea didn't begin until well into the Mesozoic Era. It was a long process, and the subsequent history of the Mesozoic can be thought of as the history of the breakup of this great landmass. Indeed, as suggested, patterns of dinosaur evolution and extinction cannot really be understood until the influence of the continents is considered.

PLATES, OCEANS, AND SEAS DURING THE TIME OF DINOSAURS. Our story begins with Pangaea. By the Late Triassic, the fragmentation of Pangaea had not yet begun. The South American, African, and European cratons were pressed against the south and east sides of the North American craton, and the equator almost perfectly bisected Pangaea.

The initial rifting of Pangaea took place in the Middle Jurassic (Figure 2.11). The effect has been likened to an "unzipping" of the Gondwana–Laurasia connection from south to north. Sediments in the Gulf Coast region of North America and Venezuela, as well as in western Africa, record the opening and widening of a seaway between the two supercontinents. At this time, some of the earliest epicontinental (or **epeiric**) seas of the Mesozoic Era first make their appearance. Epicontinental seas are shallow marine waters that cover parts of the continents, as opposed to simply residing within ocean basins. An example of a modern epicontinental sea is the North Sea, which covers part of the European craton. In the past, epicontinental seas were considerably more widespread than they are today; they were rather common during Paleozoic times and during the latter half of the Mesozoic.

The presence or absence of epicontinental seas is based upon global sea levels. High global sea levels flood the lower regions of the continents with marine waters. Sea level is controlled by many factors, but two of the most common are ice at the earth's poles and tectonism. If there is a great deal of polar ice (or glaciation, such as took place during the last one or so million years), much sea water may be bound up in ice, lowering sea levels worldwide. Likewise, during tectonically active intervals, mid-oceanic spreading centers tend to be topographically elevated, decreasing the volume of the ocean basins and thus displacing ocean water up onto the continents. During the Middle Jurassic, fluctuating epicontinental seas covered large parts of what is now western North America, eastern Greenland, eastern Africa, and Europe, where a complex system of islands and seaways developed.

Although the Middle Jurassic was an important interval of time in dinosaur evolution, terrestrial sediments that record it are rare. Why this is the case is uncertain, but probably it is just preservational "luck of the draw." Regardless of why *few* terrestrial sedimentary rocks from the Middle Jurassic are preserved, this means that few terrestrial vertebrate localities are known. An exception to this is China, where within the past 10 years, major discoveries have been made in a thick, well-preserved sequence of Middle Jurassic rocks. In the case of the Late Jurassic, however, many terrestrial sequences are known worldwide, and some of the world's best-known and best-studied terrestrial fossils hail from that time interval (see Chapter 15).

In the Late Jurassic and Early Cretaceous, the separation of Laurasia and Gondwana was complete. A broad seaway, the *Tethyan Seaway* (after *Tethys*, a Greek goddess of the sea), ran between the two supercontinents. Global sea level was relatively low, and hence epicontinental seas were not very predominant. Interestingly, the then western (present-day northern) coast of Australia seems to have been bathed in a broad, shallow, epeiric sea, as were, periodically, western North America and parts of Europe and Asia.

The mid-Cretaceous was a time of active tectonism, high global sea levels, and broad epeiric seas (Figure 2.12). The Tethyan Seaway remained a dominant geographic feature, at the same time as rifting between the European and North

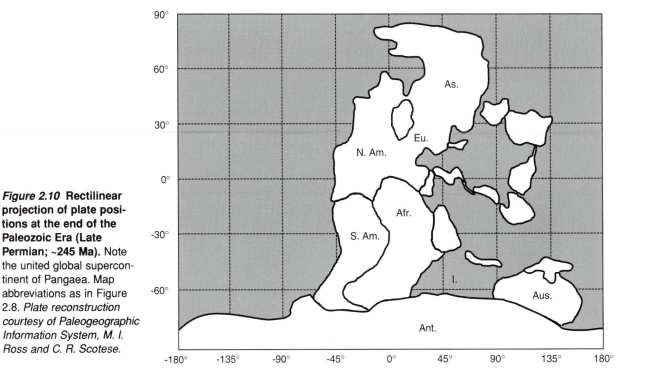

Figure 2.10 **Rectilinear projection of plate positions at the end of the Paleozoic Era (Late Permian; ~245 Ma).** Note the united global supercontinent of Pangaea. Map abbreviations as in Figure 2.8. *Plate reconstruction courtesy of Paleogeographic Information System, M. I. Ross and C. R. Scotese.*

Figure 2.11 **Rectilinear projection of plate positions in middle Mesozoic time (Early–Middle Jurassic; ~178 Ma).** Pangaea has begun to undergo its dismemberment. Map abbreviations as in Figure 2.8. *Plate reconstruction courtesy of Paleogeographic Information System, M. I. Ross and C. R. Scotese.*

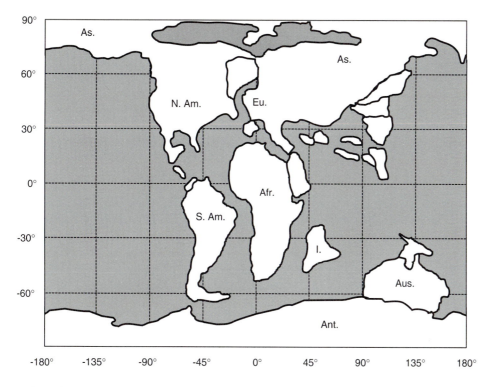

Figure 2.12 Rectilinear projection of plate positions near the end of the Mesozoic Era (mid-Cretaceous; ~90 Ma). The distribution of plates and continents is starting to take on a modern appearance. Map abbreviations as in Figure 2.8. *Plate reconstruction courtesy of Paleogeographic Information System, M. I. Ross and C. R. Scotese.*

American cratons was initiated from the south to the north. The effects of active tectonism were even more marked in Gondwana. Here, the stable supercontinent that had assembled in Early Cretaceous time fragmented into two large constituents – the African and South American cratons – as well as a smaller constituent, the Indian craton. That craton spent the next 50 million years motoring at breakneck speed (for a continent) across what became the Indian Ocean, to smash head-on into southern Asia to produce the Himalayan Mountains. But this is getting ahead of our story. While India and Madagascar were in the first bloom of unconfinement, Australia and Antarctica remained firmly united (a cratonic marriage that would not end until 50 Ma), and it is believed that a land connection remained, as it *almost* does today, between South America and Antarctica.

Europe was an **archipelago** (group of islands), dissected by small epeiric seaways. Extensive seas developed over both northern Africa and western South America. Finally, a series of North American Western Interior Seas bisected the North American craton, deriving their waters from the Arctic Ocean to the north and from the newly formed Gulf of Mexico to the south. The Western Interior Seas came and went, sometimes completely dividing the craton in half, and at other times only partially dividing it.

The global geography of the Late Cretaceous is almost familiar to us. North America became nearly isolated, connected only by a newly emergent land bridge across the modern Bering Straits to the Siberian craton (the eastern Asiatic continent). This land bridge has come and gone several times since the Cretaceous, but it was clearly a significant feature at the time. Africa and South America were fully separated, the former retaining its satellite, Madagascar, and the latter retaining a land bridge to the Antarctica–Australia continent. India was by now completely isolated and well on its hayride toward southern Asia.

The Late Cretaceous was also a time of major epeiric seas. The Western Interior Sea was an important feature of North American geography up until the very latest Cretaceous, when it receded (although it reemerged briefly, one last time in the early Cenozoic). Europe remained a complex of islands and seaways. Africa is understood to have had extensive internal seaways, as did South America and Asia. At the end of the Cretaceous, there was apparently a drop in global sea level, causing exposure on all continents.

CLIMATES THROUGH THE MESOZOIC ERA

Our picture of the Mesozoic would not be complete without some general sense of Mesozoic **paleoclimates**, or ancient climates. **Climate** is the sum of all weather conditions in a particular region and involves factors such as temperature, humidity, precipitation, and winds. In our discussion of the Mesozoic, therefore, we cannot ignore climate; it plays a critical role in where organisms live, their adaptations, and how they behave.

How can we understand past climates? Climate, after all, is a bit like performance art: If you are not there to experience it, it's gone. It turns out, though, that just as one can record the sound of a musical performance, so the earth has recorded traces that allow us to infer at least *aspects* of past climates. And, as we shall see, the record suggests that climates have not remained constant through earth history. They, like everything else, have changed through time.

Potential Effects of Plate Motions on Climate

Why should we even suspect that past climates differed from those that we experience today? Distributions of the landmasses, as well as the volume and distribution of the oceans on the globe, will drastically modify, for example, temperatures, humidity, and patterns of precipitation. In the following, we explore some of the issues, comparing the extreme case of the continents coalesced into a single landmass with that of the continents more diffusely distributed around the globe.

HEAT RETENTION IN CONTINENTS AND OCEANS. Continents and oceans respond very differently to heat from the sun. We know this from our own experience: In the dead of summer, we cannot walk barefoot on cement, and yet we can cool off in a pool. At night, the cement cools off rapidly; then, the pool seems warm. The explanation of this is relatively simple: Because fluids are mobile (as we saw with the pot of boiling soup), heat can be more easily distributed through fluids than in solids. Solids quickly become hot to the touch, but just a short distance below the surface, temperatures remain cool (a property that enables desert organisms to remain buried in a cool environment during the heat of the day and that keeps basements cool in the summer). Heat is distributed more evenly through a liquid than through a solid; thus, for the liquid to feel cool, the *entire* liquid must be cooled. In the case of a solid, the exterior can be cool while the interior remains hot; we need cool only the exterior and not the entire solid. In terms of continents and oceans, this means that oceans are slower to warm and slower to cool than continents are.

Consider how these properties of continents and oceans might modify climates, depending upon the distribution of the landmasses. In the dawn of the Mesozoic, the continents were united into the single landmass, Pangaea. Here, **continental effects** – more rapid warming and cooling of continents than of oceans – would have been more intense than in any other time in earth history. Because of its size, Pangaea should have experienced wide temperature extremes not found on earth today. It would have heated up quickly and gotten hotter, and then cooled rapidly, and then gotten colder, than modern continents, whose continental effects are mitigated by the broad, temperature-stabilizing expanses of oceans between them.

The post–Late Triassic breakup of Pangaea would have moderated the strong continental effects that we have just discussed. With the rise in global sea level, the effects of the large epeiric seas were superimposed upon the diminishing continental effects as the continents broke up. For example, the strong continentality (hot, dry summers; cold winters) observable even today in the Western Interior of North America would have been decreased by the presence in the Jurassic and Cretaceous of the Western Interior Seaways. Large bodies of water on the continents would have tended to stabilize temperatures, decreasing the magnitude and rapidity of the temperature fluctuations experienced on the continents.

OCEANIC CIRCULATION. We can be quite confident that **oceanic circulation** – the direction and patterns of oceanic currents – changed throughout the Mesozoic as well. Present-day circulation patterns, which have an important effect on continental climates, would not have been possible when all of the continents were clustered together in a single landmass. Today, **circumpolar currents**, the cold water masses that circulate at the poles, move from high to low latitudes, where they are warmed, causing the establishment of circulating cells (Figure 2.13). For example, one such cell, the Gulf Stream, takes water warmed at the equator and brings it up against northern Europe, where it warms cities near the coast to temperatures that might not have been predicted for them at such high latitudes.[5]

The circulating cells that exist today cannot have been present during the time of Pangaea. In the Southern Hemisphere, for example, Antarctica was connected to South America until Eocene times, which would have impeded a southern circumpolar current. Because as we have seen, the circumpolar currents have a marked effect on climate, the absence of a southern circumpolar current would have made for less effective cooling of water masses in southern latitudes. The breakup of Pangaea initially produced the Tethyan Seaway (Figure 2.11), but in fact, a well-developed circulation cell between North America and Europe, as well as one between South America and Africa, is not thought to have been possible until the Late Cretaceous.

[5]An example of this effect is the climate experienced in London, England. London is located at a similar latitude to Hudson's Bay, but has a milder climate.

Figure 2.13 Major circulating currents in modern oceans.

WINDS. Wind directions and speeds must have differed between the world of Pangaea and our own. To understand fully why this should be, we need to consider the concept of **high-** and **low-pressure zones**, their relationships to the oceans and continents, and, in turn, to winds and precipitation. High-pressure and low-pressure zones are caused by the *weight* of air. The vertical accumulation of large, moist, and therefore dense air masses produces a region of high pressure, whereas dry air produces a region of lower pressure.

Obviously, air masses rise when they are heated (this is how a hot air balloon works) and as air masses rise, they displace cooler air, which sinks, and a circulating or *convection cell* (much as we discussed previously) is produced.

The presence of high- and low-pressure zones marks the positions of three convection cells that ring each hemisphere. Each cell consists of air movement at the earth's surface and (because it is a convection cell) opposite circulation in the atmosphere. At the equator, air is warmed, and it rises. This creates a belt of equatorial low pressure. At about 30° latitude in each hemisphere, relatively dry, but cool, air begins to sink, piling up at that latitude. This forms a high-pressure belt at 30° latitude in each hemisphere. Circulation does not occur exactly parallel to the axis of the earth, but rather is deflected due to the spinning of the earth. **Winds** are simply moving air, and the kind of moving air that results from the convection cells produces *large-scale* winds, known as trade winds, easterlies, and westerlies. All winds are thus ultimately produced by warming due to solar energy. Figure 2.14 shows an idealized view of the wind directions in each of the three cells for each hemisphere.

CONTINENTS. Because continents heat up and cool off more rapidly than oceans, they obviously disrupt the idealized bands of high and low pressure just discussed. Continents efficiently warm air masses and produce low-pressure zones over themselves. Moreover, the speed with which continents heat up and cool off is such that the pattern of highs and lows previously described is disrupted on a seasonal basis. For example, the very large Asian continent exerts a strong continental effect, and in January, a well-known high-pressure zone (cool air) is produced over Siberia.

RAIN. Rain – or **precipitation** – has a complex, seasonally variable global pattern. Because warm winds carry more moist air than cool winds, it is not surprising that global precipitation is generally related to patterns of wind and pressure systems. Areas with high-pressure systems tend to experience drier conditions than areas

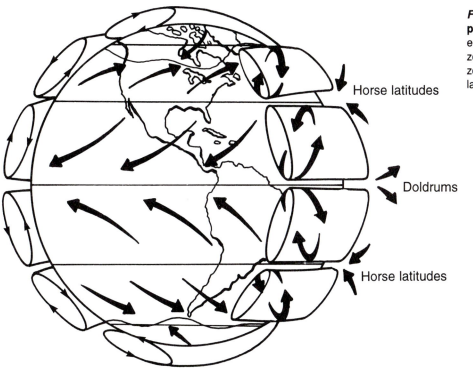

Figure 2.14 **The earth's pressure cells.** Note the equatorial low-pressure zone and the high-pressure zones at the so-called horse latitudes.

Horse latitudes

Doldrums

Horse latitudes

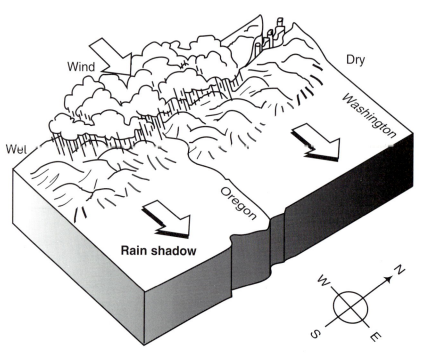

Figure 2.15 **The oro-graphic effect of the Cascade Mountains in Oregon and Washington, U.S.A.** The moist, marine climate is restricted to the west by the mountain range, whereas in the east is a desert.

with low-pressure systems. For this reason, greater precipitation occurs at the equator than around the 30° latitudes, where high-pressure systems are concentrated.

Continental effects, such as those described previously, further modify the relationship between areas of precipitation and zones of low pressure. For example, in July on the Asian continent, summer continental effects are very strong, and the Asian monsoon season is caused by a low-pressure zone that develops over northern India. This zone pulls moisture-rich oceanic air over the continent, causing the characteristic monsoonal rains on the Indian subcontinent.

Finally, mountain ranges complicate the picture by forming barriers to precipitation. Today, a famous example of such **orographic**, or mountain range, effects on climate is seen in eastern Oregon and Washington states in the United States (Figure 2.15). There, a high mountain range, the Cascades, shields the eastern interiors of these states from moist coastal air.

PANGAEA. How Pangaea might have modified atmospheric pressure, winds, and precipitation in the Late Triassic remains largely a matter of conjecture. What is not conjecture, however, is that significant differences must have prevailed. It is generally acknowledged that the idealized three-convection-cell-per-hemisphere system described previously was probably operative during the Late Triassic. As noted, however, Pangaea must have exerted spectacular continental effects that surely modified the three-cell-per-hemisphere system extensively.

Scientists reconstructing Pangaean climates from theoretical considerations have proposed that during July, a high-pressure system that was centered over the Gondwanan portion of Pangaea brought a dry winter to the southern portion of the continent (remember that in the Southern Hemisphere, July is in the dead of winter). During July in the Northern Hemisphere, winds across the broad oceanic expanse opposite from Pangaea are thought to have become very moist, possibly producing coastal monsoons in western Laurasia. As Pangaea dismembered, it is reasonable to suppose that many of these kinds of continental effects diminished or changed, sometimes radically.

Summary of Climates through the Mesozoic

TRACES ARE LEFT BY ANCIENT CLIMATES. Climates modify the bulk of sedimentary and biological processes on earth. But what sorts of traces are left by climates? To use a comparatively recent example, we know that a few thousand years ago, climates were considerably colder than they are today. We infer this from evidence that glaciers covered Canada and much of the United States. But the glaciers aren't there now; how is *their* presence inferred? The answer to this – and to how geologists reconstruct all past climates – is found in physical, biological, and geochemical evidence. In the case of glaciers, there is *physical* evidence: Lakes have been left behind, deep valleys have been carved and filled by streams of glacial meltwater, material has been deposited where it is not expected. Biological evidence includes finding the remains of plants and animals adapted for cold weather where no such weather conditions now exist. Finally, there is geochemical evidence, or evidence from the chemistry of earth materials. In soils formed during glacial times, we might observe geochemical features indicating year-round water saturation, where an abundance of water does not now occur.

LATE TRIASSIC–EARLY JURASSIC. The Late Triassic and Early Jurassic seem to have been times of heat and aridity (dryness) in comparison to today. They also were times of strong **seasonality** – that is, highly marked seasons. Paleontologists studying the Late Triassic–Early Jurassic time interval have recognized three broad climate regimes: an equatorial, year-round dry belt; narrow belts of strongly seasonal rainfall north and south of the dry belt; and middle- and upper-latitude humid belts. These concepts were developed theoretically, but they were also based upon the distribution of Late Triassic deposits of sedimentary rocks that reflect dry conditions. Such rocks include sand dune deposits (the remains of ancient deserts) and **evaporites** (rocks that are made up of minerals formed during desiccation). Triassic evaporites and sand dune deposits do not today form continuous belts around the globe; scientists have reconstructed the locations of the continents into the positions they are inferred to have occupied during the time of Pangaea in order to obtain the banding pattern.

Other climatic indicators that are keys to Late Triassic–Early Jurassic paleoclimates have been found. First, continental rocks from this time interval are very commonly **red beds**, rocks of an orange or red color due to an abundance of iron oxide. Such rocks today generally form in climates that are relatively warm. Secondly, a variety of **paleosols** – that is, ancient soil profiles – have been found with **caliche** (calcium carbonate) nodules in them. Such nodules commonly form today in soils that are located in arid climates. Finally, evidence of warm, dry climatic conditions has been obtained from **stable isotopes**. Stable isotopes are simply those that do not spontaneously decay. It has been shown that the amount of a stable isotope of oxygen, ^{18}O, varies with temperature and salinity. Therefore, by measuring how much ^{18}O is present, one can obtain a direct measure of ancient temperatures and salinities. Obviously, this isotope has become an extremely important tool for learning about ancient climates as well as a variety of other subjects, including (as we shall see) warm-bloodedness in dinosaurs (see Box 2.1 and Chapter 14).

In summary, then, the Late Triassic–Early Jurassic interval is regarded as a time generally warmer, and perhaps drier, than the present, with strong seasonality.

MIDDLE AND LATE JURASSIC.   Although not many sedimentary rocks are preserved from the terrestrial Middle Jurassic, we can still obtain a relatively good idea about Middle and Late Jurassic climates because of a wealth of data available from oceanic sediments. Most importantly, the latter two-thirds of the Jurassic, as well as the entire Cretaceous, were apparently *without* polar ice or glaciers on the northern parts of the continents. The conclusion that there was no polar ice or glaciers is largely based upon the presence of warm-climate indicators at high latitudes and upon the absence of any evidence of continental glaciation from this time. Biological warm-climate indicators include plants and certain fish, whose

---

BOX 2.1

STABLE ISOTOPES, ANCIENT TEMPERATURES, AND DEAD OCEANS

IN 1947, HAROLD C. UREY, a geochemist at the University of Chicago, made an astounding and far-reaching prediction: that stable isotopes could be used to determine ancient temperatures. As it turns out, he had come upon one of the most important tools available to earth historians. Stable isotopes have proven invaluable in reconstructions of ancient oceans, climates, ecosystems, physiologies, and a host of other subjects. Like other scientific techniques, the principle is relatively simple; however, in practice, the methods are complex and require sophisticated techniques and equipment.

Unstable isotopes were discussed previously in this chapter. In contrast, stable isotopes do *not* spontaneously decay (hence the designation "stable"). Remembering what isotopes are (varieties of an element with the same atomic number but different atomic masses), it is not surprising to discover that many elements have several stable isotopes. For example, carbon has two *stable* isotopes, ^{12}C and ^{13}C, as well as the more familiar *unstable* isotope, ^{14}C.

Like unstable isotopes, stable isotopes occur naturally. In the case of oxygen, although 99.763% of all oxygen is ^{16}O, 0.191% of all oxygen is the stable isotope ^{18}O. In the case of carbon, 98.89% of all carbon is the stable isotope ^{12}C; 1.11% of all carbon is the other isotopes of carbon: ^{13}C and ^{14}C.

The critical point is that the minor weight differences in the isotopes cause slight differences in their behavior during chemical reactions or during natural physical processes such as evaporation, precipitation, or dissolution (the process of being dissolved in a solution). Such differences in behavior, called **fractionation**, control how the isotopes sort themselves as a chemical reaction or physical process takes place.

Urey was aware of the fractionation of stable isotopes and observed that, in the case of oxygen, fractionation varied with temperature. For example, in a situation in which calcium carbonate ($CaCO_3$) was precipitated from sea water, if all other variables (for example, concentration of salts in the solution) were held constant and *only* the

distributions are thought to have been as high as 75° N and 63° S. This would put them beyond the polar fronts, which in turn suggests that the poles must not have been as cold then as we find them today.

The absence of polar ice had an important consequence for climates: Water that would have been bound up in ice and glaciers was located in ocean basins. This in turn meant higher global sea levels than we presently experience, which led to extensive epeiric seas. Indeed, there is good evidence for high sea levels and extensive epeiric seas. The increased abundance of water on the continents, as well as in the ocean basins, had a stabilizing effect on temperatures (because it decreased continental effects) and would have tended to decrease the degree of seasonality experienced on the continents.

Continental climates are enormously variable, and as we have seen, short distances can encompass huge climatic differences. Indeed, there is some evidence for Late Jurassic aridity in the form of various evaporite deposits. Likewise, Late

---

temperature at which precipitation took place was varied, the ratio of $^{18}O/^{16}O$ would vary predictably. Thus, by studying that ratio, an estimate of the temperature of precipitation could be obtained. The miniscule amounts of stable isotope could be weighed on an instrument called a **mass spectrometer**, and a paleotemperature could be calculated.

In the past 30 years, a great deal of experimental research with stable isotopes and mass spectrometers has enabled scientists to predict temperatures of precipitation from $^{18}O/^{16}O$ ratios. The technique is of extreme interest, because, for example, stable oxygen isotopes in a $CaCO_3$ shell secreted by a clam that lived on the sea floor millions of years ago can theoretically be used to provide a clue to temperatures of the water in which the clam was living. Indeed, because clams grow shells throughout the warm seasons, it is possible in certain instances to deduce seasonal temperature fluctuations that occurred millions of years ago: The fossilized shell of the clam records through its stable isotope composition the ancient seasonal temperature fluctuations that it experienced during its life! The method is potentially applicable to any organism that secretes a $CaCO_3$ shell, as well as to vertebrates, which have stable isotopic oxygen incorporated in their bones in the form of phos-

phate. For example, if indeed dinosaurs were "warm-blooded," their stable isotopic oxygen ratios should show this. As you would expect, this subject is treated in greater detail in Chapter 14, in which dinosaur "warm-bloodedness" is discussed.

In the intervening years since the original stable oxygen isotope fractionation–temperature relationship was uncovered, stable isotopes have been put to a variety of uses. As noted in this chapter, fluctuations in $^{13}C/^{12}C$ have been used to record intervals of increased atmospheric CO_2. Also, they have been used to record **productivity** – the amount of biological activity in an ecosystem – by serving as an indicator of the amount of organic carbon moving through an ecosystem. The **flux** – or cycling – of organic carbon through an ecosystem is a measure of its activity. Thus, it was by studying the $^{13}C/^{12}C$ ratios from ocean sediments deposited at the very end of the Cretaceous that oceanographers discovered that the oceans went virtually dead at that time: Isotopic carbon recorded an astounding plunge in the flux of organic carbon, which was interpreted as a severe drop in the total productivity of the world's oceans. This apocalyptic event, termed a "Strangelove Ocean" after the postnuclear world of Dr. Strangelove, is covered in greater detail in Chapter 17, when the extinction of the dinosaurs is examined.

Jurassic terrestrial oxidized sediments and calcretes in North America suggest that there, to be sure, the Late Jurassic was marked by at least seasonally arid conditions.

In the main, however, our best evidence indicates that the Jurassic was a time of warm equable climates, with less seasonality than we now experience. It appears that this was in large part a function of high global sea levels and vast flooded areas on the continents.

CRETACEOUS Paleoclimates in the Cretaceous are somewhat better understood than those of the preceding periods. It seems clear that, during the first half of the Cretaceous at least, global temperatures remained warm and equable. The poles continued to be ice free, and the first half of the Cretaceous generally saw far less seasonality than we see today. This means that although equatorial temperatures were approximately equivalent to modern ones, the temperatures at the poles were somewhat warmer. Temperatures at the Cretaceous poles have been estimated at 0°–15°C, which means that the temperature difference between the poles and the equator was only between 17°C and 26°C, considerably less than the 41°C spread that we experience at present.

It is probable that more than one culprit bears the responsibility for this climate. The Cretaceous was a time of increased global tectonic activity and associated high volcanic activity. As we have seen, an increase in tectonic activity is commonly associated with increased rates of oceanic spreading, which in turn involves elevated spreading ridges. Raised spreading ridges would have displaced more oceanic water onto the continents, and indeed, there is good evidence for extensive epeiric seas during Cretaceous times. Moreover, there was an increase in CO_2 gas in the atmosphere during Cretaceous times, a fact that has been established using ^{13}C, a stable isotope of carbon. The amount of CO_2 in the atmosphere can be correlated with the amount of ^{13}C isotope present in carbon-bearing material of Cretaceous age. The CO_2 has been attributed to volcanism, itself related to increased tectonic activity. Increased amounts of CO_2 in the Cretaceous atmosphere meant that the Cretaceous atmosphere tended to absorb more heat (long-wavelength radiation from the earth), generally warming global climates.[6]

So the first half of the Cretaceous was almost synergistic: Tectonism caused an increase in atmospheric CO_2 and a decrease in the volume of the ocean basins, which in turn increased the area covered by epeiric seas. The seas thus stabilized climates already warmed by enhanced absorption of heat in the atmosphere. Indeed, Cretaceous levels of mean global absorbed radiation are thought to have been 2.3% greater than today's. This means that the Cretaceous earth, because of its "greenhouse" atmosphere and abundance of water, retained 2.3% more heat than does the modern earth. And, because of the extensive water masses, heat was not so quickly released during cold seasons; indeed, the seasons themselves were modified. Tropical and subtropical climates have been reconstructed for latitudes as high as 70° S and 45° N.

[6]This situation is occurring today and is termed the "greenhouse effect." The increase in CO_2 in the modern atmosphere (and consequent global warming), however, is believed attributable to anthropogenic (human-originated) combustion of all types, especially automobile exhausts, and not volcanism. Study of Cretaceous climates shows that the earth has *already* undergone an experimental flirtation with greenhouse conditions, and the point has been made that the Cretaceous will help provide insights into how the modern earth will respond to such conditions.

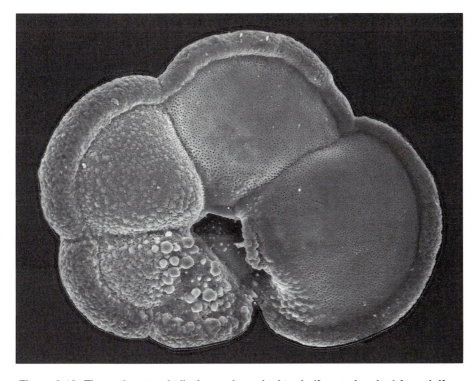

Figure 2.16. **The carbonate shell of a modern planktonic (free-swimming) foraminifer,** ***Globorotalia menardii.*** The long dimension is 0.750 mm. *Photograph courtesy of S. L. D'Hondt.*

The last 30 million years of the Cretaceous produced a deterioration of these equable conditions of the mid-Cretaceous. A pronounced withdrawal of the seas took place, and evidence exists of increased seasonality. Stable isotopes again provide important evidence of greater fluctuations in temperatures; however, this time they are aided by information from an unexpected source: the patterns of leaf margins of Late Cretaceous foliage, as well as the patterns of veins in their leaves. In the modern world, such patterns can be closely correlated with temperature and moisture. Once this indicator was "calibrated" in the present – that is, once the patterns were correlated with modern temperature and moisture levels – leaf margin and venation patterns could be used to infer previous temperatures and amounts of moisture.

Another important indicator of temperature is single-celled, shell-bearing foraminifera (Figure 2.16). The shapes of the shells of foraminifera can be correlated with a relatively narrow range of temperatures, and thus ancient representatives of the group can provide an indication of water temperatures in the past. Foraminifera serve double duty, however; because their shells are made of calcium carbonate, the shells are suitable for stable carbon and stable oxygen isotopic analyses.[7]

[7] Free-swimming (planktonic) foraminifera first made an appearance in Cretaceous oceans. Because they were (and are) geographically widespread but have relatively short species' durations, they are also superb biostratigraphic indicators for late Mesozoic and Cenozoic marine sediments.

The Cretaceous, therefore, was a world much different from our own. In its first half, warm, equable climates dominated the period. The second half, however, was marked by well-documented climatic changes, in which seasonality was increased, and in which the equator-to-pole temperature gradient became more like that which we now experience.

Important Readings

Arthur, M. A., T. F. Anderson, I. R. Kaplan, J. Velzer, and L. S. Land. 1983. Stable Isotopes in Sedimentary Geology: SEPM Short Course no. 10, 432 pp.

Barron, E. J. 1983. A warm, equable Cretaceous: The nature of the problem. Earth Science Reviews 19:305–338.

Barron, E. J 1987. Cretaceous plate tectonic reconstructions. Palaeogeography, Palaeoclimatology, Palaeoecology 59:3–29.

Berry, W. B. N. 1987. Growth of a Prehistoric Time Scale Based on Organic Evolution. Blackwell Scientific Publications, Boston, 202 pp.

Crowley, T. J., and G. R. North. 1991. Paleoclimatology. Oxford Monographs on Geology and Geophysics no. 18. Oxford University Press, Oxford, 339 pp.

Dott, R. H., Jr., and R. L. Batten. 1988. Evolution of the Earth. McGraw–Hill, New York, 120 pp.

Frakes, L. A. 1979. Climates through Geologic Time. Elsevier Scientific Publishing Company, New York, 310 pp.

Faure, G. 1991. Principles and Applications of Inorganic Geochemistry. Macmillan, New York, 626 pp.

Frazier, W. J., and D. R. Schwimmer. 1987. Regional Stratigraphy of North America. Plenum Press, New York, 719 pp.

Harland, W. B., R. L. Armstrong, A. V. Cox, L. E. Craig, A. G. Smith, and D. G. Smith. 1989. A Geological Time Scale. Cambridge University Press, Cambridge, 262 pp.

Lillegraven, J. A., M. J. Kraus, and T. M. Bown. 1979. Paleogeography of the world of the Mesozoic; pp. 277–308 in J. A. Lillegraven, Z. Kielan-Jaworoska, and W. A. Clemens, Jr. (eds.), Mesozoic Mammals, the First Two-Thirds of Mammalian History. University of California Press, Berkeley, Calif.

Lutgens, F. K., and E. J. Tarbuck. 1989. The Atmosphere: An Introduction to Meteorology. Prentice-Hall, Englewood Cliffs, N.J., 491 pp.

Robinson, P. L. 1973. Palaeoclimatology and continental drift; pp. 449–474 in D. H. Tarling and S. K. Runcorn (eds.), Implications of Continental Drift to the Earth Sciences, Vol. 1. NATO Advanced Study Institute, Academic Press, New York.

Ross, M. I. 1992. Paleogeographic Information System/Mac Version 1.3: Paleomap Project Progress Report no. 9. University of Texas at Arlington, 32 pp.

Walker, R. G., and N. P. James. 1992. Facies Models: Response to Sea Level Change. Geological Association of Canada, St. John's, Newfoundland, 409 pp.

Wilson, J. T., ed. 1970. Continents Adrift: Readings from Scientific American. W. H. Freeman Company, San Francisco, 172 pp.

CHAPTER 3

DISCOVERING ORDER IN THE NATURAL WORLD

AN IMPORTANT GOAL OF SCIENCE is to recognize order in the natural world. Understanding that order is a first step toward identifying processes that govern our world. In this chapter, we will attempt to identify patterns in the natural world and will learn rigorous methods by which those patterns are recognized. Our interest here will be focused on the **biota**, which may be considered the sum total of all living organisms that have populated the earth. There are many different kinds of organisms; our question is, is there order in all this? That is, are there patterns to be found in the apparently infinite diversity of life? And can these patterns tell us something that takes us closer to understanding the processes that account for that diversity?

HIERARCHY

As we look around us, we can see consistent patterns that contribute to natural order in the world. To cite just two obvious examples, all plants with flowers have leaves, and all birds have feathers. Indeed, the correlation between birds and feathers is so consistent that we might go so far as to identify a bird as such *because* it has feathers. Going further, can we discover patterns of organization among all organisms?

Perhaps the most significant pattern pertaining to all living organisms is the fact that their attributes – that is, all their features, from eyes, to hair, to chromosomes, to bones – can be organized into a **hierarchy**. Hierarchy refers to the rank

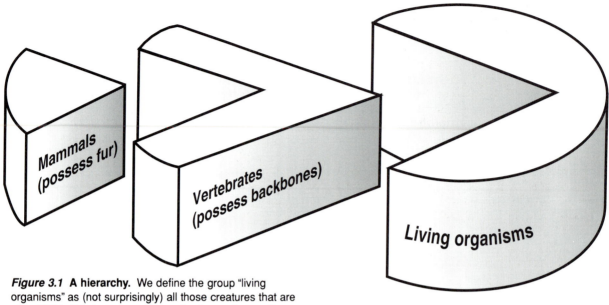

Figure 3.1 **A hierarchy.** We define the group "living organisms" as (not surprisingly) all those creatures that are alive. Within this large group is a subset: creatures with backbones. Within the subset of vertebrates are mammals, identified by the possession of fur.

or order of features. Indeed, the hierarchical distribution of features within the biota is its most fundamental property.

If we take the group that includes all living organisms, one subset possesses a backbone. We call this subset "vertebrates." Within vertebrates, some possess fur, whereas most do not. All members of the group that possesses fur are called "mammals." We choose features that will define smaller and smaller groups within larger groups. The arrangement is hierarchical, because those creatures possessing fur are a subset of all animals possessing a backbone, which are in turn a subset of all living organisms (Figure 3.1). Although so far we have been discussing only backbones and fur, *all* features of living organisms, from the possession of DNA – which is ubiquitous – to highly restricted features such as the possession of a brain capable of producing a written record of culture, can be arranged hierarchically.

Although life is commonly referred to as infinitely diverse (indeed, we earlier used such a phrase), the concept is not accurate; life really is *not* infinitely diverse. Rather, it is connected by the sharing of features in a hierarchy. Life's diversity actually takes the form of many variations on a primitive body plan. Always, however, unmodified or only slightly modified vestiges of the original plan remain, and these provide the keys to revealing the fundamental hierarchical interrelationships of the biota.

Characters

To recognize the hierarchy, we must identify features of organisms. These are termed **characters**. A character is an isolated or abstracted characteristic of an organism. The character acquires its meaning not when it appears as a single feature on a particular organism, but when its *distribution* among a selected group of organisms is considered. For example, suppose that we are interested in the character "possession of claws." A single individual can possess claws, but it is the *distribution* of claws among different organisms that is interesting and informative. The group Felidae – cats – is generally linked on the basis of the possession of claws that retract. Thus, not only is that household hairball Garfield a felid, but also *all* cats – including bobcats, lions, jaguars, and the terror of the tar pits, the saber-toothed tiger – are felids. And by the same token, if someone tells us that a given mammal is a felid, we could expect it to have retractile claws.

Because characters are distributed hierarchically, their *position* in the hierarchy is obviously of great significance. Consider the simple example of feathers on birds. Since all birds have feathers, it follows that if one wanted to tell a bird from a nonbird (any other organism), he or she need only observe that the bird is the one that has the feathers. On the other hand, the character "possession of feathers" is not useful for distinguishing, say, an eagle from a duck; both have feathers. To distinguish one bird from another, characters other than feathers must be used to identify subsets within the group that includes all birds.

These distinctions are extremely important in establishing the hierarchy appropriately, and for this reason, characters may function in two ways: as **general** characters and as **specific** characters. Simply put, a character is specific when it characterizes (or is diagnostic of) all members of a group, whereas a character is general when it diagnoses a larger group. A character may be specific in one group, but general in a smaller subset of that group (because it is now being applied at a different position in the hierarchy).

Suppose as a description to help you find someone you had never met, you were told, "He has two eyes." This would obviously be of little help, since all humans have two eyes. The character of possession of two eyes is a general character among humans. Indeed, it is a general character among all vertebrates and, alone, the character of two eyes would not distinguish a person from a frog or a guppy. But at an even more general level in the hierarchy, the character would be specific: Possession of two eyes would distinguish a vertebrate (two eyes) from an earthworm (no eyes) or a spider (four eyes). Likewise, consider again the example of feathers in birds. The presence of feathers is a specific character when birds and nonbirds are considered together, because the presence of feathers characterizes birds, and it is specific to birds as a whole.

Cladograms

We commonly use a **cladogram** to visualize the hierarchies of characters in the biota. A cladogram (pronounced clay-doh-gram; *clados* – branch; *gramma* – letter) is a branching diagram that is used to depict the hierarchical distribution of shared characters.

To understand how a cladogram is constructed, we begin with two things to group – say a car and a pickup truck. Notice that anything can be grouped; it does not necessarily apply only to living (or to extinct) organisms. Cars and trucks may be linked by any number of characters. The characters must, of course, be observable features of each. For example, "used for hauling lumber" could not be used,

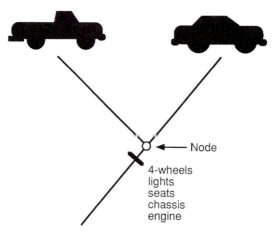

← Node

4-wheels
lights
seats
chassis
engine

Figure 3.2 A cladogram. The car and pickup are linked by the characters listed at the hatch mark, just below the node. The node itself defines the things to be united; commonly a name is attached to the node that designates the group. Here, such a name might be "four-wheeled vehicles."

because hauling lumber is a *function*, not a character. A cladogram of a car and a truck is shown in Figure 3.2.

For characters, we choose the presence of (1) four wheels, (2) lights, (3) seats, (4) a chassis, and (5) an engine. The cladogram simply connects these two separate objects (the car and the pickup truck) based upon the characters that they *share*. The features are identified (and itemized) on the cladogram adjacent to the **node**, which is a bifurcation (or two-way splitting) point in the diagram. Figure 3.2 shows this relationship.

The issue becomes more complicated (and more interesting) when a third item is added to the group (Figure 3.3). Consider a motorcycle. Now, for the first time, because no two of the three items are identical, two will share more in common with each other than either does with the third. It is in this step that the hierarchy is established. The group that contains all three vehicles is diagnosed by certain features shared by all three. Now, however, we have also established a subset containing two vehicles. Because the two are linked together on the cladogram, not only do they share the characters that link all three, but above and beyond these they share further characters that link them to the exclusion of the third vehicle. Characters such as lights and seats would be general when discussing the subset composed of two vehicles, since those characters are diagnostic only at a more general level in the hierarchy.

How these characters, and even the vehicles, are arranged on the cladogram is controlled by the *choice* of characters. Suppose that instead of "seats" we had specified bucket seats, and instead of "four wheels" we had simply specified wheels. Bucket seats would then no longer be a general character defining all vehicles but instead would unite only the car and the truck. Likewise, the presence of wheels would be a general condition pertaining to all three, instead of uniting trucks and cars. The choice of characters and how we express them have a lot to do with the cladograms that result from our analysis.

Now suppose that instead of the characters that we listed earlier, we had chosen wheels, engine, lights, seat, and non–passenger space < passenger space. These char-

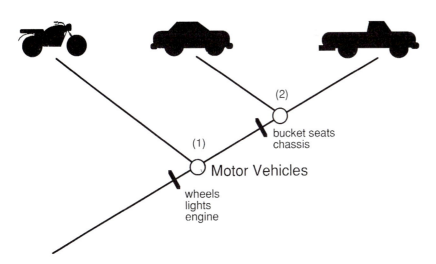

Figure 3.3 **One possible distribution of three motor vehicles.** Members of the group designated by node 1 are united by the possession of wheels, lights, and an engine; that group could be called Motor Vehicles. Within the group Motor Vehicles is a subset united by possession of bucket seats and a chassis. That subset is designated at node 2.

acters produce a quite different cladogram. The new cladogram, Figure 3.4, shows cars and motorcycles linked more closely to each other than to a pickup truck. This arrangement is counterintuitive and contradicts the cladogram in Figure 3.3.

How do we choose? Most of the characters one might think of tend to support the cladogram in Figure 3.3 and not that in Figure 3.4; they suggest that in its design, a car shares much more in common with a pickup truck than it does with a motorcycle. And, those features that it shares with a motorcycle generally are also present in the pickup truck; they are general features for all three, rather

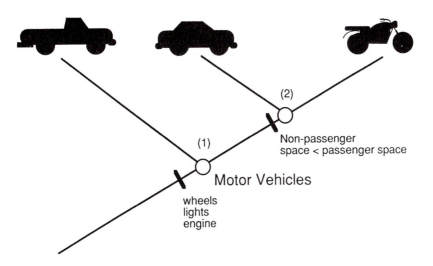

Figure 3.4 **An alternative distribution of three motor vehicles.** The characters selected suggest that the car and motorcycle share more in common with each other than either does with the truck.

than specific features that clearly diagnose a car–motorcycle subset within the three vehicles.

So far, we have presented the cladogram only as a graphic method of showing hierarchies. In fact, cladograms are powerful tools for studying the evolutionary relationships among organisms. As we shall see in the next chapter, their use in the past 15 or so years has completely revolutionized our understanding of the interrelationships of organisms and has led to some remarkable changes in how scientists view the biota.[1] Terms as fundamental as "reptile," "dinosaur," and "bird" have come to have startling new meanings, as a result of **cladistic analysis**, or analysis using a cladogram. An important goal of this book and of our study of dinosaurs is to acquire a strong grasp of these new approaches.

ORGANIC EVOLUTION

Before we can understand the great events and rhythms that have occurred through the 3.5 billion years of biotic history, we need a way to discover the pattern of descent of the earth's creatures. Considered individually, fossil and Recent organisms are a myriad of apparently disconnected data points in time, but considered as evolving groups (lineages), striking patterns emerge that enrich our understanding of ourselves and the world around us. Accordingly, we are interested in **phylogeny**, the history of the descent of organisms.

It is in this respect that cladistic analysis makes a fundamental, profound contribution. Using character hierarchies portrayed on cladograms, we can establish **clades** or **monophyletic groups** (to add to the nomenclature, these are sometimes termed **natural groups** as well; these terms are all synonymous). These are groups of organisms that have evolutionary significance because the members of each group are more closely related to each other than they are to any other creature. For example, it is probably no surprise that humans are a monophyletic group: All members of *Homo sapiens* are more closely related to each other than they are to anything else. The idea that a group is monophyletic has a second, more subtle ramification: It implies that all members of that group share a more recent common ancestor with each other than with any other organism.

Evolution

ORGANIC EVOLUTION IS A FACT.   By this, we mean that if one accepts that the human mind, with its strengths and limitations, is capable of understanding aspects of the natural world and that scientific methods are appropriate tools for that type of inquiry, the biota has undergone **evolution**. In the biological context, evolution refers to *descent with modification*: Organisms have changed and modified through time, and each new generation is the most recent bearer of the genetic thread that

---

[1]Cladistic methods were first developed by an entymologist, Willi Hennig, in a book titled *Grundzüge einer Theorie der phylogenetischen Systematik*. Hennig's work had a minor impact on European biologists, but it was not until the 1966 publication of an English translation of a revised version of the 1950 work (entitled simply *Phylogenetic Systematics*) that cladistic methods became relatively well known. During the late 1960s and throughout much of the 1970s and early 1980s, it became a kind of *cause célèbre* (see Box 4.3). The real strengths of the method eventually triumphed, and today, virtually all phylogenetic reconstruction is done by means of cladograms.

connects life. In this sense, each new generation is forward looking, in that its members potentially contain changes relevant for the future, and is connected to the past by features that its members retain from their heritage.

That evolution has occurred is not a particularly new idea; it was articulated by a variety of Enlightenment and post-Enlightenment philosophers and natural historians.[2] The unique contribution of Charles Darwin and Alfred Russel Wallace (who jointly presented their ideas at an 1858 meeting of the Linnaean Society of London) was that the driving force behind evolution is natural selection. Here, however, we are most concerned with the *record* of evolution – an observable pattern of descent with modification – regardless of the processes (such as natural selection) that may be responsible for it. It is important, however, to decouple mentally the idea of evolution (itself fundamentally a pattern) from the idea of natural selection (an important process driving the pattern), because our efforts will largely be directed to the pattern and not to processes such as natural selection.

Evolution amounts to modifications (in morphology, in genetic make-up, in behavior, etc.), so that although some changes are developed in the descendants, many of the ancestral features are retained. Implicit in this are the relationships between anatomical structures. For example, we postulate a relationship between the five "fingers" in the human "hand" and the five "toes" in the front "foot" of a lizard. Here, the English language is confusing; we are really talking about the digits of the forelimbs, a particular feature that happens to have been conserved (or maintained) through time in these two lineages (humans and lizards). The digits on the forelimbs of lizards and humans can be traced back in time to digits on the forelimb of the common ancestor of humans and lizards. We call these anatomical structures **homologues**, and two anatomical structures are said to be **homologous** when they can, by deduction from their distributional patterns, be traced back to a single original structure in a common ancestor (Figure 3.5). Thus, it is reasonable to suppose that the digits of the forelimbs of all mammals are homologous with those of, for example, dinosaurs. That is because these digits can be traced back to the digits in the forelimbs of the common vertebrate ancestor of mammals and dinosaurs. The wings of a fly, however, are not homologous with those of a bird, because both pairs of wings cannot be traced to a single structure on a common ancestor. Because the wings of a fly and the wings of a bird perform in similar fashion (they allow flight to take place), they are considered **analogues** and are said to be **analogous** (Figure 3.6). Obviously, recognition of evolution necessitates the recognition of homologous anatomical structures.

An obvious yet often ignored clue to the fact that evolution has taken place is the hierarchical distribution of characters in nature. If descent with modification has taken place, what patterns of character distributions might one expect to find? Modification of ancestral body plans through time would produce the distribution of characters that we observe: a hierarchical arrangement in which some homologous characters, or groups of characters, are present in all organisms, in which some characters are found in somewhat smaller groups, and in which some characters have a very restricted distribution and are found in only a few organisms.

---

[2]Although scientific debate about the so-called "theory" of evolution is seized upon by creationists as in some way undermining the fact of evolution, the discussion among scientists is really about the relative importance of underlying causal mechanisms and rates of the evolutionary process, rather than whether evolution occurs.

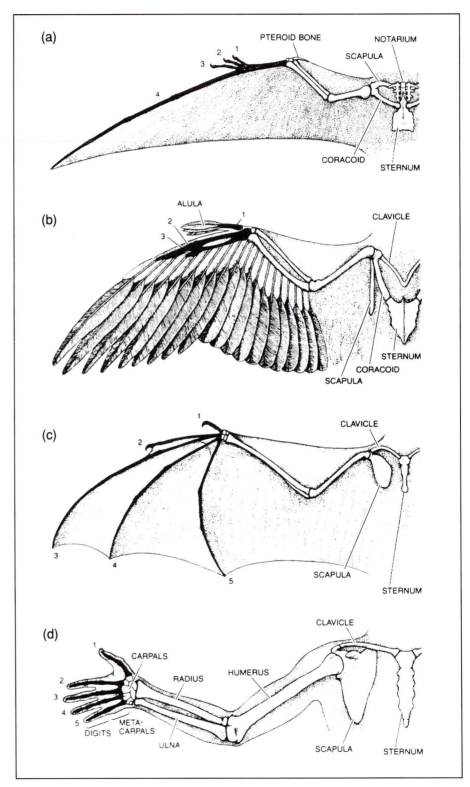

Figure 3.5 **Homologues.** Homologues are anatomical structures that can be traced back to a single structure in a common ancestor. The front limbs of all tetrapods – including pterosaurs, birds, bats, and humans – are all homologous and retain the same basic structure and bone relationships even though the appearance of these forelimbs may be outwardly different. Homology forms the basis for hypotheses of evolutionary relationships. Source: W. Langston, Jr., 1981,"Pterosaurs," *Scientific American,* 244:122–136.

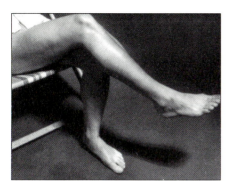

human being
Homo sapiens

desert locust
Schistocerca gregaria

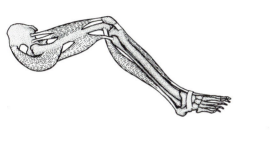

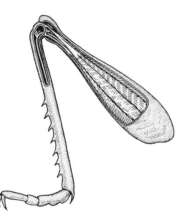

Figure 3.6 **Analogues.** Analogues may perform similar functions, and even look outwardly similar, but internally they can be very different. Here a human leg is contrasted with that of a locust. Although both are legs, the two structures are different. Human muscles are on the outside of the skeleton, whereas locusts' muscles are on the inside of their skeletons.

Cladograms and the Reconstruction of Phylogeny

Accepting the fact of evolution, there must be a single phylogeny – a single genealogy – that documents the interrelatedness or connectedness of all life. This is not an unfamiliar concept, because most of us have seen "trees" that purport to document who came when and from whom. Such "trees of life" are common in textbooks and museum displays and deeply influence most people's ideas about the pattern of evolution (Figure 3.7). These trees commonly show an ancestral fossil creature rising out of the primordial sludge and giving rise to a host of other fossil and living creatures. But how does one make a tree of life? How do we figure out who gave rise to whom? After all, no human was present to observe the appearance of the first dinosaur on the face of the earth. And how do we know that, with the fossil record as incomplete as it is, one fossil that we happen to find just fortuitously turns out to be the very ancestor of some other fossils that we happened to find? Because of their rarity, the chance of that occurring, especially among vertebrates, is vanishingly small. In this book, therefore, we avoid trees of life and instead use cladistic analysis to reconstruct evolutionary patterns.

Cladistic analysis is a tool particularly suited to reconstructing phylogeny. Because cladograms are hierarchical branching diagrams, they allow us to depict shared morphology. Moreover, a cladogram is testable in a way that a tree can

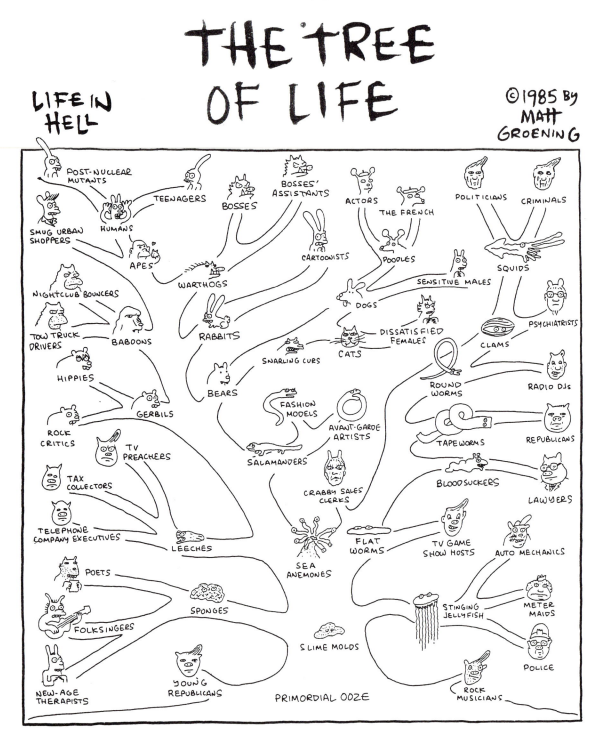

Figure 3.7 A tree of life. This particular one is a satire. The image of evolution as a tree, however, is completely familiar. From the *Big Book of Hell,* copyright 1990 by Matt Groening. All rights reserved. *Reprinted by permission of Pantheon Books, a division of Random House, Inc., New York.*

never be. We can never be sure that we can recognize the exact ancestor of a particular group, nor can we ever be sure that once we have found a particular fossil, we can identify its exact descendants. How could one ever demonstrate such a thing? On the other hand, a cladogram specifies particular characters, which either are, or are not, present in the organisms being grouped.

To reconstruct phylogeny, we need a way to recognize how closely two creatures are related. How this is done is superficially very simple: Things that are closely related share specific features. We have seen how the word "specific" is applied to characters that uniquely diagnose (identify) a particular group. Now, in the context of phylogeny, we observe that closely related organisms tend to share unique features. This is not startling; we have seen the results of breeding, in which the offspring look, and sometimes act, very much like their parents.

Cladograms were initially presented here without their being placed within an evolutionary context. Considered in an evolutionary context, the specific characters that we said characterize groups can be treated as homologous among the groups that they link. Feathers in birds again provide a good example. If all birds have feathers (and birds are monophyletic), the implication is that the feathers found in eagles and those found in ducks can in fact be traced back to feathers that must have been present in the most recent common ancestor of eagles and ducks. This is in fact putting the cart before the horse, however. It is the distribution of characters that helps us determine which groups are monophyletic and which are not, and in the case of birds, the conclusion that they are monophyletic is in part based upon the fact that birds all share the distinctively specific character of feathers (among other characters). In an evolutionary context, specific characters are termed **derived**, and general characters are termed **ancestral**.[3] Only derived characters provide evidence of natural (monophyletic) groups, because, as newly evolved features, they are potentially transferable from the first organism to acquire them to all its descendants. Ancestral characters – those with a much more ancient history – provide no such evidence of unique natural relationships. To illustrate this, we resort yet again to birds and feathers. Feathers are among the shared, derived characters that unite the birds as a monophyletic group. On a cladogram, therefore, we look for characters that unite a bifurcation point in the diagram. As befits the hierarchy in nature, the cladogram documents monophyletic groups within monophyletic groups. In Figure 3.8, a small part of the hierarchy is shown: Humans (a monophyletic group, possessing shared, derived characters) are nested within the mammals (another monophyletic group possessing shared, derived characters). Notice that the character of warm-bloodedness is ancestral for *Homo sapiens*, but derived for mammals. As we have seen, features can be derived or ancestral, all depending upon what part of the hierarchy one is investigating.

The cladogram need not depict every organism within a monophyletic group. If we are talking about humans and carnivores, we can put them on a cladogram and show the derived characters that diagnose them, but we might (or might not)

---

[3] Occasionally the term *advanced* is used instead of *derived*, and *primitive* is used instead of *ancestral* or *general*. The terms *primitive* and *advanced* might be easily misunderstood, because here, *primitive* certainly does not mean worse or inferior, and *advanced* certainly does not mean better or superior. In an evolutionary sense, these simply refer to the timing of evolutionary change; advanced characters evolved later than primitive characters.

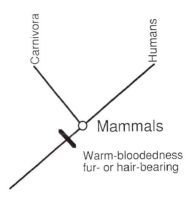

***Figure 3.8* A cladogram showing humans within the larger group, Mammalia**. Mammalia is diagnosed by warm-bloodedness and possession of fur (or hair); many other characters unite the group as well. Carnivora, a group of mammals that includes bears and dogs (among others) is shown to complete the cladogram. Carnivores all uniquely share a special tooth, the carnassial, and humans all uniquely share various gracile features of the skeleton. Note that all mammals (including humans and carnivores) are warm-blooded and have fur (or hair), but only humans have the gracile skeletal features, and only members of Carnivora have the carnassial tooth.

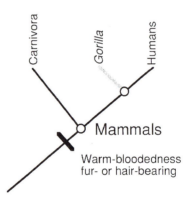

***Figure 3.9* Addition of the genus *Gorilla*.** The addition of gorillas to the cladogram does not alter the basic relationships outlined on the cladogram.

include other mammals (for example, a gorilla). Nonetheless, if the hierarchical relationships that we have established are correct, the addition of other organisms into the cladogram should not alter its basic hierarchical arrangement. Figure 3.9 shows the addition of one other group into the cladogram from Figure 3.8. The basic relationships established in Figure 3.8 still obtain even with the new organisms added.

Parsimony

From the foregoing, it should be apparent that, put in an evolutionary context, a cladogram becomes a **hypothesis of relationship** – that is, a hypothesis about how closely (or distantly) organisms are related. With a given set of characters, it

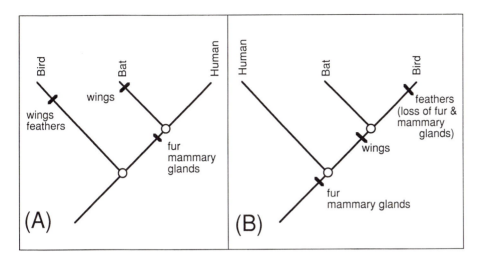

Figure 3.10 Two possible arrangements for the relationships of birds, bats, and humans. (A) requires wings to have evolved two times; (B) requires birds to have lost fur and mammary glands. These, as well as many other characters, suggest that (A) is the more parsimonious of the two cladograms.

may be possible to construct several possible cladograms (as we saw that there were in the example of the pickup truck, the car, and the motorcycle). We can distinguish among different hypotheses of relationship using the principle of **parsimony**. Parsimony, a sophisticated philosophic concept first articulated by the 14th-century English theologian William of Ockham, dictates in this case that the *simplest* cladogram is the point of departure from which further consideration of the hypothesis can occur. There is no law that states that natural systems must behave simply; indeed, one may reasonably expect complexity of them. Still, why presume a priori more steps if fewer can provide the same information? Methodologically, therefore, one begins with the simplest hypothesis and considers it in the context of new or independent evidence.

A bird, a human, and a bat will serve as a simplified example. We start with the following characters: wings, fur, feathers, and mammary glands. Figure 3.10 shows two cladograms that are possible from these animals and their characters. In the one in which the bird is most closely linked with the bat, the bird has to lose mammary glands and it has to lose fur. In the cladogram in which the bat and the human share a most recent common ancestor, wings must be invented by evolution two times. The cladogram in which the human and bat are linked to each other more closely than either is to the bird, is the simpler of the two; that is, it requires fewer evolutionary events or steps. The cladogram linking the human and bat together remains uncomplicated by the addition of more characters and, crucially, we have no independent evidence suggesting that this would be unlikely; however, the addition of virtually all other characters that pertain to the creatures in question (e.g., the arrangement, shape, and number of bones, particularly those in the skull and wings; the structure of the teeth; the biochemistry of each organism, etc.) only further complicates the cladogram that links birds and bats. Based upon parsimony, therefore, the cladogram that shows bats and humans to have more in common with each other than either does with a bird, is preferred. And indeed, as a statement about the *evolution* of these vertebrates, it is extremely likely that bats and humans share a more recent common ancestor with each other than either does with a bird (which, obviously, is why they are classified together as mammals). In this case, the use of shared, derived characters has led us to the most parsimonious working hypothesis with regard to the evolution of these three creatures.

BOX 3.1

THE EVOLUTION OF WRIST WATCHES

DO "WRIST WATCHES" HAVE a common heritage beyond merely postdating a sundial? In this example we can use a cladogram to infer the evolutionary relationships of the group of things we call "wrist watches."

Consider three types of wrist watches: a wind-up watch, a digital watch, and a watch with a quartz movement. Six cladograms are possible for these instruments, but it can be seen that by the definition of a cladogram, (a) and (b) for each type are identical. This is because the groups at a node share the characters listed at that node, regardless of order. For this reason, we really have only three cladograms to consider.

One might at first wish to place the digital watch in the smallest subset, in the most derived position (as in Types I and II), since it is the most modern, technologically advanced, and sophisticated of the three. Remember, however, how the cladogram is established: on the basis of *shared, derived* (or a*dvanced*) characters. Cladograms Types I and II say that the digital watch shares the most characters with either a wind-up watch (Type I) or a quartz watch (Type II). A look at the characters themselves suggests that this is not correct: Wind-up and quartz watches are both analogue watches (have a dial with moving, mechanical hands), and their internal mechanisms consist of complex gears and cogs to drive the hands at an appropriate speed. The digital watch, on the other hand, consists of microcircuitry and

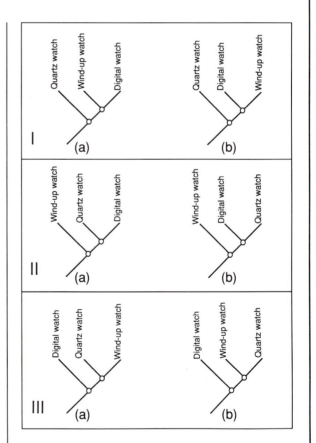

Figure Box 3.1A Six possible arrangements of three timepieces on cladograms. Note that each pair is redundant; the order in which the objects on each "V" are presented is irrelevant. Each pair is said to be "commutatively equivalent."

Science and Testing Hypotheses

The tabloid literature contains headlines such as "Scientists discover two-headed twins from alien mother." The implication of this is that somehow the claim of a two-headed twin from an alien mother has credibility if it comes from a scientist, as opposed to, say, your neighbor. Students often preface questions in general science classes with remarks such as "Well, I'm not very good in science, but . . . ." Both of these kinds of statements seem to imply that science is something special

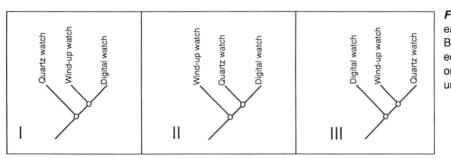

Figure Box 3.1B Because each pair of cladograms in Box 3.1A is commutatively equivalent, there are really only three cladograms under consideration.

a microchip, with essentially no moving parts. It is apparently something very different, and from its characters, bears little relationship to the other "watches."

What *is* the digital watch? In an evolutionary sense, it is a really a computer masquerading (or functioning) as a timepiece. In our hypothesis of relationship, the watchbands and cases of watches have evolved independently two times (once in computers and once in watches), rather than the guts of the instrument, itself, having evolved two times. That the watchbands and cases evolved independently two times is a more parsimonious hypothesis than arguing that the distinctive and complex internal mechanisms (themselves consisting of many hundreds of characters) of the watches evolved independently two times.

What, then, is a watch? If the term "watch" includes digital watches as well as the other two more conventional varieties, then it should also include computers, since a digital watch has the shared, derived (advanced) characters of computers. The cladogram suggests that the term "watch" does not describe an evolutionarily meaningful (monophyletic) group, in the sense that a

cladogram that includes digital watches, wind-up watches, and quartz watches must also include computers, as well as a variety of more conventional mechanical timing devices (such as stopwatches). Rather, the term "watch" may be thought of as some other kind of category: It describes a particular function (timekeeping) in conjunction with size (relatively small).

In this example, we are fortunate in that should we so choose, we can check the cladogram-based conclusions by studying the historical record to find out about the evolution of wrist watches, digital watches, and quartz watches. Obviously this is not possible to do with the record of the biota, because there is no written or historical record with which to compare our results. The characters of each new fossil find, however, can be added to existing cladograms and a hypothesis of relationship selected according to the principle of parsimony. In our discussions of the biota, we attempt to establish categories that are evolutionarily significant (monophyletic groups) and to avoid groups that have less in common with each other than with anything else.

and perhaps more significant than other realms of human inquiry.

Although science may be thought of as many things, at its base it is only an approach to certain kinds of issues rooted in a particular type of logic. In this sense it is nothing more than a tool for solving a restricted series of problems. Although science can be a problem-solving tool, there is in fact a variety of potentially significant problems that are not particularly amenable to a scientific solution. Examples of such questions might be, Is there a God?; Does she love me?; Why don't I like hairy men?; and, Is this beautiful music? It is not by means of science that the answers are discovered.

Other questions, however, are more amenable to scientific inquiry. For example, a scientific hypothesis (although a terribly simplistic one) is, The sun will rise tomorrow. This statement can be thought of as a hypothesis with specific predictions. The hypothesis (that the sun will rise tomorrow) is **testable**; that is, it makes a prediction that can be assessed. The test is straightforward: We wait until tomorrow morning and either the sun rises or it does not. If the sun does not rise, the statement has been *falsified*, and the hypothesis can be rejected. If the sun does rise, the statement has *not been falsified*, and the hypothesis cannot be rejected. For a variety of sophisticated philosophical reasons, scientists do not usually claim that they have *proven* the statement to be true; rather, the statement has simply been tested and not falsified. *One of the basic tenets of science is that it consists of hypotheses that have predictions that can be tested.* We will see many examples of hypotheses in the coming chapters; all must involve predictions that can be tested. Our ability to test these will determine the importance of these hypotheses as scientific contributions.

As previously noted, in an evolutionary context, cladograms can be considered hypotheses of phylogenetic relationship. Using hierarchical distributions of characters, we can construct hypotheses of the relationships among the various organisms, both living and extinct. There are obviously no proofs, for how can one "prove" that a lion is more closely related to a tiger than to a wolf? Since no one was around to observe the many speciation events that took place in the evolution of these three forms, "proof" of their relationships is impossible. At best, we can *test hypotheses* of their genealogical connections using shared characters. The cladograms that we construct are said to be most **robust**, that is, strongest, if they survive falsification attempts. The cladogram that is most resistant to falsification embodies the most robust hypothesis of relationships.

We have remarked that a cladogram is testable in a way that a tree of life is not. It is now clear how a cladogram can be tested. The addition of characters can cause the rejection of a given cladogram by demonstrating that it is not the most parsimonious character distribution. The tree of life, however, presents more difficulties. How can we ever really be sure who gave rise to whom? Can we really be sure that some organism, which lived 345 million years ago and just happened to be fossilized, was the real ancestor of some other organism, which lived 344 million years ago and also just happened to be fossilized? Aside from being an improbable coincidence of preservation, it is not easily testable. A tree of life, although perhaps initially more satisfying than a cladogram, is thus really more of a story, or **scenario**, than a testable, scientific hypothesis. For this reason, in this text we shall have to content ourselves with cladograms, and we must not confuse them with trees of life. As will become evident, much can be learned from cladograms that will contribute to our desire to know what occurred in ages long past.

Important Readings

Cracraft, J., and N. Eldredge, eds. 1981. Phylogenetic Analysis and Paleontology. Columbia University Press, New York, 233 pp.

Eldredge, N., and J. Cracraft. 1980. Phylogenetic Patterns and the Evolutionary Process: Method and Theory in Comparative Biology. Columbia University Press, New York, 349 pp.

Funk, V. A., and D. R. Brooks. 1990. Phylogenetic Systematics as the Basis of Comparative Biology. Smithsonian Institution Press, Washington, D.C., 45 pp.

Hennig, W. 1979. Phylogenetic Systematics (translation by D. D. Davis and R. Zangerl). University of Illinois Press, Urbana, Ill., 263 pp.

Jepsen, G. L., G. G. Simpson, and E. Mayr. 1949. Genetics, Paleontology, and Evolution. Princeton University Press, Princeton, N.J., 445 pp.

Nelson, G., and N. Platnick. 1981. Systematics and Biogeography, Cladistics and Vicariance. Columbia University Press, New York, 567 pp.

Moore, J. A. 1993. Science as a Way of Knowing. Harvard University Press, Cambridge, 530 pp.

Ridley, M. 1992. Evolution. Blackwell Scientific Publications, Cambridge, Mass., 670 pp.

Stanley, S. M. 1979. Macroevolution. W. C. Freeman and Company, San Francisco, 332 pp.

Wiley, E. O., D. Siegel-Causey, D. Brooks, and V. A Funk. 1991. The Compleat Cladist: A Primer of Phylogenetic Procedures. University of Kansas Museum of Natural History Special Publication no. 19, Lawrence, Kan., 158 pp.

CHAPTER 4

INTERRELATIONSHIPS OF THE VERTEBRATES

WHAT IS A DINOSAUR and where does it fit among other vertebrates? In this chapter, we use cladograms to discover fundamental interrelationships among the larger groups of vertebrates, focusing on the groups that include the dinosaurs. In doing so, we will uncover some remarkable things not only about dinosaurs, but also about many of the living vertebrates. For example, we will address questions like, How many times has warm-bloodedness evolved in the vertebrates? (answer: at least two, possibly three times); How many times has powered flight been invented by vertebrates? (answer: three independent times); Is a cow a fish? (answer: in an evolutionary sense, clearly!); Did all the dinosaurs become extinct? (answer: definitely not); and, Which is closer to a crocodile – a lizard or a bird? (answer: a bird).

THE EARLY BIOTA AND ITS HISTORY

Life is generally understood to be monophyletic. This intuitively comforting conclusion should not be taken for granted, for who can say how many early forms of molecular "life" arose, proliferated, and died out in the primordial oceans of 3.5 billion years ago? Regardless, all modern life is united by the possession of RNA, DNA, cell membranes with distinctive chemical structure, a variety of amino acids (proteins), the metabolic pathways (i.e., chemical reaction steps) for their processing, and the ability to replicate itself (not simply grow). These are all shared, derived characters of life.

A variety of lines of evidence indicate that the earliest atmosphere on earth was virtually without oxygen (**anaerobic**). The earliest cells, 3.5 billion years ago,

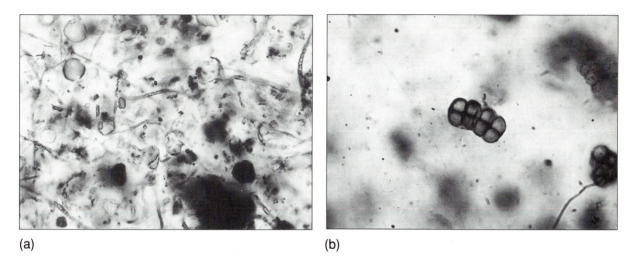

(a)   (b)

Figure 4.1 **Ancient single-celled organisms. (a)** A 2000-million-year-old fossil of a prokaryotic cell from the Gunflint Formation, Ontario, Canada. The spheres to the upper left of the photograph are 10 microns (10^{-5} m) in diameter; **(b)** 750-million-year-old prokaryotic cell from the Draken Formation, Svalbard. Each cell unit is 4 microns (4×10^{-6} m) in diameter. *Photographs courtesy of A. H. Knoll, Harvard University.*

are thought to have been **heterotrophs**; that is, they simply obtained organic molecules (for sustenance) from the substrate in which they lived. More than two billion years ago, however, the earliest **autotrophs** evolved – organisms that were capable, through **photosynthesis**, of utilizing inorganic nutrients and energy from the sun to produce complex molecules for nutrition. Photosynthesis is a reaction that combines carbon dioxide and water and energy (through light) to produce carbohydrates (sugar) and oxygen. This process takes place with the aid of the molecule chlorophyll, the well-known source of green color in plants. The oxygen produced as a result of photosynthesis was originally, and continues to be, the main source of the oxygen in the earth's atmosphere. The world is still divided into heterotrophs and autotrophs; obviously (because we don't photosynthesize), all vertebrates are heterotrophs.

The oldest fossils known on earth, found in South Africa, Australia, and Siberia, indicate that the earliest forms of life were individual (not colonial), single-celled, **prokaryotic** (*pro* – before; *karyo* – kernel, a reference to the nucleus, or central "kernel" of the cell) organisms (Figure 4.1). Prokaryotic cells are small, have no nucleus or other internal cell partitions, and are the type of cell found in bacteria. Today, along with prokaryotes, however, there is a second very different kind of cell on earth: **eukaryotes** (*eu* – true). Eukaryotic cells are larger and more complex, and they have a nucleus and a variety of internal chambers called **organelles**, each with specific functions. All other single-celled organisms, as well as multicellular life, are eukaryotic. The evolution of eukaryotes from prokaryotes, arguably the single most significant event in the history of life, probably occurred about 1.5 billion years ago. An evolutionary step of this magnitude is hard to explain, but one very plausible scenario, proposed in the late 1960s by L. Margulis of the University of Massachusetts, suggests that eukaryotic cells are in fact the products of prokaryotic cells' initially having been swallowed by other prokary-

(a)   (b)

Figure 4.2 **Two 565-million-year-old examples of the Ediacaran fauna, the earliest multicellular life known on earth.** The fossils are preserved as impressions in fine-grained sandstones. **(a)** *Charnia masoni*, from the White Sea, Russia; and **(b)** *Charniodiscus arboreus*, from the Pound Quartzite, South Australia. The ruler is marked in centimeters. *Photographs courtesy of A. H. Knoll, Harvard University.*

otes. This theory, called the **endosymbiotic theory** of the origin of eukaryotes (*endo* – inside; *symbios* – living together), proposes that the ingested prokaryotes continued to live within their hosts in a mutually beneficial arrangement, adopting specialized functions as organelles. A large body of evidence supports this theory. For example, the DNA and RNA of certain organelles is like that of prokaryotic cells and different from that of the nucleus; certain organelles have a separate cell membrane; and organelles have separate reproductive mechanisms. Moreover, chemical reactions within the tissue of eukaryotic cells are not like those of certain organelles, whose own chemical reactions more closely resemble those found in prokaryotes. Clearly, evolution does not necessarily proceed in tiny, mutation-based steps occurring over millions of years!

At least two other evolutionary events had to occur before life as we know it today could have evolved. The first was the origin of sex, and the second was the evolution of multicellular organisms. Although we don't often think of it this way, sex is a means by which the genetic material of two individuals can be reshuffled and a third combination of genetic material can be produced. With asexual reproduction, variety can occur only over an extremely long period of time, by small-scale mutations in the genetic material that somehow did not irreparably damage the affected organism. By contrast, with **recombination** through sex (the production of new *combinations* of DNA in the offspring with each reproduction event) a tremendous *variety* of genetic material is possible in a relatively short period of time.

The fossil record is largely silent about the evolution of multicellularity until about 670 million years ago, at which time an assemblage of multicellular organisms is first recorded. This **fauna**, or group of animals, was found in the Ediacara Hills of Western Australia. All that is left of these animals are impressions in sandstone (Figure 4.2).

Quite a different story is recorded about 570 million years ago, when the first **skeletons** – that is, tissue hardened by mineral deposits – appear in the fossil record. We begin to find small shells at the base of the Paleozoic Era, many of which are attributable to groups that have modern descendants. After a second wave of evolution some 50 millions of years into the Cambrian – that is, during the middle Cambrian (530 million years ago) – a fantastic radiation of forms, in which a tremendous diversity of *body plans* (including vertebrates) materialized, took place relatively rapidly.

Body Plans

All organisms are subject to design constraints (or restrictions): They live in a fluid medium (usually air or water), they are acted upon by gravity, and their ancestry limits the structures they can evolve. In other words, evolution is the modification of what one has inherited from more ancestral forms. It is decidedly not the business of inventing wholesale structures like, for example, turbofans in place of the flapping wings of birds (even though jet engines might be a better means of propelling a body through the air). Design constraints, fixed by ancestry and the limits of the physical world, clearly set a finite range of possible body plans.

Many body plans, and in all likelihood their design constraints, are predictable and repeated with some variations throughout the biota. Such repetitions do not occur as the result of a single evolutionary event. Rather, design constraints are such that different lineages of organisms reinvent each structure. When the reinvention of a structure independently takes place in two lineages, the evolution is said to be **convergent**. This, in fact, means that the structures are analogous, rather than homologous.

High levels of activity dictate a number of convergent, analogous (independently evolved) structures that are repeated in one form or another throughout a variety of different organisms. Because muscles can only contract, opposing – or **antagonistic** – **muscle masses** are found in most **motile** (moving; as opposed to stationary, or **sessile**) organisms. With antagonistic muscle masses, the contraction of one muscle extends the muscle opposing it. In vertebrates, antagonistic muscles are distributed *around* the rigid support provided by the jointed *internal* skeleton. In arthropods, these are enclosed *within* the rigid support provided by the jointed *external* skeleton (see Figure 3.6). Complex musculature requires sophisticated coordination, and this is usually accomplished by a centralized cluster of nerves and neural material called the **brain**. In highly active creatures, the brain is usually located at the front of the creature in a head structure that is distinct from the rest of the body. This condition is called **encephalization** and of course pertains to all vertebrates. Most arthropods, being motile, are also highly encephalized creatures (barnacles, with their sessile lifestyle, are a notable exception), as are some mollusks, such as squids and octopuses. With motility and encephalization commonly comes **bilateral symmetry**, a kind of symmetry in which the right and left halves of the body are mirror images of each other.

Another example of convergence in body plans is **segmentation**. Segmentation simply involves division of the body into repeating units, which in turn permits the isolation of parts of the body. Segmentation is a common way that organisms have solved the need to produce coordinated movement.

BOX 4.1

THE BIOLOGICAL CLASSIFICATION

ORGANISMS ARE COMMONLY CLASSIFIED according to the biological classification system, the categories of which have been dutifully memorized by generations of nascent biologists. The classification was first developed by the Swedish naturalist Carolus Linnaeus (1707–1778), who, a full one hundred years before Darwin first published *The Origin of Species*, created a formalized hierarchical nomenclature to be applied to the biota. His system of increasingly smaller categories, the Linnaean system, has been used since Linnaeus's time to classify organisms. The oft-memorized categories that he established are, in order of decreasing size, Kingdom, Phylum, Class, Order, Family, Genus, and Species. Thus, for example, a cat would be classified in the following way. Kingdom: Animalia (a cat is an animal); Phylum: Chordata (the embryo has a notochord); Class: Mammalia (it has fur and mammary glands); Order: Carnivora (it bears a distinctive pair of shearing teeth, the carnassials); Family: Felidae (it has certain features in the number of teeth and in its skull); Genus: *Felis*, and Species: *domesticus* (it is the variety that became domesticated). Individuals are generally referred to by italicized generic (genus) *and* specific (species) names; that is, *Felis domesticus*. Commonly in this binomial scheme, the generic name is abbreviated, as in *F. domesticus*, or *T. rex*. Each of these names (Mammalia, *Felis*) is termed a taxon (plural: taxa).

Organisms get names applied to them by the scientists who first recognize them as new or distinct. Hence, paleontologists have created names like *Deinonychus* ("terrible claw") to celebrate the formidable claws on the feet of this carnivorous dinosaur. Occasionally, a name can be quite whimsical, such as *Cuttysarkus*, a lizard whose name celebrates a well-known scotch whiskey.

Any classification must have a purpose, and the biological classification is no exception. We classify by many things; our movies are commonly classified by subject (horror) or by suitability for viewing (PG-13; R), but in the case of the biota, implicit in the classification is the *degree of relatedness*. Thus, all members of a single species are said to be more closely related to each other than any is to anything else. Because a true reflection of degree of relatedness is sought in the biological classification, the rule presumably applies to all levels of the hierarchy. That is, at any level in the classification, it is assumed that all members of a group are more closely related to each other than they are to members of an equivalent, but different, group. We have seen how this applies in *Homo sapiens*. In the case of the Linnaean order Carnivora, for example, it is assumed that all carnivores are more closely related to each other than they are to taxa in any other order.

Here we will avoid using most Linnaean terms except generic and specific ones. The reason for this is that the level in the hierarchy that is designated by any particular Linnaean term is largely arbitrary: There is no particular type of character difference or amount of morphological divergence that is equated with a particular Linnaean term. For example, some people might designate the dinosaur group Ornithischia an order, and some people might consider it a suborder, but in neither case have we learned more about Ornithischia itself. For us, the important facts about Ornithischia are its shared, derived characters, and where (relative to other groups) it is situated in the hierarchy. Moreover, if a formal name were introduced to designate every rank of monophyletic group revealed by the cladogram, this book, already filled with a formidable nomenclature, would be nothing but an apparently endless list of formal names. The cladogram plainly indicates the hierarchical relationships without resorting to formalized Linnaean designations.

Sequential motion becomes possible, because one part of the body can respond independently of another. Moreover, the segments themselves can be modified through evolution into specialized organs for a variety of functions, most commonly for locomotion or for obtaining food. We all know that arthropods (including the familiar insects, spiders, crabs, shrimp, and lobsters) are segmented animals; less well known is that vertebrates are segmented animals. The segmentation of the body plan is still seen in the repeated structure of the vertebrae, in ribs, and in muscles. Segmentation and encephalization are, in part, an effective solution to the problems of size and mobility; the fact that arthropods and vertebrates share these characters does not mean that arthropods and vertebrates are phylogenetically close.[1] Vertebrates and arthropods evolved quite different body plans (the most obvious difference is the internal skeleton of vertebrates and the external skeleton of arthropods), with convergent structures related to their active life strategies.

In the Linnaean biological classification system (see Box 4.1), the term **phylum** is a grouping of organisms whose make-up is supposed to connote a basic level of organization that is shared by all of its members. The idea is that the members of a phylum may modify or adjust aspects of their morphology, but the fundamental body plan of the phylum remains constant. For example, a whale is a rather different creature from a salamander, but few would deny the basic shared similarities of their body plans. Both are included within the Chordata, a Linnaean phylum that includes creatures – almost all of them vertebrates – that possess a dorsal nerve cord (a nerve cord along the length of the back).

The Middle Cambrian Explosion and the Earliest Vertebrates

We are forced to interpret Cambrian life from a very sketchy record, one that gives us only tiny glimpses of what things must really have been like. One of those tiny glimpses is the Burgess Shale, an outcrop of rock in the Canadian Rockies that represents the edge of a reef that was active during the Middle Cambrian (530 million years ago). At the edge of this reef, rubble and mud periodically fell, burying and preserving thousands of small creatures. These ancient creatures are today beautifully fossilized (if a bit flattened) and indicate that in terms of body plans, the Middle Cambrian was a time interval of remarkable diversity. This bestiary is like a parts store for segmented organisms. Among the creatures found there, are *Opabinia*, a five-eyed, segmented, presumed carnivore equipped with a single vacuum cleaner–style hose at the front end; *Odontogriphus*, a flattened, segmented swimmer with an oblong mouth surrounded by a ring of tentacles; *Odaria*, which looks like a segmented cucumber with a triangular tail, living in a muscle shell; *Wiwaxia*, an odd, plated mound of a creature with large bladelike spines sticking out of its back; and *Anomalocaris*, a large (up to 2 m!), free-swimming, segmented predator with a ring of circular mouthplates and a pair of feeding arms. Several of these Burgess

---

[1]Because all life is monophyletic, we can be sure that vertebrates and arthropods share some common ancestor, however distant. In this case, we surmise that the common ancestor of arthropods and vertebrates must have been some multicellular creature that existed far back in Precambrian time.

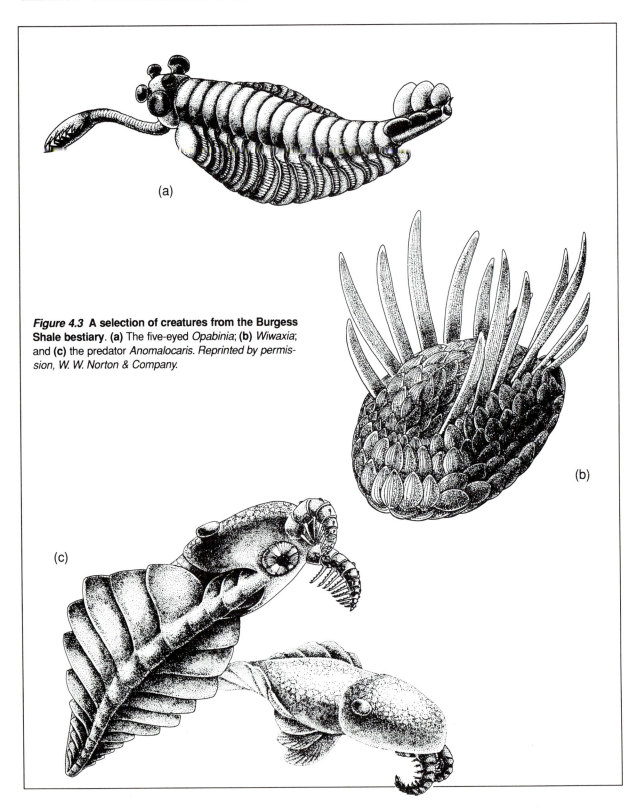

Figure 4.3 **A selection of creatures from the Burgess Shale bestiary.** **(a)** The five-eyed *Opabinia*; **(b)** *Wiwaxia*; and **(c)** the predator *Anomalocaris*. *Reprinted by permission, W. W. Norton & Company.*

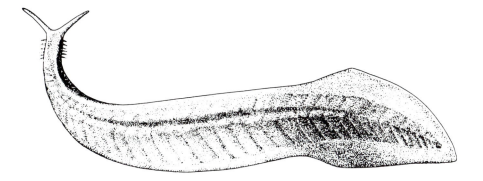

Figure 4.4 Pikaia, a presumed chordate from the Middle Cambrian Burgess Shale. *Reprinted by permission, W. W. Norton & Company.*

Shale creatures are shown in Figure 4.3. Many of the Burgess Shale animals, although extinct, are constructed along body plans that are still seen in living organisms. Some, however, had body plans that are unlike anything currently alive.

Buried among the many specimens and vast array of body plans is a modest-looking creature called *Pikaia gracilens* (Figure 4.4). *Pikaia* is about 5 cm in length and, in its flattened condition, looks a bit like a strip of ribbon. It was initially described in 1911 as a "worm," but with closer examination, *Pikaia* seems to show the characters that are diagnostic of Chordata: a nerve cord running down the length of its back, a stiffening rod that gives the nerve cord support (the notochord), and V-shaped muscles (composed of an upper and a lower part) with repeated segments, a character that is familiar to most of us because it is present in modern fish.[2] We – and the dinosaurs – would appear to have chordate relatives as far back as the Cambrian!

CHORDATES

Although *Pikaia* provides an inkling about our primitive relatives, what we know about the early evolution of vertebrates and their forebears among Chordata comes principally from living organisms, with sometimes a goodly mixture of information from relevant fossils (particularly for true vertebrates).

The chordates consist of *Pikaia* from the Burgess Shale (but see Footnote 2), urochordates, cephalochordates, and, most important for our story, vertebrates, all of which can be united on the basis of (1) features of the throat (pharyngeal gill slits), (2) the presence of a **notochord** at some stage in their life histories, and (3) the presence of a dorsal, hollow nerve cord. A notochord is a stiffening rod running down the back of the animal. Above the notochord in

[2] The claim that *Pikaia* is some kind of chordate has been contested by N. Butterfield of Harvard University. Butterfield believes that chordate tissues cannot be preserved in the way those of *Pikaia* are. Therefore, he is unwilling to place *Pikaia* in any modern group.

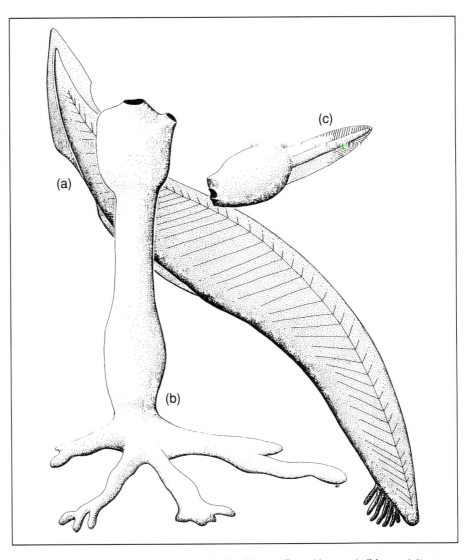

Figure 4.5 **Two primitive chordates. (a)** *Amphioxus* (*Branchiostoma*); **(b)** an adult sea squirt (*Clavelina*); and **(c)** the juvenile form of the sea squirt (*Clavelina*). Note that the juvenile form is free-swimming and has a notochord running down its tail. It metamorphoses into the stationary adult by planting itself on its nose and rearranging its internal and external structures.

chordates is the nerve cord, encompassed within a distinct tail region behind the gut. This distinctive suite of characters – pharyngeal gills, notochord, and nerve cord – appears to have evolved only once, thus uniting these animals as a monophyletic group

Urochordates, commonly called "sea squirts," have a sessile adult form, and the evidence of chordate ancestry is found only in their free-swimming larvae (in which the notochord is evident). The larvae eventually give up their roving ways, park themselves on their noses, and develop a filter to trap food particles from water that they pump through their bodies (Figure 4.5).

Cephalochordates, the closest relative to the vertebrates, are best known to us in the form of the small, water-dwelling lancet (*Amphioxus* or *Branchiostoma*). Again, known basically from present-day forms (one specimen was recovered from the Carboniferous), these creatures share with the vertebrates a host of derived features, including segmentation of the muscles of the body wall, separation of upper and lower nerve and blood vessel branches, and many newly evolved hormone and enzyme systems. As our first cousins among chordates, cephalochordates themselves are united by important derived characters of the head region. Figure 4.6 is a cladogram of the Chordata, taking us through the major groups of chordates to Tetrapoda.

Vertebrata

We have now reached Vertebrata. The features that unite vertebrates, with one exception, include calcified skeletal tissue (i.e., **bone**) divided into discrete parts called **elements**, and a variety of other characters (Figure 4.6).

Gnathostomata

What we might unthinkingly regard as the body plan for all vertebrates is really the plan for a subset of the vertebrates called Gnathostomata (*gnathos* – jaw; *stome* – mouth). Gnathostomes are vertebrates with true jaws and a variety of other features (Figure 4.6). In fact, the absence of jaws is primitive for vertebrates, and the evolution of jaws was an important innovation.

So, who are these gnathostomes? Well, *we* certainly are, as are sharks (starring in *Jaws*), ray-finned fishes (familiar gracing our aquariums and filleted on our dinner plates), and lobe-finned fish (fleshy-finned creatures that tinkered with muscular fins and, possibly, air breathing). Other, less familiar, extinct gnathostome fishes that do not concern us here also cruised the ancient seas.

Sharks and their relatives belong to a clade within Gnathostomata called **Chondrichthyes** (*chondros* – cartilage; *ichthys* – fishes). The remaining fishes are grouped together within **Osteichthyes** (*os* – bone; *ichthys* – fish), bony fishes that include the ray-finned and lobe-finned gnathostome groups. The ray-finned branch of osteichthyans is called **Actinopterygii** (*actino* – ray; *ptero* – wing) and is one of the most diverse groups (more than 20,000 living species) ever to have evolved among vertebrates (Figure 4.6).

The other osteichthyan branch, the lobe-fins or **Sarcopterygii** (*sarco* – flesh), includes *lungfish* (of which only three types are alive today); *coelocanths* (of which only one type is alive today); extinct barracuda-like carnivorous forms; and, surprisingly, *tetrapods* (which share the derived characters of the group; see Figure 4.6 and Box 4.3). Indeed, bones homologous to those of the limbs, pelvis, vertebral column, and skull of tetrapods are all found within nontetrapod members of the lobe-finned clade, strongly uniting the tetrapods to other members of this group (Figure 4.7). Those homologies – and many others – indicate that it is here, among the lobe-fins, that the ancestry of Dinosauria – as well as our own ancestry – is to be found.

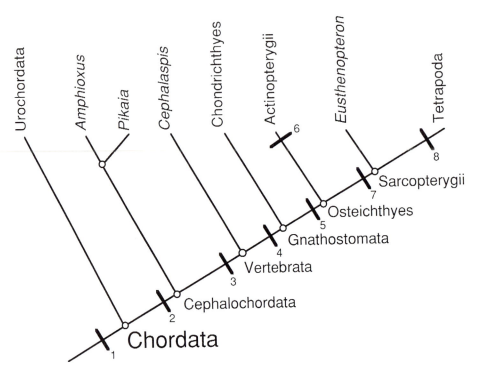

Figure 4.6 Cladogram of Chordata. Because this is a book about dinosaurs (and not all chordates), we have provided diagnoses for only some of the groups on the cladogram. Hatches denote the shared, derived characters of the groups. The characters are the following: (1) pharyngeal gill slits, a notochord, and a nerve cord running above the notochord along its length; (2) segmentation of the muscles of the body wall, separation of upper and lower nerve and blood vessel branches, and new hormone and enzyme systems; (3) bone organized into elements, neural crest cells, the differentiation of the cranial nerves, the development of eyes, the presence of kidneys, new hormonal systems, and mouthparts; (4) true jaws; (5) bone in the endochondral skeleton; (6) ray fins; (7) distinctive arrangement of bones in fleshy pectoral and pelvic fins (Figure 4.7); and (8) skeletal features relating to mobility on land – in particular, four limbs. Consistent with a cladistic approach, only monophyletic groups are presented on the cladogram. Some of the groups may not be familiar: For example, *Cephalaspis* and *Eusthenopteron* are not discussed in the text. *Cephalaspis* was a primitive, jawless, bottom-dwelling, swimming vertebrate, and *Eusthenopteron* was a predaceous lobe-fin, bearing many characters present in the earliest tetrapods. *Cephalaspis* and *Eusthenopteron* are included here to complete the cladogram as monophyletic representatives of jawless vertebrates (the nonmonophyletic "Agnatha") and lobe-finned fishes, respectively.

TETRAPODA

The gnathostomes central to our story are **tetrapods**. Tetrapoda (*tetra* – four; *pod* – foot) connotes the appearance of limbs, an adaptation that is strongly associated with land. According to the traditional classification, there are four classes of tetrapods:

TETRAPODA
1. Amphibia
2. Reptilia
3. Aves
4. Mammalia

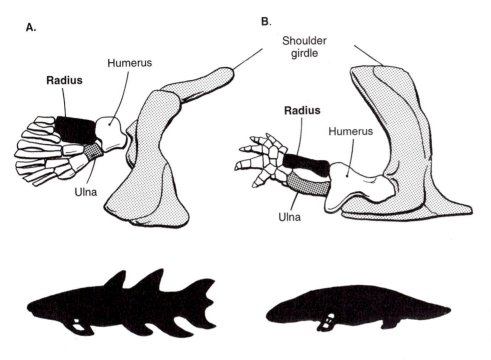

Figure 4.7 **The pectoral girdle of (A) a lobe-fin fish and (B) a tetrapod.** Note homologous bones.

The living amphibians are frogs, salamanders (and newts), and a group of rare, tropical, limbless creatures called caecilians. The living reptiles are crocodiles, turtles, snakes and lizards, and the tuatara, an unusual lizard-like creature that lives only in New Zealand. Mammals and birds (Aves), obviously, are common forms that are familiar to all of us.

The traditional classification is actually a very inadequate way to reflect the interrelationships of the tetrapods. Figure 4.8 is a cladogram showing the major groups of tetrapods. It shows the phylogenetic relations of the major groups of tetrapods and leads to a very different understanding of vertebrate interrelations from that implied by the traditional classification.

The Tetrapod Skeleton

The tetrapod skeleton is a modification of the skeletal component of the fundamental vertebrate body plan. We shall see in succeeding chapters how, through evolution, dinosaurs have modified this basic skeleton in a variety of ways. Of course, all tetrapods possess the diagnostic characters of chordates (Figure 4.9). The backbone is composed of distinct, repeated structures (the **vertebrae**), which consist of a lower spool (the **centrum**), above which, in a groove, lies the spinal cord. This relationship was first developed in gnathostomes and modified in tetrapods. Planted on the centrum and straddling the spinal cord is a vertically oriented splint of bone called the neural arch. Various **processes**, parts of bone that are commonly ridge-, knob-, or blade-shaped, may stick out from each vertebra. These can be for muscle and/or ligament attachment, or they can be sites against which the ends of ribs can abut. The repetition of vertebral structures, a relic of the segmented condition that is primitive for vertebrates, allows flexibility in the

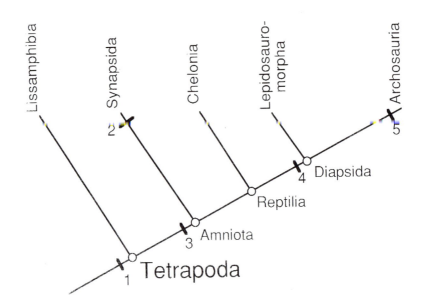

Figure 4.8 **Cladogram of Tetrapoda.** Diagnostic characters include (1) the tetrapod skeleton (Figures 4.9 and 4.10); (2) a lower temporal fenestra (see Figure 4.14); (3) presence of an amnion; (4) lower and upper temporal fenestrae (Figure 4.14); and (5) an antorbital fenestra (Figure 4.16). Lepidosauromorpha is a robust monophyletic group, the living members of which are snakes and lizards. Chelonia – turtles – are reptiles whose primitive, completely roofed skulls place them near the base of Reptilia.

backbone. In general, however, the tetrapod backbone is considerably more complex than the backbones that preceded it, because in tetrapods the backbone acts not only to facilitate locomotion, but also to support the body out of water.

Sandwiching the backbone are the **pelvic** and **pectoral girdles** (Figure 4.10). These are each sheets of bone (or bones) against which the limbs attach for the support of the body. The pelvic girdle – which includes a block of vertebrae called the **sacrum** – is the attachment site of the hindlimbs; the pectoral girdle is the attachment site of the forelimbs. Each side of the pelvic girdle is made up of three bones: a flat sheet of bone, called the **ilium**, that is fused onto processes from the backbone; a piece that points forward and down, called the **pubis**; and a piece that points backward and down, called the **ischium**. Primitively, the three bones of the hip come together in a depressed area of the pelvis called the **hip socket**.

The pectoral girdle consists of a flat sheet of bone on each side of the body, the **scapulae** (singular – scapula), or shoulder blades. These are attached to the outside of the ribs by ligaments and muscles.

A couple of other important elements deserve mention. The breastbone (**sternum**) is generally a flat or nearly flat sheet of bone that is locked into its position on the chest by the tips of the thoracic (or chest) ribs. The rib cage is supported at its front edge by bones in the shoulder (the **coracoid** bones).

Limbs in tetrapods show a consistency of form, an arrangement that was pioneered in their sarcopterygian ancestors (Figure 4.7). All limbs, whether fore or hind, have a single upper bone, a joint, and then a pair of lower bones. In a forelimb, the upper arm bone is the **humerus**, and the paired lower bones (forearms) are the **radius** and **ulna**. The joint in between is the **elbow.** In a hindlimb, the upper bone (thigh bone) is the **femur**, the joint is the **knee**, and the paired lower bones (shins) are the **tibia** and **fibula**. Note that knees and elbows bend in opposite directions.

Beyond the paired lower bones of the limbs are the wrist and ankle bones, termed **carpals** and **tarsals**, respectively. The bones in the palm of the hand are called **metacarpals**, the corresponding bones in the foot are called **metatarsals**, and collectively they are termed **metapodials**. Finally, the small bones that allow flexibility in the digits of both the hands (fingers) and the feet (toes) are called **phalanges** (singular – **phalanx**). At the tip of each digit, beyond the last joint, are the **ungual phalanges**. Until very recently, the primitive condition in tetrapods was considered to be the possession of five digits on each limb. Hence, in terms of the numbers of digits they possess, humans, for example, were thought to retain the primitive condition. Now it is known that tetrapods primitively had as many as eight digits on each limb. Early on in the evolutionary history of tetrapods, this number rapidly reduced to, and stabilized at, five digits on each limb, although many groups of tetrapods subsequently reduced that number even further (Figure 4.10).

At the **anterior** or **cranial** (head) end of the vertebral column of chordates are the bones of the head, composed, as we have seen, of the **skull** and **mandible** (lower jaws). Primitively, the skull has a distinctive arrangement: Central and toward the back of the skull is a bone-covered box containing the brain, the **braincase**. At the back of the brain case is the **occipital condyle**, the knob of bone that connects the braincase (and hence the skull) to the vertebral column. The opening in the braincase that allows the spinal cord to attach to the brain is called the **foramen magnum** (*foramen* – opening; *magnum* – big). Located on each side

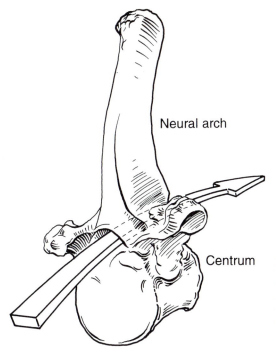

Neural arch

Centrum

Figure 4.9. **A vertebra from a rhinoceros.** The nerve cord (indicated by arrow) lies in a groove at the top of the centrum and is straddled by the neural arch.

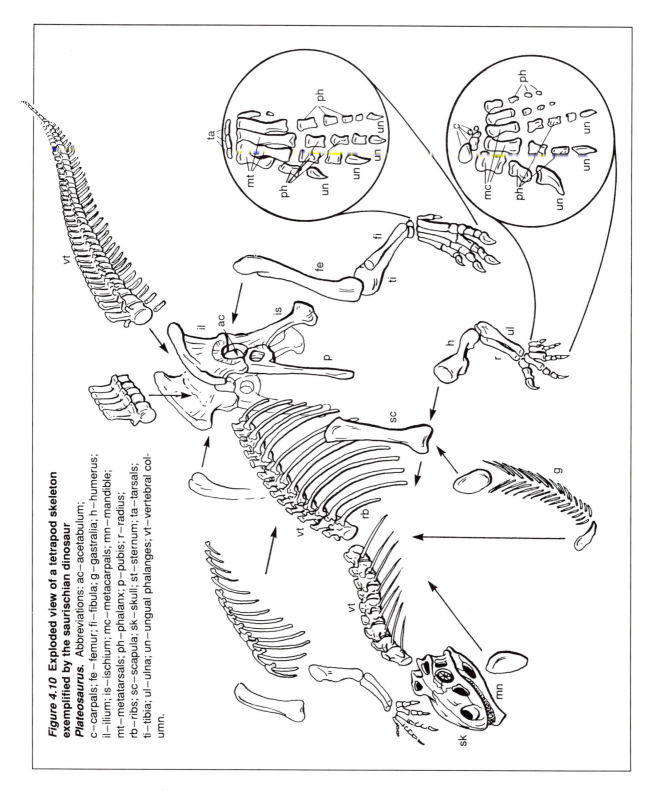

Figure 4.10 Exploded view of a tetrapod skeleton exemplified by the saurischian dinosaur *Plateosaurus*. Abbreviations: ac–acetabulum; c–carpals; fe–femur; fi–fibula; g–gastralia; h–humerus; il–ilium; is–ischium; mc–metacarpals; mn–mandible; mt–metatarsals; ph–phalanx; p–pubis; r–radius; rb–ribs; sc–scapula; sk–skull; st–sternum; ta–tarsals; ti–tibia; ul–ulna; un–ungual phalanges; vt–vertebral column.

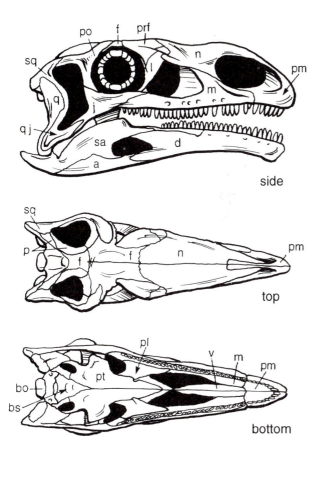

Figure 4.11 Skull and mandible. Abbreviations: a, angular; bo, basioccipital; bs, basisphenoid; d, dentary; f, frontal; fm, foramen magnum; j, jugal; l, lacrimal; m, maxilla; n, nasal; oc, occipital condyle; p, parietal; pl, palatine; pm, premaxilla; po, postorbital; prf, prefrontal; pt, pterygoid; q, quadrate; qj, quadratojugal; sa, surangular; sq, squamosal; v, vomer.

of the braincase are openings for the **stapes**, the bone that transmits sound from the **tympanic membrane** (ear drum) to the brain. Covering the braincase and forming much of the dorsal (upper) part of the skull is a curved sheet of interlocking bones, the **skull roof**. Among tetrapods, the bones comprising the skull roof have a distinctive pattern with clearly recognizable positions (Figure 4.11).

Primitively, the skull roof has several important openings. Located midway along each side of the skull is a large, round opening – the eye socket, or **orbit**. At the anterior tip, or **rostral**, part of the skull is another pair of openings – the **nares**, or nostril openings. Finally, located dorsally and centrally in the skull is a small opening called the **pineal**, or "third eye." This is a light-sensitive window to the braincase that has been lost in most living tetrapods and thus is not terribly familiar to us.

Flooring the skull, above the mandible, is a paired series of bones, organized in a flat sheet, which forms the **palate**.[3]

Tetrapods share a variety of derived features. We have seen many of these in

[3] In mammals the palate forms the floor of the nasal cavity, a chamber leading from the nostrils and the roof of the oral cavity (mouth), so that air breathed in through the nostrils is guided to the back of the throat. As a result, it is possible for chewing and breathing to occur at the same time. Similar kinds of palates (called secondary palates) are known in other tetrapods besides mammals, but primitively the nares lead directly to the oral cavity through tubes called choanae. So, if food were to be extensively chewed in the mouth, it would quickly get mixed up with the air that is breathed in. For this reason, extensive chewing is not a behavior of primitive tetrapods!

the tetrapod skeleton: the distinctive morphologies of the girdles and limbs, as well as the fixed patterns of skull roofing bones. The likelihood of all of these shared similarities' evolving convergently is remote; for this reason, these characters establish Tetrapoda as a monophyletic group.

Amniota

Only reptiles, birds, and primitive mammals possess in their eggs a membrane, the **amnion**, that retains moisture and allows the embryo to be continuously bathed in liquid (Figure 4.12). The amnion occurs in conjunction with several other features: a shell, a large yolk for the nutrition of the developing embryo, and a special bladder for the management of embryonic waste. These features have important ramifications for the organisms that possess them. Foremost among these, eggs can be laid on land without their drying out; therefore, the presence

---

BOX 4.2

POSTURE:
It's Both Who You Are *and* What You Do

TETRAPODS THAT ARE MOST highly adapted for land locomotion tend to have an erect posture. This clearly maximizes the efficiency of the animal's movements on land, and it is not surprising that, for example, all mammals are characterized by an erect posture. Tetrapods, such as salamanders, that are adapted for aquatic life display a **sprawling posture**, in which the legs splay out from the body in the same plane as the torso. The sprawling posture appears to have been inherited from the original position of the limbs in early tetrapods, whose sinuous trunk movements (inherited from swimming locomotion) aided the limbs in land locomotion.

Some tetrapods, such as crocodiles, have a **semi-erect posture**, in which the legs are directed at something like 45° away from the body, toward the ground (Figure Box 4.2). Does this mean that the semi-erect posture is an adaptation for an aquatic *and* a terrestrial existence? Clearly not, because a semi-erect posture is present in the large, fully terrestrial monitor lizards of Australia (goanna) and Indonesia (Komodo dragon). If adaptation is driving the evolution of features, why don't the fully terrestrial lizards have a fully erect

posture, and why don't the aquatic crocodiles have a fully sprawling posture? The issue is more complex and is best understood through adaptation to a particular environment or behavior, *as well as* through inheritance.

If we consider posture simply in terms of the ancestral and derived characters, the ancestral condition in tetrapods is a sprawling posture. An erect posture represents the most highly evolved (advanced) state of this character. Are animals with sprawling postures simply more primitive than those with erect postures?

In 1987, D. R. Carrier of Brown University hypothesized that the adoption of an erect posture represents the commitment to an entirely different mode of respiration (breathing) as well as locomotion (see Chapter 14 on "warm-bloodedness" in dinosaurs). Those organisms that possess a semi-erect posture may reflect the modification of a primitive character (sprawling) for greater efficiency on land, but they may also reflect a retention of the less-derived type of respiration. Ornithodirans (Figure 4.15) and mammals both have fully erect postures, which represent a full commitment to a terrestrial existence as well as to a more derived type

of an amnion marks the commitment to a land-based existence. Moreover, the embryos can develop within the egg more completely, rather than having to go through a larval stage during which development takes place while the animal is foraging. In contrast, amphibians, in retaining the primitive condition of an egg without an amnion (**anamniotic** egg), must lay their eggs in the water or in a moist environment on land, as well as maintain a free-swimming, foraging larval state (e.g., in frogs, the tadpole). The amnion, therefore, is an important character that fundamentally unites the clade Amniota within the tetrapods. Many other characters (literally, hundreds) also unite Amniota, including the number and position of the skull, trunk, and limb bones, as well as – among the living amniotes – aspects of the soft (nonskeletal) tissues and molecular systems. These features occur throughout the group and, taken together, powerfully suggest that the amniotes are monophyletic. Dinosaurs, too, share the skeletal anatomy that characterizes Amniota, and for this reason they fall within this group.

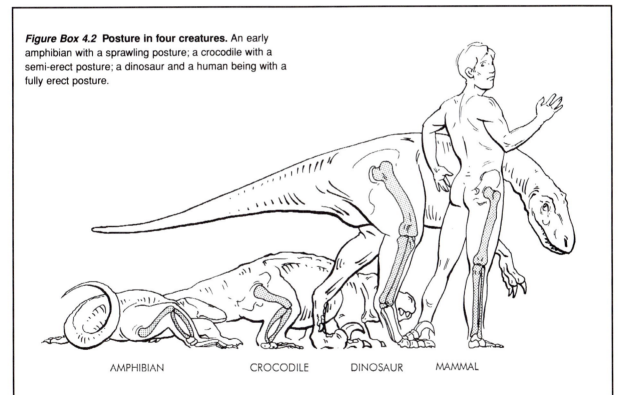

Figure Box 4.2 **Posture in four creatures.** An early amphibian with a sprawling posture; a crocodile with a semi-erect posture; a dinosaur and a human being with a fully erect posture.

AMPHIBIAN   CROCODILE   DINOSAUR   MAMMAL

of respiration. The designs of all these organisms are thus reflections of their inheritance, of their habits, and of their mode of respiration. Who can say what other influences are controlling morphology?

Interestingly, the cladogram (Figure 4.8) shows that the most recent common ancestor of ornithodirans and mammals – some primitive amniote – was itself an organism with a sprawling posture. Because ornithodirans and mammals (or their precursors) have been evolving independently since that most recent common ancestor (which is, after all, the implication of the expression "most recent common ancestor"), an erect posture must have evolved twice in Amniota: once in the synapsid lineage and once in the most recent common ancestor of Ornithodira.

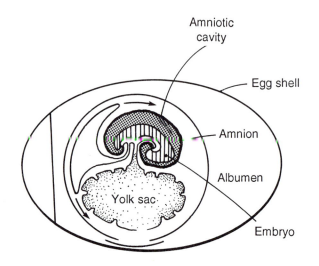

Figure 4.12 Amniote egg.

And what of the non-amniotes? The living non-amniotic tetrapods are the modern Amphibia (or **Lissamphibia**), but in the past a bewildering variety of different non-amniotic tetrapods existed. With the exception of the living amphibians, most non-amniotes were extinct by the middle of the Jurassic, not long after the first true dinosaurs appeared on earth.[4]

Synapsida

Synapsids are one of the two great lineages of amniote tetrapods. *All* mammals (including ourselves) are synapsids, as are a host of extinct forms, traditionally (and, as we shall see, misleadingly) called "mammal-like reptiles." The split between the earliest synapsids and the earliest representatives of the other great lineage, the reptiles (which includes the dinosaurs), occurred between 310 and 320 million years ago. Since then, therefore, the synapsid lineage has been evolving independently, genetically unconnected to any other group.

Synapsids are united by their common possession of the distinctive synapsid skull type, which is a departure from the primitive tetrapod skull type. As we noted earlier, the primitive condition in tetrapods consists of a sheet of interlocking bone covering the braincase. In synapsids, however, the skull roof has developed a low opening behind the eye – the **lower temporal fenestra** (*fenestra* – window) – whose position is such that, near its base, the skull roof seems to form an arch over each side of the braincase region (see Figure 4.14). From this comes the name synapsid (*syn* – with; *apsid* – arch). Jaw muscles pass through this opening to attach to the upper part of the skull roof. Many other features that we won't discuss here unite Synapsida as well.

[4]Although most of these archaic anamniotes were extinct by the Middle Jurassic, recent discoveries in Australia suggest that a relict few lingered on there until the Early Cretaceous!

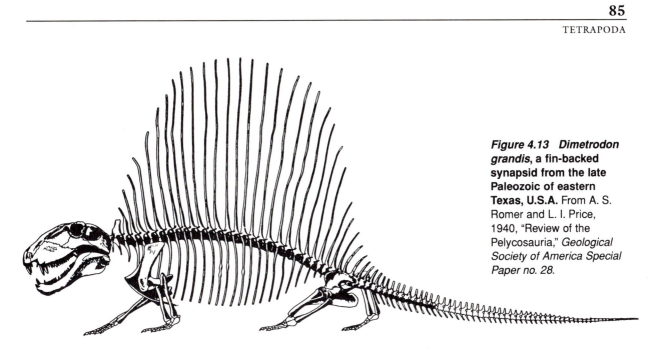

Figure 4.13 *Dimetrodon grandis*, **a fin-backed synapsid from the late Paleozoic of eastern Texas, U.S.A.** From A. S. Romer and L. I. Price, 1940, "Review of the Pelycosauria," *Geological Society of America Special Paper no. 28.*

Synapsids are an important group, and their variety and complexity cannot be done justice in this book. The most famous of the synapsids with elongate neural spines is the late Paleozoic *Dimetrodon*, from the southwestern United States (Figure 4.13). *Dimetrodon* was a 2-m-long, powerful quadruped with a deep skull full of nasty, carnivorous teeth, and, most likely, a malevolent personality to match. Although passed off on the cereal-box circuit as a dinosaur, *Dimetrodon* is a much closer relative of humans than it is of any dinosaur that ever lived!

Synapsids radiated during the late Paleozoic and, by the Middle Triassic, were the dominant terrestrial vertebrates. In the early Mesozoic, they developed a worldwide distribution and had diversified into a variety of herbivorous and car-

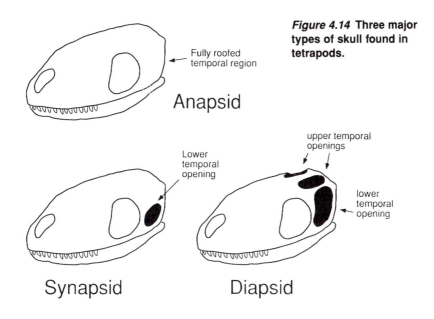

Figure 4.14 Three major types of skull found in tetrapods.

Fully roofed temporal region

Anapsid

Lower temporal opening

Synapsid

upper temporal openings

lower temporal opening

Diapsid

nivorous habits. By the Late Jurassic, all that was left was a clade of tiny, scrappy, furry night-dwellers: mammals! What happened to synapsids in the Late Triassic and Early Jurassic remains a mystery (see Chapters 5 and 14) that may have relevance to, of all subjects, dinosaur physiology!

REPTILIA

The other great clade of amniotes is **Reptilia** (*reptere* – to creep). Reptilia refers to Anapsida and Diapsida, and to all the forms down to their most recent common ancestor (Figure 4.8). From the modern world, we know these animals best as turtles, snakes, lizards, crocodiles, the tuatara, and birds, and from the past, as dinosaurs and their close relatives, the pterosaurs. Today there are an estimated 15,000 species of Reptilia; who knows how many other members of this great clade have come and gone? Reptilia is diagnosed by a braincase and skull roof that are uniquely constructed and by unique features of the neck vertebrae. Figure 4.11 shows the typical reptilian arrangement of bones in the skull roof and braincase.

Notice that we included birds among the living members of Reptilia. How can a bird be a reptile? Clearly we have a decidedly different Reptilia from the traditional motley crew of crawling, scaly, nonmammal, nonbird, nonamphibian creatures that most of us think of when we think of reptiles. If it is true that crocodiles and birds are more closely related to each other than either is to snakes and lizards, then a monophyletic group that includes snakes, lizards, and crocodiles *must* also include birds. The implication of calling a bird a reptile is that birds share the derived characters of Reptilia, as well as have unique characters of their own.

Anapsid

Within Reptilia are two equally important clades. The first – Anapsida (*a* – without; *apsid* – arch) – itself consists of Chelonia (turtles) and some extinct, bulky, quadrupeds that do not concern us here. Legendary stalwarts of the world, turtles are unique: These venerable creatures with their portable houses, in existence since the Late Triassic (210 million years ago), will surely survive another 200 million years if we let them.

Diapsida

Diapsida (*di* – two; *apsid* – arch) is united by a suite of shared, derived features, including having two temporal openings in the skull roof, an **upper** (or *supra*) and a **lower** (or *infra*) temporal fenestra. The upper and lower temporal openings are thought to have provided space for bulging jaw muscles as well as to have increased the surface area for their attachment. Figure 4.14 diagrams the three skull types found in amniotes: anapsid (the inherited, ancestral condition), synapsid, and diapsid.

There are two major clades of diapsid reptiles. The first, **Lepidosauromorpha** (*lepido* – scaly; *sauros* – lizard; *morpho* – shape; note that the suffix *sauros* means "lizard" but is commonly used to denote anything "reptilian"), is com-

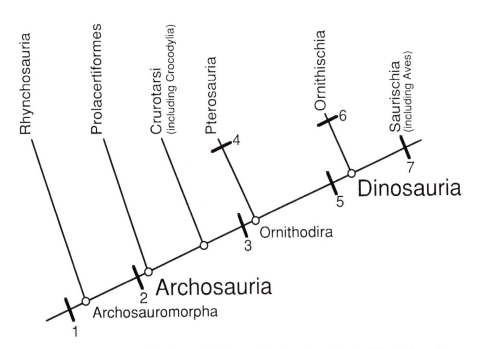

Figure 4.15 Cladogram of Archosauria. Diagnostic characters include (1) teeth in sockets, elongate nostril, high skull, and vertebrae now showing evidence of embryonic notochord; (2) antorbital fenestra (see Figure 4.16), loss of teeth on palate, and new shape of articulating surface of ankle (calcaneum); (3) erect posture (shaft of femur is 90° to the head; the upper part of the hip socket is thickened or has a ridge; the ankle has a modified mesotarsal joint); (4) a variety of extraordinary specializations for flight, including an elongate digit IV; (5) perforate acetabulum (see Chapter 5 for greater detail); (6) predentary and rearward projection of pubic process (see Part II: Ornithischia); and (7) asymmetrical hand with distinctive thumb, elongation of neck vertebrae, and changes in chewing musculature (see Part III: Saurischia).

posed of snakes and lizards and of the tuatara (among the living), as well as a number of extinct lizard-like diapsids.[5]

Archosauria

Finally, we come to the other great clade of diapsids, **Archosauromorpha** (*archo* – ruling; *sauros* – lizard; *morpho* – shape). Of particular interest to us, however, is the famous clade within the archosauromorphs, **Archosauria** (*archo* – ruling; *sauros* – lizard) among whom are included crocodiles, dinosaurs, pterosaurs, and birds (Figure 4.15).

Archosauromorpha is supported by many important, shared, derived characters, which are included on the cladogram in Figure 4.15. Within the archosauromorphs are a series of basal members that are known mostly from the Triassic. Some bear a superficial resemblance to large lizards (remember, however, that they cannot be true lizards, which are lepidosaurs), whereas others look like reptilian pigs.

The last of the aforementioned – prolacertiforms – possess a number of sig-

[5]Two marine groups, ichthyosaurs and plesiosaurs, have also been placed among the diapsids, but these are obviously not germane to our story.

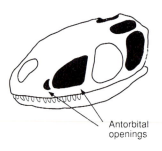

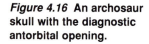
Antorbital
openings

Figure 4.16 **An archosaur skull with the diagnostic antorbital opening.**

nificant evolutionary innovations (Figure 4.15), most notably an opening on the side of the snout, just ahead of the eye, called the **antorbital fenestra** (Figure 4.16). These are the characters that unite the Archosauria, the group that contains crocodilians, birds, and dinosaurs. Crocodilians and their close relatives belong to a clade called **Crurotarsi**[6] (*cruro* – shank; *tarsus* – ankle); birds and their close relatives constitute a clade called **Ornithodira** (*ornith* – bird; *dira* – neck). Crocodiles have had a glorious and diverse history, and in the past, there have been seagoing crocodiles with flippers instead of legs, crocodiles with teeth that look more mammalian than crocodilian, and crocodiles that appear to have been well adapted for running on land. Other crurotarsans included a variety of carnivorous, piscivorous (fish-eating), and herbivorous forms (Figure 4.17).[7]

Ornithodira brings us quite close to the ancestry of dinosaurs. This group is composed of two major clades, Dinosauria (*deinos* – terrible; *sauros* – lizard) and Pterosauria (*ptero* – winged; *sauros* – lizard). Ornithodirans are united by the fact that within the archosaurs, they alone possess an **erect**, or **parasagittal**, **posture** – that is, a posture in which the plane of the legs is perpendicular to the plane of the torso and is tucked under the body (Figure 4.18; see also Box 4.2).

In ornithodirans, an erect posture consists of a suite of anatomical features with important behavioral implications. In particular, the head of the femur (thighbone) is distinctly offset from the shaft. The head of the femur itself is barrel-shaped (unlike the familiar ball seen in a human femur), so that motion in the thigh is largely restricted to forward and back, within a plane parallel to that of the body. The ankle joint is modified to become a single, *linear* articulation. This type of joint, termed a modified **mesotarsal** joint, allows movement of the foot only in a plane parallel to that of the body (forward and back). Note that again this situation differs from that in humans, in which the foot is capable of rotating movement. The upshot of these postural adaptations is that all ornithodirans are highly specialized for **cursorial** (i.e., running, as in the "cursor" on a computer screen!) **locomotion**; that is,

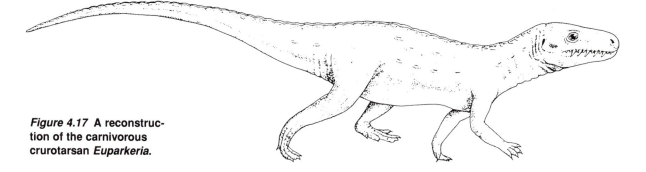

Figure 4.17 **A reconstruction of the carnivorous crurotarsan *Euparkeria*.**

---

[6]Some paleontologists prefer J. A. Gauthier's (1986) use of the term Pseudosuchia for this group. Membership in Pseudosuchia is very similar to that of Crurotarsi, and the differences are really not within the scope of this text to explore.

[7]Historically, basal archosauromorphs have all been jumbled together under the name **Thecodontia** (theco – socket; dont – tooth; see Chapter 13) because their teeth are set in sockets (much as our own are). But teeth in sockets applies to all archosauromorphs (Figure 4.15); how, therefore, can this character be used to distinguish one archosauromorph from another?

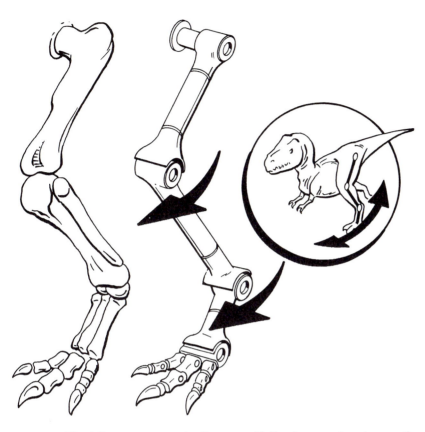

Figure 4.18 **The fully erect posture in dinosaurs.** Unlike, for example, a human, the bones of the leg permit it to move in only one plane of motion: forward and backward.

they are highly specialized for efficient locomotion on land. Ornithodirans are *fundamentally*, or at least primitively, terrestrial beasts (see Box 4.2).

Pterosaurs, the brainy, impressive "flying reptiles" from the Mesozoic, are highly modified ornithodirans, with as many as 40 derived features that unite them as a natural group. Their smallest members were sparrow-sized, and their largest members had wingspans as large as 12 m, making them the largest flying organisms that have ever lived (for reference, the wingspan of the two-person Piper "Cub" is also about 12 m). These extraordinary Mesozoic beasts demand a detailed treatment unfortunately not possible here.

Dinosaurs

This leaves us at long last with the subject of our book, **Dinosauria**. Dinosaurs can be diagnosed by a host of shared, derived characters, many of which are elaborated in Chapter 5 (Figures 5.4 and 5.5). Within dinosaurs, two groups are commonly recognized, **Saurischia** (*sauros* – lizard; *ischia* – hip) and **Ornithischia** (*ornith* – bird; *ischia* – hip). Treatment of these two groups constitutes much of the remainder of this book.

BOX 4.3

FISH AND CHIPS

AS 1978 TURNED TO 1979, a provocative and entertaining letter and reply were published in the scientific journal *Nature,* discussing the relationships of three gnathostomes: the salmon, the cow, and the lungfish.* English paleontologist L. B. Halstead argued that, obviously, the two fish must be more closely related to each other than either is to a cow. A group of English cladists disagreed, pointing out that in an evolutionary sense, a lungfish is more closely related to a cow than to a salmon. Who was right?

The vast majority of vertebrates are what we call "fishes." They all make a living in either salt or fresh water and, consequently, have many features in common that relate to the business of getting around, feeding, and reproducing in a fluid environment more viscous than air. But as it turns out, even if it describes creatures with gills and scales that swim, "fishes" is not an *evolutionarily* meaningful term; there are no shared, derived characters that unite *all* fishes that cannot also be applied to all nonfish gnathostomes. The characters that pertain to fishes are either characters that are present in all gnathostomes (i.e., primitive in gnathostomes) or characters that evolved independently.

The cladogram in Figure Box 4.3 is universally regarded as correct for the salmon, the cow, and the lungfish. In light of what we have discussed, this cladogram might look more familiar using the groups to which these creatures belong: Salmon are ray-finned fishes, cows are tetrapods, and lungfish are lobe-finned fishes. Clearly, lobe-finned fishes share more *derived* characters in common with tetrapods than they do with ray-finned fishes. Thus, there are two clades on the cladogram: (1) lobe-finned fishes + tetrapods; and

*L. B. Halstead, 1978, "The cladistic revolution – can it make the grade?" *Nature* 276:759–760; B. G. Gardiner, P. Janvier, C. Patterson, P. L. Forey, P. H. Greenwood, R. S. Mills, and R. P. S. Jeffries, 1979, "The salmon, the lungfish and the cow: a reply," *Nature* 277:175–176.

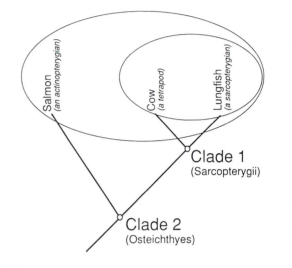

Figure Box 4.3. **The cladistic relationships of a salmon, a cow, and a lungfish.** The lungfish and the cow are more closely related to each other than either is to the salmon.

(2) lobe-finned fishes + tetrapods + ray-finned fishes. Clade 1 is familiar as the sarcopterygians. Clade 2 occurs at the level of Osteichthyes and looks like part of the cladogram presented in Figure 4.6 for gnathostome relationships. If only the organisms in question are considered, the only two natural groups on the cladogram must be (1) lungfish + cow; and (2) lungfish + cow + salmon (i.e., representatives of the sarcopterygians and osteichthyes, respectively).

Which are the "fishes?" Clearly the lungfish and the salmon. But, the lungfish and salmon do not in themselves form a natural group unless the cow is also included. Hence, if "fishes" are monophyletic, they must also include cows. The cladogram is telling us that the term "fishes" has *phylogenetic* significance only at the level of Osteichthyes (or even below)! But we can (and do) use the term "fishes" *informally.* Fish and chips will never be a burger and fries!

This, then, completes our long trek through Vertebrata to find Dinosauria. Having constructed many cladograms, we can now answer more completely the questions set forth at the beginning of this chapter. In response to, How many times has warm-bloodedness evolved in the vertebrates? it would appear that "warm-bloodedness" (an unfortunate term that we will abandon, with good reason, in Chapter 14) has evolved as many as three times: once in the synapsid lineage (in mammals or their near ancestors) and twice in the ornithodiran lineage (in birds and in pterosaurs). It is possible, however, that it only evolved two times and that the basal ornithodiran, that is, the organism that was the ancestor of *all* ornithodirans, was itself "warm-blooded." If so, then all ornithodirans were prim-

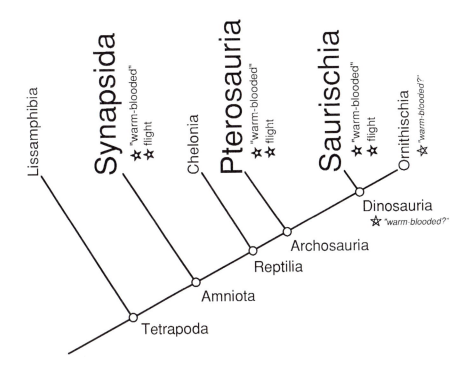

Figure 4.19 **The distribution of warm-bloodedness and flight in Tetrapoda, presented on a cladogram constructed from figures 4.8 and 4.15.** Filled stars denote known instances of warm-bloodedness and flight; unfilled stars and italics denote the *possibility* that warm-bloodedness and/or flight characterized a group. The cladogram shows that warm-bloodedness and flight evolved at least three times in Tetrapoda. Within Synapsida, warm-bloodedness occurs in all mammals, and flight occurs in bats. Within Archosauria, warm-bloodedness and flight occurred in pterosaurs (which are now known to have been insulated), and in birds (Saurischia). Warm-bloodedness has been proposed for Ornithischia, suggesting to some that all Dinosauria might be characterized by warm-bloodedness (see Chapter 14). Warm-bloodedness and flight, however, are not fundamental characteristics of Archosauria; Figure 4.15 shows that archosaurs (e.g., crocodiles) were certainly primitively non-flying, and almost certainly cold-blooded animals. The cladogram thus shows that warm-bloodedness and flight evolved three independent times: once in bats, once in pterosaurs, and once in birds (or their dinosaurian near-relatives).

itively "warm-blooded," which means that all dinosaurs must have been "warm-blooded," too. The evidence for and against this possibility will be explored in Chapter 14. As to, How many times has powered flight been invented by vertebrates? we know that flight occurred once in the synapsids (in bats) and twice in Ornithodira (in birds and in pterosaurs). If flight had evolved in the ancestor of all ornithodirans, then flight would be primitive among the ornithodirans, and *all* ornithodirans should be flying creatures. Obviously, many among Dinosauria were not (a flying *Stegosaurus* is hard to imagine), and so we can be relatively certain that powered flight evolved independently in three lineages of vertebrates. These relationships are shown in Figure 4.19.

Is a cow a fish? This is discussed in Box 4.3, and the "take-home" message from this question is that many of the features that we might intuitively use to group organisms are primitive characters and not suitable for recognizing evolutionarily significant groups. With regard to whether all the dinosaurs went extinct, the negative answer to this question requires fuller elaboration in Chapter 13. Finally, it should now be clear why a bird is closer to a crocodile than to a lizard. Birds and crocodiles are archosaurs, whereas lizards are not. The common ancestor of birds and crocodiles was some early archosaur, and thus it is only at the level of Diapsida that lizards, crocodiles, and birds are related. This, as noted previously, bodes ill for the traditional Reptilia; crocodiles, snakes, lizards, turtles, and the tuatara do *not* form a monophyletic group, unless one also includes birds. Our Reptilia includes birds!

Important Readings

Benton, M. J., ed. 1988. The Phylogeny and Classification of the Tetrapods, Vol I. Systematic Association Special Volume 35A, Oxford, 377 pp.

Butterfield, N. J. 1990. Organic preservation of non-mineralizing organisms and the taphonomy of the Burgess Shale. Paleobiology 16:272–286.

Carpenter, K., and P. J. Currie, eds. 1990. Dinosaur Systematics. Cambridge University Press, New York, 318 pp.

Carrier, D. R. 1987. The evolution of locomotor stamina in tetrapods: circumventing a mechanical constraint. Paleobiology 13:326–341.

Cowen, R. 1995. History of Life. Blackwell Scientific Publications, Boston, 462 pp.

Gauthier, J. A. 1986. Saurischian monophyly and the origin of birds; pp. 1–56 *in* K. Padian (ed.), The Origin of Birds and the Evolution of Flight. Memoirs of the California Academy of Sciences, no. 8, San Francisco.

Gauthier, J. A., A. G. Kluge, and T. Rowe. 1988. Amniote phylogeny and the importance of fossils. Cladistics 4:105–209.

Gould, S. J. 1989. Wonderful Life: The Burgess Shale and the Nature of History. W. W. Norton, New York, 347 pp.

LaPorte, L. F., ed. 1974. Evolution and the Fossil Record. Scientific American Offprint Series, W. H. Freeman and Company, San Francisco, 222 pp.

Moy-Thomas, J. A., and R. S. Miles. 1971 Palaeozoic Fishes. W. B. Saunders Company, Philadelphia, 259 pp.

Padian, K., and D. J. Chure, eds. 1989. The Age of Dinosaurs. Short Courses in Paleontology no. 2, The Paleontological Society, University of Tennessee Press, Knoxville, Tenn., 210 pp.

Prothero, D. R., and R. M. Schoch, eds. 1994. Major Features of Vertebrate Evolution. Short Courses in Paleontology no. 7, The Paleontological Society, University of Tennessee Press, Knoxville, Tenn., 270 pp.

Schopf, J. W., 1983. The Earth's Earliest Biosphere: Its Origin and Evolution. Princeton University Press, Princeton, N.J., 543 pp.

Stahl, B. J. 1985. Vertebrate History: Problems in Evolution. Dover Publications, New York, 604 pp.

CHAPTER 5

THE ORIGIN OF DINOSAURIA

WHO WERE THE DINOSAURS? Where did they come from? In a phylogenetic context, these two questions are really one and the same, because *who* an organism is, is really the key to its ancestry. Our goal in this chapter will be to consider precisely what a dinosaur is and, in doing so, to explore the origins of Dinosauria.

HISTORY OF DINOSAURIA

What dinosaurs are and how they came to be are questions long pondered by scientists since the creation of the name by Sir Richard Owen just over 150 years ago. Owen created "Dinosauria" to encompass a group of exceedingly large, pachydermous reptiles from what he referred to as the Secondary Age (the Mesozoic Era).[1] *Iguanodon* (an ornithopod), *Megalosaurus* (a theropod), and *Hylaeosaurus* (an ankylosaur) were its first members.

---

[1]The term "Secondary" came from a now-outdated concept of how rocks were formed. In 1759, G. Arduino, an Italian mining expert, developed a history of the rocks in northern Italy. He viewed these rocks as having been deposited by a retreating ocean. For him, the oldest rocks were designated Primary, and as the seas receded, Secondary, and Tertiary rocks were consecutively deposited. Thus Primary, Secondary, and Tertiary strata represented sequentially younger rocks. The scheme was elaborated by the eminent German naturalist A. G. Werner, who in 1787 published a highly influential history of the earth in which these terms were applied. Although in both schemes, rock types *and* age relations were embodied in the terms, the terms "Tertiary" and another, "Quaternary," remain with us today as age designations (the notion of a characteristic rock type has been abandoned, because most rock types can be produced and deposited at any time).

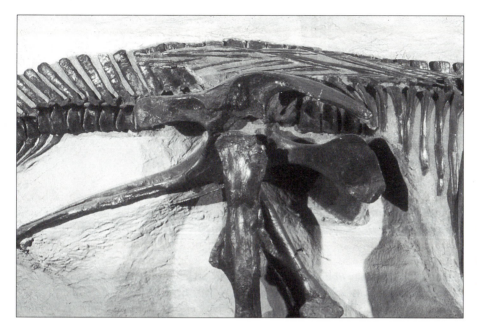

Figure 5.1 **The pelvis of the hadrosaurid *Prosaurolophus.*** A splint of the pubis lies along the base of the ischium, exemplifying the ornithischian condition. *Photograph courtesy of the Royal Ontario Museum.*

Dinosaurs – The Mesozoic Unnaturals

In 1887, Cambridge paleontologist H. G. Seeley concluded that there were two kinds of dinosaurs and that Dinosauria, as conceived by Owen, was not monophyletic. Ornithischia (*ornith* – bird; *ischia* – hip), he suggested, were all those dinosaurs that had a birdlike pelvis, in which at least a part of the pubis runs posteriorly, along the lower rim of the ischium (Figure 5.1). Saurischia (*sauros* – lizard) were those that had a pelvis more like a lizard, in which the pubis is directed anteriorly and slightly downward (Figure 5.2). This pelvic distinction has held sway ever since[2] and has been bound up in the debate of the origin of birds (see Chapter 13).

That dinosaurs had one or the other kind of pelvis has been of great importance to understanding the shape of the evolutionary tree of these animals, but in Seeley's hands, it went considerably further. For it implied to him that the ancestry of Ornithischia and Saurischia was to be found *separately* within what was then being called "Thecodontia" (a name coined by none other than Sir Richard Owen in 1859 and one that we met in the previous chapter and will meet again in Chapter 13), the heterogeneous nondinosaurian basal archosaurs. Dinosauria to Seeley was not a natural group (Figure 5.3).

This major theme, that dinosaurs were **diphyletic** (i.e., having two separate origins), was continued into this century, principally by F. von Huene. Throughout his remarkably long career (he published actively from the early 1900s to the 1960s), von Huene's studies came to epitomize independent dinosaur origins. According to him and many who came after, dinosaurs had at

---

[2]Virtually any discussion of dinosaurs, from the back of cereal boxes, to museum displays, to books devoted solely to dinosaurs, *begins* with this important distinction.

Figure 5.2 **The pelvis of the ornithomimid *Dromicieomimus***. The pubis is directed forward only, exemplifying the saurischian condition. *Photograph courtesy of the Royal Ontario Museum.*

least two, and more probably three or four, separate origins from different "thecodontian" groups. Certainly, saurischians and ornithischians had separate origins; after all, their hip structures were different. And among saurischians, surely sauropods and theropods had separate origins. After all, they *look* so different! And finally, among ornithischians, ankylosaur ancestry was also often sought separately within some other "thecodontian" group.

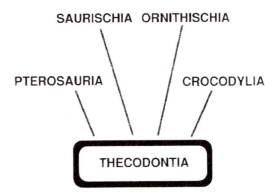

Figure 5.3 **Seeley's evolutionary scenario of the origin of dinosaurs.**

Following in the tradition of von Huene, British paleontologist A. J. Charig (now retired from the Natural History Museum, London) was a major student of dinosaur systematics in the 1960s and 1970s. It was through his studies of basal archosaurs that Charig traced the various dinosaur lineages to their separate origins among various "thecodontian" archosaurs. Dinosaurs to Charig and to many of his co-workers remained an unnatural group.

Dinosauria – It's a Natural!

The first inkling that things were changing as regards dinosaur relationships began in 1974 with a short publication in the British science journal *Nature*. In it, R. T. Bakker (then at Harvard University) and P. M. Galton (University of Bridgeport, Connecticut) attempted to resurrect Dinosauria as a monophyletic taxon, using a number of skeletal features and speculations about dinosaurian physiology. Their analyses – which reunited saurischians and ornithischians, and also included birds within Dinosauria – met a great deal of resistance and, in a few cases, open hostility (see Chapter 13). Similar resistance was met by Argentinian paleontologist J. F. Bonaparte, who speculated in 1976 that Dinosauria is a clade.

Starting in 1984, a variety of cladistic analyses were applied to Archosauria. The thrust of these studies was along four fronts: the disbanding of "Thecodontia," the origin of dinosaurs, the internal pattern of relationships within ornithischian and saurischian clades, and the relationships of birds to dinosaurs. We will discuss birds as dinosaurs in Chapter 13. Dinosaurian interrelationships will be treated within each of the successive taxonomic chapters. We have already discussed the dismemberment of "Thecodontia" in Chapter 4 and will do so further in Chapter 13. As it turns out, there are no unifying diagnostic

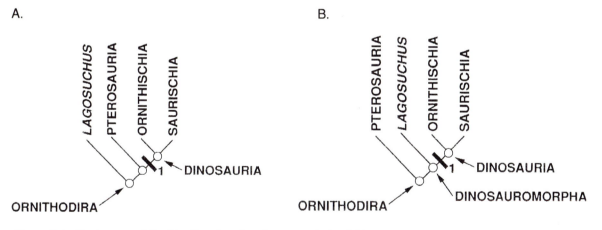

Figure 5.4 **Cladogram of Ornithodira showing the monophyly of Dinosauria.** In **(A)**, Pterosauria is the closest relative to Dinosauria, whereas in **(B)**, *Lagosuchus* is the closest relative. Derived characters at 1: loss of postfrontal bone, lapping of the ectopterygoid over the pterygoid in the palate, exposure of the head of the quadrate bone in lateral view, reduction of the post-temporal opening, three sacral vertebrae, rear-facing shoulder joint, asymmetric hand with small outer two digits with few finger bones, open hip socket, ascending process on the astragalus, projection of a flange on the upper end of the front surface of the tibia (cnemial crest), inturned head of femur, and S-shaped (sigmoidal) third metatarsal.

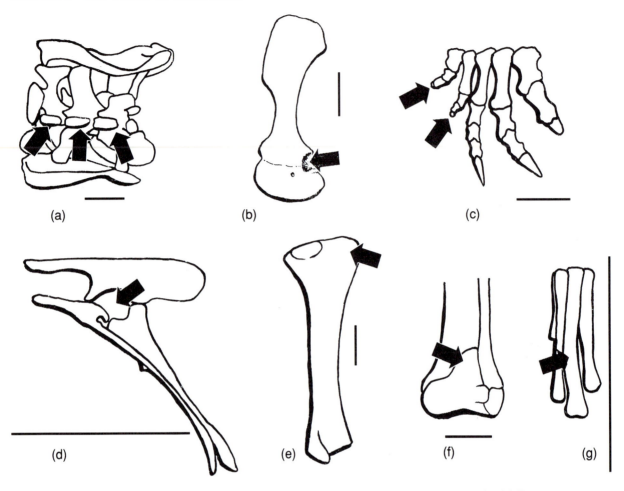

Figure 5.5 **Some of the derived characters uniting Dinosauria. (a)** Three or more sacral vertebrae as seen from above (shown in *Plateosaurus*; note arrows); **(b)** rear-facing shoulder joint (scapulacoracoid; shown in *Dilophosaurus*; note arrow); **(c)** asymmetric hand with two small outer digits (shown in *Plateosaurus;* note arrows); **(d)** open hip socket (shown in *Scutellosaurus*; note arrow); **(e)** cnemial crest on tibia (shown on *Plateosaurus*; note arrow); **(f)** ascending process on the astragalus (shown on *Allosaurus*; note arrow); and **(g)** sigmoidal third metatarsal (shown in *Lesothosaurus*; note arrow). Scale = 10 cm.

features that identify "Thecodontia" that do not also apply to Archosauromorpha. "Thecodontia" is not monophyletic.

In 1986, J. A. Gauthier of the California Academy of Sciences identified more than 10 derived features that established the clade Dinosauria (Figure 5.4). With Gauthier's research into dinosaur origins came the recognition that some animals – *Herrerasaurus* and *Staurikosaurus*, which we will meet in more detail in Chapter 12 – weren't members of either Saurischia or Ornithischia. Gauthier considered them the most basal members of Dinosauria, but *outside* the mega-clade of Ornithischia plus Saurischia.

Newer and not yet fully published research by P. C. Sereno (University of Chicago) and a number of co-workers differs from Gauthier's earlier assessment of basal dinosaur relationships. This research positions *Herrerasaurus* and *Staurikosaurus*, as well as a new dinosaur called *Eoraptor* (see Chapter 12), not only within the clade of saurischians and ornithischians, but within Theropoda.

DINOSAURIAN MONOPHYLY. If *Staurikosaurus* and *Herrerasaurus* (and *Eoraptor*) turn out to be theropods, then Dinosauria reverts to being defined as all descendants of the most recent common ancestor of Saurischia and Ornithischia (Figure 5.4). Such a definition is quite close to that provided by Owen, even if Sir Richard was not necessarily thinking along the lines of evolution and common descent. Some of the derived characters uniting Dinosauria are illustrated in Figure 5.5.

In all, Dinosauria appears to be a well-supported monophyletic clade. Its constituent members – Ornithischia and Saurischia – each have a monophyletic origin (see Introductions to Parts II and III) and the pattern of internal relationships of taxa within these clades is also reasonably well understood. Still, there are places, as we shall see, where controversy – and therefore intense research activity – still reigns.

ORIGINS

How does one find the ancestor of a clade? Simply put, the hierarchy of characters in the cladogram defines for us what features ought to be present in an ancestor, what features ought to be absent in an ancestor, and what the less evolved (general) and more evolved (derived) states of a particular character might be. As we have seen, the likelihood of the very progenitor of a lineage being fossilized is very low; however, we can commonly find representatives of closely related lineages that embody most of the features of the hypothetical ancestor.

According to Gauthier, as well as to K. Padian of the University of California at Berkeley, pterosaurs – otherwise highly modified for flight (Figures 5.4A and 5.6) – may be the closest archosaurian relatives to dinosaurs, together sharing four derived features (see Chapter 4). The clade of pterosaurs + dinosaurs (Ornithodira) then shares a close relationship with a slender, long-limbed animal from the Middle Triassic of Argentina called *Lagosuchus* (*lago* – rabbit; *suchus* – crocodile) to form what Gauthier termed Ornithosuchia. *Lagosuchus* (Figure 5.7) was a very small (less than 1 m), long-legged (hence the name) bipedal carnivore or insectivore. The head of this creature is very poorly known. Nevertheless, few paleontologists would disagree that this tiny creature embodies many of the features that were ancestral for all Dinosauria; the diminutive *Lagosuchus* is probably close to the ancestry of all the spectacular vertebrates encompassed within Dinosauria.

Sereno, in contrast, goes further, placing *Lagosuchus*, as well as two other small, contemporary archosaurs (*Lagerpeton* [*erpet* – a creeper] and *Pseudolagosuchus* [*pseudo* – false]) as the closest dinosaurian relatives (Figure 5.4B). This clade of dinosaurs and *Lagosuchus*, called Dinosauromorpha, shares a sigmoidal vertebral column in the neck region, a shortening of the forelimb, and several modifications of the ankle bones and of the metatarsals. On the strength of these features, it appears that Dinosauria shares closest relationship with archosaurs like *Lagosuchus*, *Lagerpeton*, and *Pseudolagosuchus*. Broader relationships of dinosauromorphs are with pterosaurs, which together with a few other taxa make up Ornithodira (see Chapter 4).

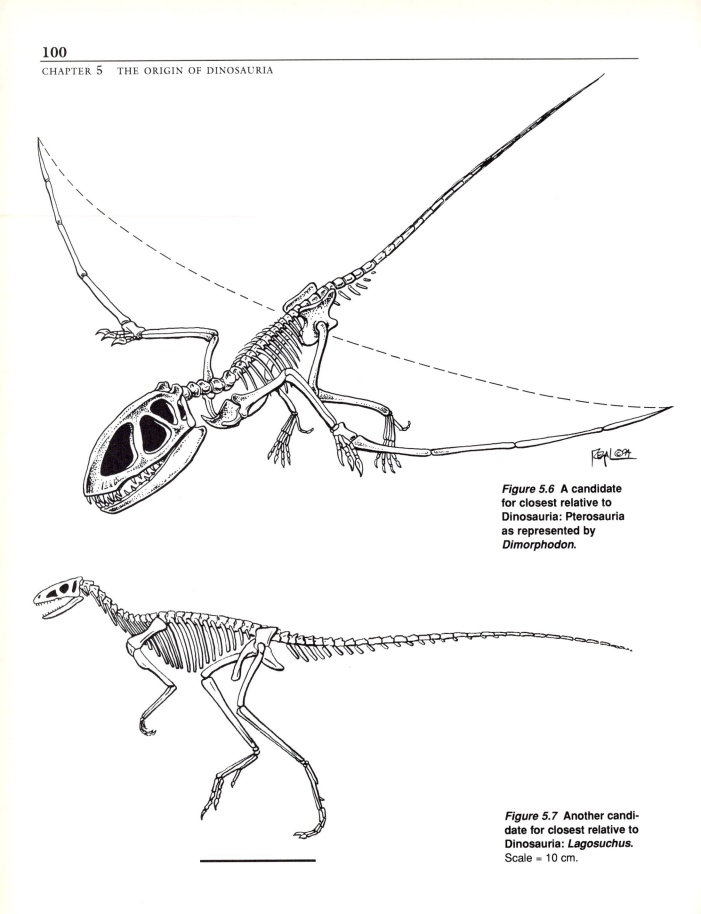

Figure 5.6 **A candidate for closest relative to Dinosauria: Pterosauria as represented by** *Dimorphodon.*

Figure 5.7 **Another candidate for closest relative to Dinosauria:** *Lagosuchus.* Scale = 10 cm.

There is an interesting and important consequence of this phylogeny. With archosaurs like *Lagosuchus* closest to dinosaurian ancestry, apparently dinosaurs were primitively **obligate bipeds**. This means that the earliest dinosaurs were creatures that were completely and irrevocably bipedal. Because the primitive stance for archosaurs is quadrupedal, and because Dinosauria is monophyletic, it follows that creatures like *Apatosaurus, Triceratops, Ankylosaurus,* and *Stegosaurus* – in fact, *all* quadrupedal dinosaurs – must have **secondarily evolved** (or reevolved) their quadrupedal stance. They must have (phylogenetically) gotten back down on four legs, as it were, after having been up on two. In fact, you can see the remnant of bipedal ancestry when you look at a stegosaur or a ceratopsian, in which the back legs are quite a bit longer than the front!

THE RISE OF DINOSAURS:
Superiority or Luck?

To see how the features that uniquely diagnose Dinosauria might affect evolutionary success, let's first set the stage for the emergence of dinosaurs in the Triassic. From its outset some 245 million years ago, the Triassic was dominated on land by synapsids. Among these were sleek, dog-like predators and rotund, beaked and tusked herbivores. From the middle, and toward the end, of the Triassic, these synapsids shared the scene with herbivorous archosauromorphs and a few carnivorous crocodile-like archosaurs. Yet toward the Late Triassic, approximately 225 million years ago, there was a great change of fortunes for these animals. The majority of synapsids went extinct (one highly evolved group of synapsids, the mammals, of course survived), and only the dinosaurs, pterosaurs, crocodiles, and a few other taxa among archosaurs survived. And it was only dinosaurs that rose to dominance, by which it is meant that they became the most abundant, diverse, and probably the most visible group of vertebrates on land.

We have noted that all dinosaurs had a fully erect posture. Charig envisioned the development of erect posture as occurring approximately coincidently with the appearance of dinosaurs.[3] For him, the fully erect posture provided for longer strides and more effective walking and running ability. The flashy new archosaurs (the emerging dinosaurs) that had these new, "improved" features were then able to compete effectively against contemporary predatory therapsids for their food sources (the herbivorous therapsids and archosauromorphs, both of which lacked such fully erect limb posture).

We can call this – and any other evolutionary advantage – a **competitive edge**. An implication in the competitive edge concept is that *all other things are equal*; hence, it is the new evolutionary innovation (in this case, the fully erect posture) that confers the advantage on the group that possesses it. We call the pattern of waxing and waning dominance (as one group supersedes another in evolutionary time) the **wedge** (Figure 5.8).

Beginning in the late 1960s, Bakker was making similar arguments about the competitive superiority of "warm-bloodedness" in dinosaurs (see Chapter 14 for

---

[3]Things are complicated by the fact that Charig also viewed dinosaurs as being at least diphyletic; nevertheless, the scenario we present is reasonably close to his general views on the evolution of dinosaur locomotion.

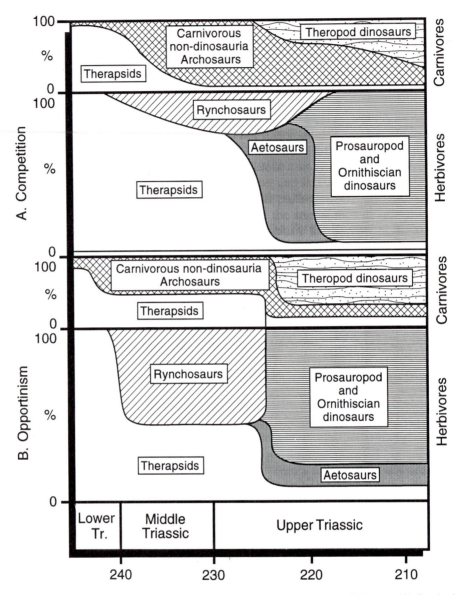

Figure 5.8 Two views of the origin of dinosaurs during the Late Triassic. **(A)** Gradual competitive replacement of synapsids, primitive archosaurs, and rhynchosaurs (both herbivores and carnivores) by herbivorous and carnivorous dinosaurs. **(B)** Rapid opportunistic replacement mediated by extinction.

an elaboration of Bakker's view of the competitive superiority of dinosaurs). As we shall see, he believed that instead of limbs, it was the achievement of internally produced heat that gave dinosaurs (or their immediate ancestors) a competitive edge over contemporary and supposedly "cold-blooded" therapsids and archosauromorphs. The same conclusions apply: Dinosaurs win, therapsids lose, and the truth of the competitive superiority of "warm-blooded" over "cold-blooded" can be read directly from the pattern of faunal succession at the end of the Triassic. Again, we have a hypothesis that comes down to the competitive edge producing the wedge.

Both ideas – locomotion and thermoregulation – sound like plausible arguments tied to the notion of competitive superiority of members of one clade over another. Both, however, have some serious disadvantages. Remember that all other things must be equal for the competitive edge concept to apply. But can we really be sure that all things are held equal? In an ecosystem with its immensely complex web of species interactions, it is very unlikely that all things are equal and that any competitive advantage comes down to one evolutionary innovation. It's not like a natural-history drag race, in which two virtually identical cars are pitted against each other under identical conditions, and the one with better tires wins. Any number of factors beyond simple predation can and do influence the waxing and waning of populations. The idea that one factor caused large-scale changes in an ancient ecosystem is probably simplistic.

In the case of fully erect posture, it is particularly problematical. For though it is true that the herbivorous synapsids and archosauromorphs did not have fully improved postures, the carnivorous synapsids (who presumably fed upon them) did. Thus, the edge that dinosaurs are supposed to have possessed during the Middle to Late Triassic was shared by the carnivorous members of the synapsid clade.

The competitive edge, thus, is not a completely satisfactory explanation of the wedge pattern. Still, the *pattern* itself is valid. Must the "wedge" be produced by some competitive edge?

No, says M. J. Benton (University of Bristol). Edges and wedges can't be documented in the Middle to Late Triassic fossil record of the earliest dinosaurs and their predecessors. In order for edges to lead to wedges, all of the players in the game have to be present to interact with each other. And according to Benton, they were not (note Figure 5.8). Instead, work by P. E. Olsen (Columbia University) and H.-D. Sues (Royal Ontario Museum), as well as Benton, has shown that the fossil record of the last part of the Triassic is marked not by one mass extinction, but by two mass extinctions. The first appears to have been the more extreme and, ultimately, more relevant to the rise of dinosaurs. This earlier Late Triassic extinction completely decimated some archosauromorphs and nearly obliterated the synapsids, as well as several major groups of predatory archosaurs. Likewise, there was a major extinction in the plant realm. The important seed-fern floras (the so-called *Dicroidium* flora, which contained not only seed ferns, but also horsetails, ferns, cycadophytes, ginkgoes, and conifers) all but became extinct as well, to be replaced by other conifers and bennettitaleans (large cycad-like plants). Dinosaurs appeared as the dominant land vertebrates only after this great disappearance of great numbers of therapsids and of many archosaurs and archosauromorphs. Thus, the initial radiation of dinosaurs, according to Benton, was done in an ecological near-vacuum, with mass extinction followed by opportunistic replacement. No need for competitive edge, because there was no competition.

Benton has suggested that the Late Triassic extinctions may be linked with climatic changes (the regions first inhabited by dinosaurs appear to have been hotter and more arid, a change from the more moist and equable) and thence to alterations in terrestrial floras and faunas. The abrupt extinction of the *Dicroidium* flora may have caused the extinction of herbivores specialized on them and, thereby, of the predators feeding on the herbivores. According to Benton, far from being a long-term *competitive* takeover, this rapid loss of the dominant land-living vertebrates set the stage for the *opportunistic* evolution of dinosaurs (in the vacuum left by the extinction).

The end-Triassic extinctions may have been driven by climatic shifts, but Olsen and colleagues think that another, more dramatic forcing factor can be identified: asteroid impact! They even proposed a "smoking gun": an impact crater close in age to the first of the Triassic extinctions. This impact structure, the Manicouagan crater in northern Quebec, Canada, is 70 km in diameter, large enough, Olsen and colleagues concluded, to have accommodated an asteroid with enough force to have done the job. Unfortunately for this idea, the available data are not strongly supportive. The age of the Manicouagan crater is somewhat older than either of the two Late Triassic extinction events.

So the underlying processes that ultimately drove the ascendancy of dinosaurs during the Middle and Late Triassic remains a mystery. We don't even know if this event was due to intrinsic factors – that is, aspects of the animals' biology – or extrinsic factors, such as an asteroid's coming out of the sky and disrupting the ecosystem in fundamental ways. But by whatever means dinosaurs wrested the terrestrial realm from the various synapsids (including mammals), 160 million years later the tables again turned and mammals inherited an earth *this* time deserted by the very dinosaurs who had taken it from them 160 million years earlier.

Important Readings

Bakker, R. T. 1975. Dinosaur renaissance. Scientific American 232:58–78.

Bakker, R. T., and P. M. Galton. 1974. Dinosaur monophyly and a new class of vertebrates. Nature 248:168–172.

Benton, M. J. 1983. Dinosaur success in the Triassic: a noncompetitive ecological model. Quarterly Review of Biology 58:29–55.

Benton, M. J. 1984. Dinosaurs' lucky break. Natural History 93(6):54–59.

Benton, M. J. 1990. Origin and interrelationships of dinosaurs; pp. 11–30 *in* D. B. Weishampel, P. Dodson, and H. Osmólska (eds.), The Dinosauria. University of California Press, Berkeley.

Charig, A. J. 1972. The evolution of the archosaur pelvis and hindlimb: an explanation in functional terms; pp. 121–155 *in* K. A. Joysey and T. S. Kemp (eds.), Studies in Vertebrate Evolution. Winchester, New York.

Charig, A. J. 1976. "Dinosaur monophyly and a new class of vertebrates": a critical review; pp. 65–104 *in* A. d'A. Bellairs and C. B. Cox (eds.), Morphology and Biology of the Reptiles. Academic Press, New York.

Desmond, A. J. 1979. Designing the dinosaur: Richard Owen's response to Robert Edmond Grant. Isis 70:224–234.

Gauthier, J. A. 1986. Saurischian monophyly and the origin of birds. Memoirs of the California Academy of Sciences 8:1–55.

Olsen, P. E., N. H. Shubin, and M. H. Anders. 1987. New Early Jurassic tetrapod assemblages constrain Triassic-Jurassic tetrapod extinction event. Science 237:1025–1029.

Sereno, P. C. 1991. Basal archosaurs: phylogenetic relationships and functional implications. Journal of Vertebrate Paleontology (Supplement) 11:1–53.

Sereno, P. C., C. A. Forster, R. R. Rogers, and A. F. Monetta. 1993. Primitive
dinosaur skeleton from Argentina and the early evolution of Dinosauria. Nature
361:64–66.

Wellnhofer, P. 1991. The Illustrated Encyclopedia of Pterosaurs. Crescent Books,
New York, 192 pp.

PART II

ORNITHISCHIA

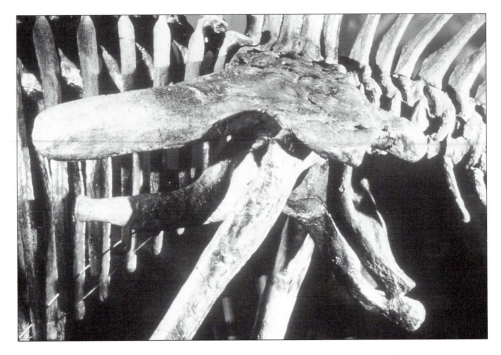

Figure II.1 Left lateral view of the ornithischian pelvis as exemplified by *Stegosaurus.* Note that the pubic bone is rotated backward to lie under the ischium in what is known as the opisthopubic condition. *Photograph courtesy of the Royal Ontario Museum.*

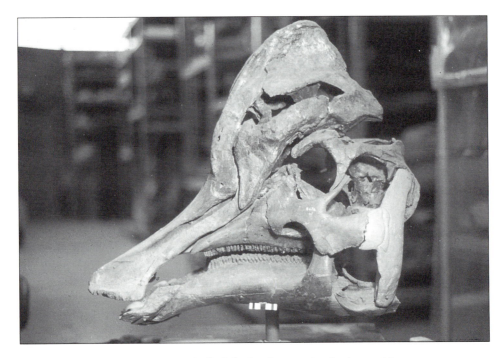

Figure II.2 Left lateral view of the skull of the lambeosaurine hadrosaurid *Corythosaurus.* Note the predentary bone capping the front of the lower jaw.

ORNITHISCHIA:
Armored, Horned, and Duckbilled Dinosaurs

LITTLE DID H. G. SEELEY guess that Ornithischia, one of the two great clades of dinosaurs, would be such a diverse group. Since 1887, we have gained an immense amount of information not only about the existence of new ornithischians (e.g., Pachycephalosauria, Heterodontosauridae, Protoceratopsidae), but also about the anatomy and diversity both of known and of newly discovered groups.

All the diversity and anatomical detail do not cloud the fact that Ornithischia is monophyletic. As is clear from the name Ornithischia, the pelvis is reminiscent of that found in birds: The original process of the pubis has rotated backward to lie close to, and parallel with, the ischium. This is called the **opisthopubic** condition (Figure II.1). The other landmark condition of ornithischians is the presence of a bone called the **predentary**, an unpaired, commonly scoop-shaped bone that caps the front of the lower jaws (Figure II.2).

Although these are the sine qua non of ornithischians, there are numerous other derived features

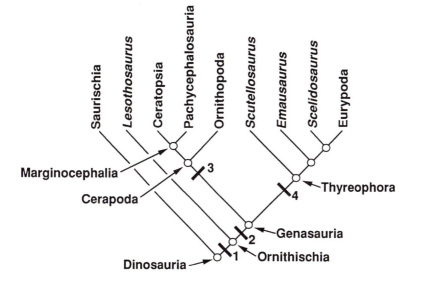

Figure II.3 Cladogram of Dinosauria, emphasizing relationships within Ornithischia – in particular those of *Lesothosaurus*, cerapodans, and basal thyreophorans. Derived characters at 1: opisthopubic pelvis, predentary bone, toothless and roughened tip of snout, reduced antorbital opening, palpebral bone, jaw joint set below level of the upper tooth row, cheek teeth with low subtriangular crowns, at least five sacral vertebrae, ossified tendons above the sacral region, small prepubic process along the pubis, and long and thin preacetabular process on the ilium; at 2: muscular cheeks (as indicated by the positioning of tooth rows away from the sides of face), spout-shaped front to the mandibles, and reduction in the size of the opening on the outside of the lower jaw (the external mandibular foramen); at 3: gap between the teeth of the premaxilla and maxilla, five or fewer premaxillary teeth, finger-like lesser trochanter; at 4: transversely broad postorbital process of the jugal, parallel rows of keeled scutes on the back surface of the body.

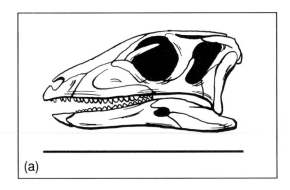

(a)

Figure II.4 **Left lateral view of the skull (a) and skeleton (b) of the basal ornithischian,** *Lesothosaurus.* Scale = (a) 10 cm; (b) 10 cm.

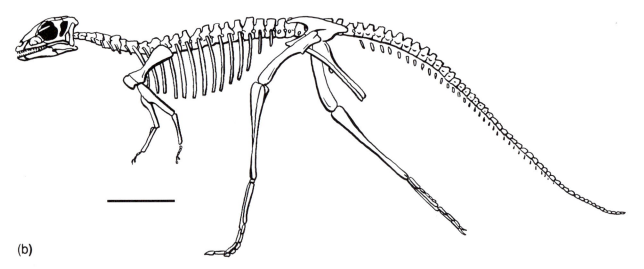

(b)

shared by these dinosaurs, including a narrow bone – the **palpebral**, that crosses the outside of the eye socket (this bone was probably mobile and set within the upper eyelid); and **ossified** (bony) tendons above the sacral region (and probably farther along the vertebral column as well), for stiffening the backbone at the pelvis (Figure II.3, node 1).

Ornithischia is made up of a number of well-diagnosed clades. For our discussion, we draw on a number of phylogenetic analyses of ornithischian clades, principal among them studies by P. C. Sereno (University of Chicago), D. B. Norman (Cambridge University), and M. R. Cooper (National Museum of Zimbabwe).

The initial split of Ornithischia is into the lone form *Lesothosaurus* and a clade termed **Genasauria** (*gena* – cheek) by Sereno (Figure II.3). The small, long-limbed herbivore *Lesothosaurus* (named for Lesotho, South Africa, where this dinosaur was discovered) was first christened by University of

Bridgeport's P. M. Galton in 1978 (Figure II.4). This Early Jurassic form had earlier been grouped with Ornithopoda, principally on the basis of primitive characters. With more recent cladistic analyses, however, it now appears to be fully ensconced as the most basal of known ornithischians.

In contrast, all remaining ornithischians – Genasauria – share the derived characters indicated in Figure II.3 (node 2), including reduction in the size of the opening on the outside of the lower jaw (the **external mandibular foramen**).

Genasaurs subsequently split into **Thyreophora** and **Cerapoda**. Taking each in turn, Thyreophora (*thyreo* – shield; *phora* – bearer; a reference to the fact that these animals have dermal armor) – a name originally proposed by Hungarian paleontologist F. B. Nopcsa in 1915 – consist of those genasaurs in which the **jugal** (one of the cheek bones) has a transversely broad process behind the eye and in which there are parallel rows of keeled dermal armor

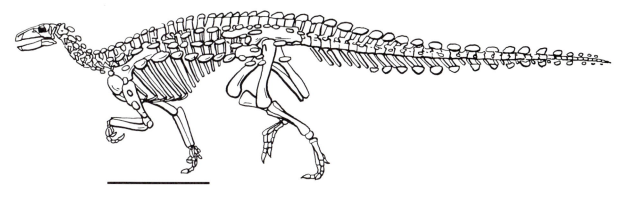

Figure II.5 **Left lateral view of the skull and skeleton of *Scelidosaurus*.** Scale = 50 cm.

scutes on the back surface of the body. The most familiar thyreophorans are stegosaurs and ankylosaurs, but there are some more primitive, yet no less important, thyreophorans. At the base of the tree is *Scutellosaurus* (*scutellum* – small shield), which was described from Lower Jurassic rocks of Arizona by E. H. Colbert (Museum of Northern Arizona, Flagstaff) in 1981. *Scutellosaurus* was a small, gracile bipedal herbivore with a back covered by small, oval-shaped plates of dermal armor. Slightly higher up the thyreophoran tree, we come to one of the newest of these armor-bearers to be discovered, *Emausaurus* (EMAU is the abbreviation for Ernst-Moritz-Arndt-Universität), also from Lower Jurassic strata (but from Germany) by H. Haubold (Geiseltalmuseum, Halle, Germany) in 1991. Finally, we have another Early Jurassic thyreophoran, the English dinosaur *Scelidosaurus* (*skelis* – limb). *Scelidosaurus* is a moderate-sized, heavily built herbivore whose limb morphology suggests that it may have been at times a quadruped and at others a biped (Figure II.5). First described by England's Sir Richard Owen in 1860, *Scelidosaurus* is the closest relative to the crowning thyreophoran clade, a taxon called **Eurypoda** (*eury* – wide; *pod* – foot).

Eurypodans consist of both stegosaurs (Chapter 6) and ankylosaurs (Chapter 7) and share as many as 20 important features not found in any of the more basal thyreophorans. These characters include short and stocky metacarpal and metatarsal bones; reduction in the **fourth trochanter** (the large process on the shaft of the femur); shortened **postacetabular process of the ilium** (the part of the ilium behind the hip socket); loss of a phalanx in digit IV of the foot; special bones that fuse to the margins of the eye socket; loss of a notch between the quadrate (the bone that buttresses the lower jaw against the skull roof) and the back of the skull; and enlargement of the forward part of the ilium (the upper bone of the pelvis).

As earlier mentioned, Thyreophora has as its sister-taxon Cerapoda (*cera* – horn), which are those genasaurs that share a **diastema** (or gap) between the teeth of the premaxilla and maxilla, five or fewer premaxillary teeth, and finger-like **lesser trochanter** (a process at the top of the femur), among other derived characters (Figure II.3, node 3). Within this large group, we encounter ornithopods (Chapter 10) and marginocephalians. **Marginocephalia** (*margin* – margin; *cephal* – head), a group united by having a narrow shelf formed from both the parietal and squamosal bones that extends over the back of the skull, a reduced contribution of the premaxillary bone to the palate (this bone primitively forms an important part of the front of the palate), and a relatively short pubis, is formed of two well-known ornithischian taxa, pachycephalosaurs (Chapter 8) and ceratopsians (Chapter 9).

Important Readings

Haubold, H. 1991. Ein neuer Dinosaurier (Ornithischia, Thyreophora) aus dem unteren Jura des nördlichen Mitteleuropa. Revue de Paléobiologie 9:149–177.

Sereno, P. C. 1986. Phylogeny of the bird-hipped dinosaurs (Order Ornithischia). National Geographic Research 2:234–256.

CHAPTER 6

STEGOSAURIA:
Hot Plates

DUMB AS A DODO. Dumb as a dinosaur! Both expressions link small brain-size and irredeemable stupidity to extinction. And in the pantheon of Dumb and Extinct, who is more celebrated than *Stegosaurus*?

Stegosaurs (or at least, *Stegosaurus*) rank today among the most familiar of all dinosaurs (Figure 6.1). As their name implies, stegosaurs (*stego* – roof; *sauros* – lizards) all had rows of bones, called **osteoderms** (*osteo* – bone; *derm* – skin), that sometimes developed into spines and plates along the neck, back, and tail. A particular type of these osteoderms – the **parascapular spines** – developed over the shoulder blades. Beyond such derived characters, stegosaurs are further characterized by their quadrupedal limb posture, and long, thin, yet relatively small heads (remember those small brains!) with simple teeth that suggest herbivory. The forelimbs in all stegosaurs are short and massive; the hindlimbs are long and columnar. The toes of both fore- and hindlimbs ended in broad hooves. The disparity in limb proportions gives these animals a profile that sloped strongly forward and downward toward the ground when they walked. Stegosaurs spanned 3 to 9 m in length and weighed in at an estimated 300 to 6,500 kg. By just about any standard, these were large animals, but still only of modest size when compared with their herbivorous contemporaries, the gigantic sauropods!

In this chapter, we will take up stegosaur intelligence (or at least its measurable surrogates), along with several other aspects of stegosaur paleobiology (among them, thermoregulation and display). But let us here lay to rest the nonsense of lack of evolutionary success among stegosaurs. These dinosaurs did very well. They rose in diversity from their origin sometime in or before the Middle Jurassic to an acme of at least seven species in the Late Jurassic. From then on,

Figure 6.1 **The best known of all plated dinosaurs, the North American *Stegosaurus*, from the Late Jurassic.** *Photograph courtesy of the Royal Ontario Museum.*

stegosaurs declined to one or two in the Early Cretaceous, and to one by the early Late Cretaceous. During their time on earth, stegosaurs spawned upward of a dozen species and stood their ground, both behaviorally and evolutionarily, with unique and complex style. And, although never particularly diverse at any one time, stegosaurs nevertheless accumulated a somewhat cosmopolitan geographic distribution (Figure 6.2).

HISTORY OF STEGOSAUR DISCOVERIES

19th Century

The earliest discoveries of stegosaurs were made in England in the early 1870s, but even the fact that they were something new (stegosaurs) went unappreciated because the bones were too fragmentary. Some were studied by the great anatomist and vertebrate paleontologist Sir Richard Owen of the British Museum (Natural History); others became the subject of papers by H. G. Seeley of Cambridge University and F. B. Nopcsa of Vienna in the early 20th century. Originally given names that have long since been changed for a variety of reasons, these first European stegosaurs now bear the monikers *Dacentrurus* (*da* – very; *kentron* – spiny; *oura* – tail; named in 1902 by F. A. Lucas of the National Museum of Natural History [Smithsonian Institution]; Figure 6.3b) and *Lexovisaurus* (after the Lexovii people of Europe; named by R. Hoffstetter of the Muséum National d'Histoire Naturelle in Paris in 1957; Figure 6.3c).[1]

---

[1] *Lexovisaurus* was originally named *Omosaurus* by Owen; however, it was found that the name was pre-occupied by another animal.

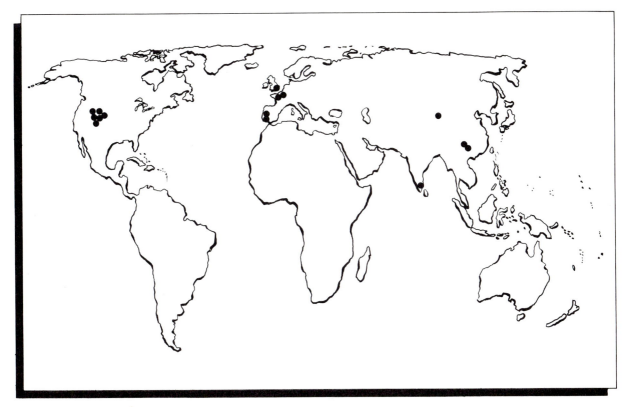

Figure 6.2 **Global distribution of Stegosauria.**

Meanwhile, as these European stegosaurs were being discovered, across the Atlantic Ocean, things dinosaurian began to brew – better yet, boil! For it was toward the end of the 1870s that a veritable wealth of dinosaur remains – including many virtually complete skeletons – were discovered at Como Bluff, Wyoming, and at many other localities in the western United States. This was the time of the Great Dinosaur Rush (see Box 6.1), and *Stegosaurus* (Figures 6.1, 6.4a, 6.4e, and 6.5) was one of its products.

Stegosaurus is the largest member of the clade (up to 9 m in length) and is known from a wealth of material, all of it from the famous Upper Jurassic Morrison Formation of North America (see Chapter 15). *Stegosaurus* has an elongate, narrow snout, which bears a roughened margin. In life, this was covered by a horn-covered beak called a **rhamphotheca** (*rhampho* – beak; *theca* – cup or sheath). Because the teeth (Figure 6.6) are deeply set in from the sides of the face, *Stegosaurus* probably had cheeks.

20th Century

From the 1870s and 1880s until the early 20th century, little was heard about stegosaurs. This changed with the discovery of *Kentrosaurus* (*kentro* – prickly,

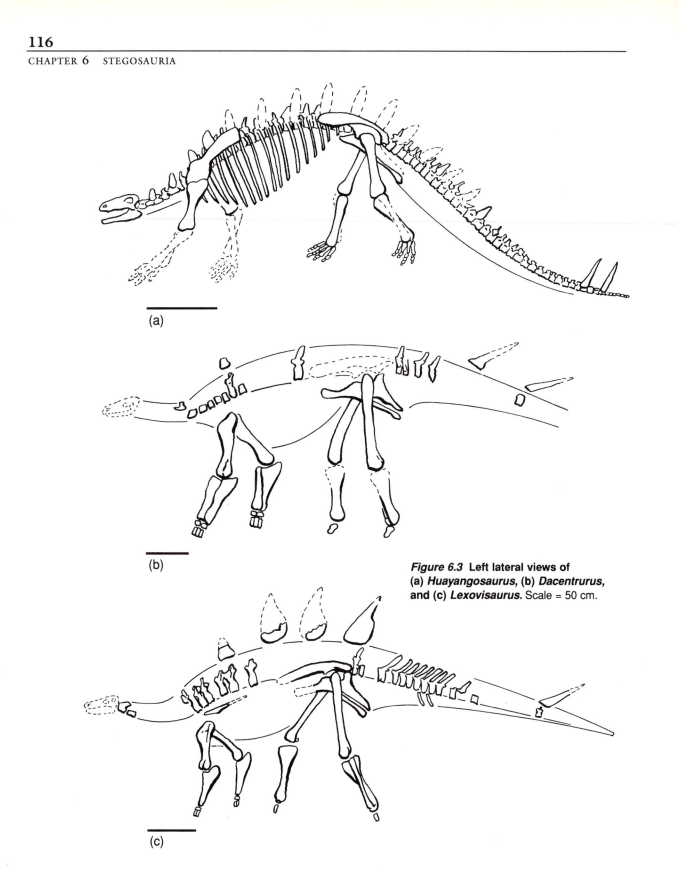

Figure 6.3 Left lateral views of
(a) *Huayangosaurus,* (b) *Dacentrurus,*
and (c) *Lexovisaurus.* Scale = 50 cm.

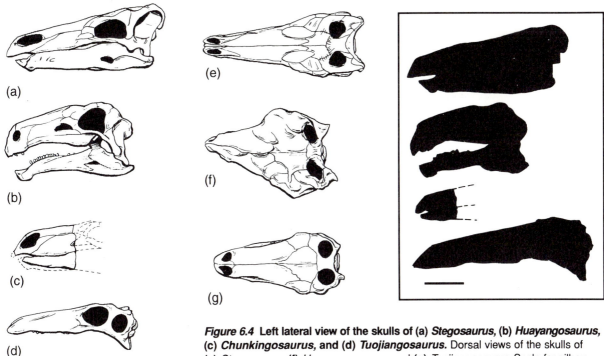

Figure 6.4 Left lateral view of the skulls of (a) *Stegosaurus*, (b) *Huayangosaurus*, (c) *Chunkingosaurus*, and (d) *Tuojiangosaurus*. Dorsal views of the skulls of (e) *Stegosaurus*, (f) *Huayangosaurus*, and (g) *Tuojiangosaurus*. Scale for silhouettes = 10 cm.

referring to its abundant spines; Figure 6.7), one of the many spectacular finds from the famous Tendaguru expeditions (see Box 11.1).

Kentrosaurus is not the only stegosaur to have been discovered on the African continent. South African beds that date from the Late Jurassic–Early Cretaceous transition have yielded an elongate snout fragment, replete with very simple looking upper teeth (Figure 6.6), that – 55 years and three distinguished paleontologists later – has been called *Paranthodon* (*para* – around; *antho* – flower; *odon* – tooth, referring to the shape of the teeth).

China is a treasure-trove of stegosaurs, but it was not until 1959 that the first stegosaur, *Chialingosaurus* (from Chialing), was discovered and described by China's premier vertebrate paleontologist, C.-C. Young, of the Institute of Vertebrate Paleontology and Paleoanthropology in Beijing. In the 1970s and 1980s China yielded five different species, mostly through the considerable work of Dong Z.-M. and co-workers at the Institute of Vertebrate Paleontology and Paleoanthropology. In close succession, these paleontologists discovered and named *Wuerhosaurus* (after Wuerho) from Xinjiang Province; and *Tuojiangosaurus* (from Tuojiang; Figures 6.4d and 6.4g), *Huayangosaurus* (a reference to Hua Yang Guo Zhi, the Chin Dynasty name for Sichuan; Figures 6.3a, 6.4b, and 6.4f), and *Chunkingosaurus* (from Chungking, China; Figure 6.4c), all from Sichuan.

In the midst of this flurry of new stegosaur discoveries in China, P. Yadagiri and K. Ayyasami from the Geological Survey of India in Hyderabad described a very important early Late Cretaceous stegosaur, *Dravidosaurus* (from Dravidandu, the southern part of India), in 1979. This, the sole stegosaur from

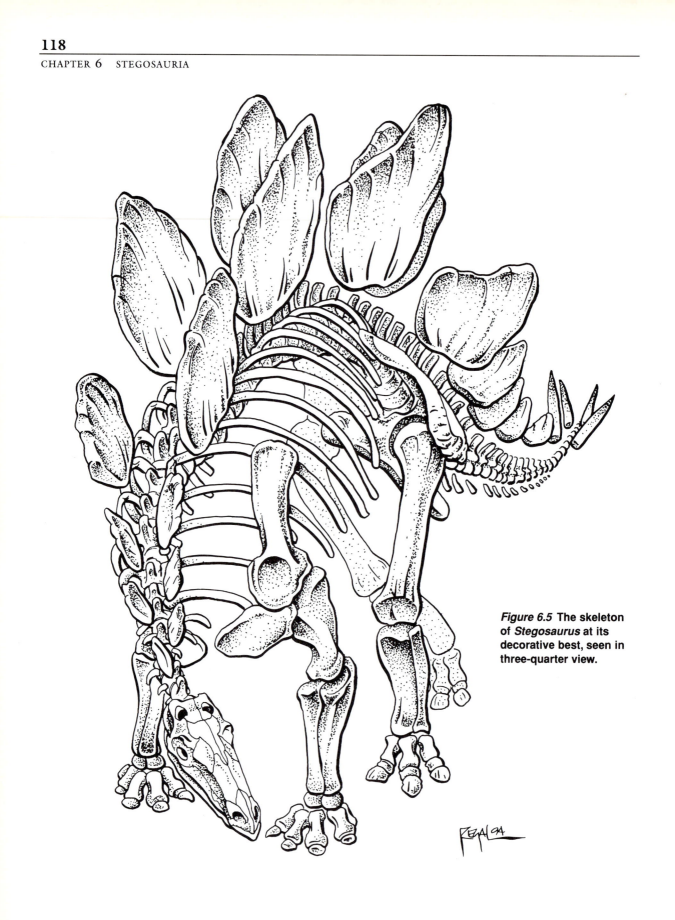

Figure 6.5 **The skeleton of *Stegosaurus* at its decorative best, seen in three-quarter view.**

India and the only stegosaur of Late Cretaceous age, is known from extremely fragmentary material. To this day, *Dravidosaurus* is the most recent stegosaur – and the most recent new stegosaur discovery.

STEGOSAURIA DEFINED AND DIAGNOSED

Stegosauria is a well defined, robustly diagnosed monophyletic clade of ornithischian dinosaurs (Figure 6.8). As will be discussed later in this chapter, its sister-group consists of ankylosaurs, the armored dinosaurs best known from the Cretaceous Period of North America and central and eastern Asia (Chapter 7). Together, stegosaurs and ankylosaurs make Eurypoda (see Part II: Ornithischia).

Returning to the group at hand, Stegosauria is that clade of ornithischian dinosaurs defined by the common ancestor of *Huayangosaurus* and *Stegosaurus* and all of the descendants of this common ancestor. As shown by P. C. Sereno and Dong, the stegosaur clade can be diagnosed on the basis of a number of important skull and postcranial features (Figure 6.8), including back vertebrae with very tall neural arches and highly angled transverse processes (Figure 6.9), and a great number of features relating to the development of osteoderms. In particular, osteoderms may form as pairs of long spines from the shoulder to the tip of the tail or may be more plate-like along the forward part of the back, angling slightly away from the center line of the animal, whereas those located more posteriorly are longer and more spine-like. Always at the end of the tail are long pairs of spines. Finally, as we have seen, virtually all stegosaurs have a parascapular spine.

STEGOSAUR DIVERSITY AND PHYLOGENY

We are beginning to recognize the diversity of stegosaurs, particularly with recent discoveries in China and India. Our understanding of phylogenetic patterns that encompass these taxa, however, has not kept pace, and what follows is a very general account of who is related to whom within Stegosauria.

BASAL SPLIT. The basal split within Stegosauria is between *Huayangosaurus* on the one hand and remaining species on the other (Figure 6.10). This divergence took place sometime before the latter half of the Middle Jurassic, the age for *Huayangosaurus*. *Huayangosaurus* itself has a number of uniquely derived features, among them the oval depression between the premaxilla and maxilla, a great number of cheek teeth, and a small horn on the top of the skull roof. Yet *Huayangosaurus* lacks a number of important derived features that are shared by a more inclusive group of stegosaurs. This group, called **Stegosauridae**, appears to be monophyletic (Figure 6.10, node 1). There may even be additional shared, derived features (not listed in Figure 6.10) that unite this clade – for example, loss of premaxillary teeth and reorientation of the quadrate bone – however, preservation of many stegosaurs is poor, and we must proceed with caution in concluding that these features have phylogenetic significance.

STEGOSAURIDAE. Within Stegosauridae, we enter the murky world of stegosaurid systematics. We'll have to wend our way through often frustratingly spotty preser-

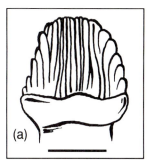

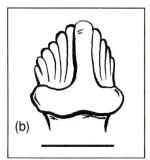

Figure 6.6 Inner view of an upper tooth of **(a)** *Stegosaurus* and **(b)** *Paranthodon*. Scale = 4 mm.

Figure 6.7 **The skeleton of Kentrosaurus, a spiny stegosaur from the Late Jurassic of Tanzania.** *Photograph courtesy of the Institut und Museum für Geologie und Paläontologie, Universität Tübingen.*

vation: Only *Tuojiangosaurus*, *Stegosaurus*, and to a degree *Chungkingosaurus* are at all well known from articulated material. Others, such as *Kentrosaurus*, *Lexovisaurus*, and *Dacentrurus*, can be reconstructed with some reliability from disarticulated but associated specimens. A few workers – among them, Dong Z.-M. and P. M. Galton – have ventured some preliminary analyses of these interrelationships. We'll hazard another view here. The cladogram we present here (Figure 6.10) may stand or fall with continued and more detailed research and discovery. It may even fall apart on the basis of your future work!

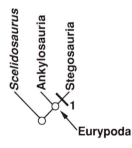

Figure 6.8. **Cladogram of "higher" Thyreophora, emphasizing the monophyly of Stegosauria.** Derived characters at 1: large oval cavity on the palatal part of the quadrate, back vertebrae with very tall neural arches with highly angled transverse processes, loss of ossified tendons down the back and tail, a broad and plate-like flange (the acromion process) on the forward surface of the shoulder blade, large and block-like wrist bones, elongation of the prepubic process on the pubic bone, loss of the first digit of the foot, loss of one of the toe bones of the second digit, formation of long spines of plates from the shoulder toward the tip of the tail, and parascapular spine.

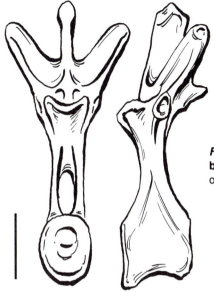

Figure 6.9 Front and left lateral views of one of the back vertebrae of *Stegosaurus.* Note the great height of the neural arch. Scale = 10 cm.

MONOPHYLETIC GROUPS WITHIN STEGOSAURIDAE.   A smaller group of stegosaurs appears to be nested within Stegosauridae. The characters that diagnose this monophyletic group – a group that excludes not only *Huayangosaurus* but also *Dacentrurus* – are listed in Figure 6.10, node 2. Although not the earliest, the most *primitive* member of this subclade is *Kentrosaurus,* which has plate-

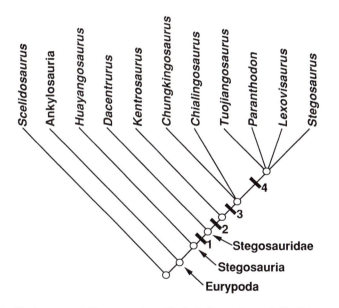

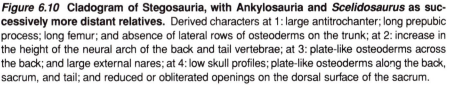

Figure 6.10 Cladogram of Stegosauria, with Ankylosauria and *Scelidosaurus* as successively more distant relatives. Derived characters at 1: large antitrochanter; long prepubic process; long femur; and absence of lateral rows of osteoderms on the trunk; at 2: increase in the height of the neural arch of the back and tail vertebrae; at 3: plate-like osteoderms across the back; and large external nares; at 4: low skull profiles; plate-like osteoderms along the back, sacrum, and tail; and reduced or obliterated openings on the dorsal surface of the sacrum.

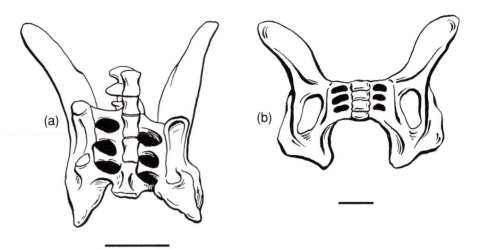

Figure 6.11 View of the underside of the pelvis of (a) *Huayangosaurus* and (b) *Tuojiangosaurus.* Scale = 20 cm.

like osteoderms over the neck and forward part of the trunk but lacks the derived feature of having them over the entire back region, one of the features shared by all remaining stegosaurs.

THE MOST-DERIVED CLADE OF STEGOSAURIDS.   The remaining stegosaurs include the Late Jurassic forms *Chungkingosaurus, Chialingosaurus, Tuojiangosaurus, Lexovisaurus, Stegosaurus,* and possibly *Paranthodon,* all of which can be grouped as a monophyletic taxon, diagnosable by having plates over all of the vertebrae of the back and also by having large external nares (?secondarily reduced in *Tuojiangosaurus*; Figure 6.10, node 3). With *Chungkingosaurus* and *Chialingosaurus* split off as basal members, we are left with *Tuojiangosaurus, Paranthodon, Lexovisaurus,* and *Stegosaurus,* all of which share derived features (Figure 6.10, node 4). The character of reduced or obliterated openings on the dorsal surface of the sacrum is shown in Figure 6.11. This diversification began by at least the beginning of the Late Jurassic, an estimate that is based on the age of the earliest known member of the clade (*Lexovisaurus*). How these derived stegosaurs relate to each other is still uncertain.

Still, we have gone from considering Stegosauria as a collection of different genera and species to understanding these forms as part of a reconstructed, branching phylogeny. Looked at from this perspective, we get a far greater sense of the pattern and dynamics of stegosaur evolution than we would treating these animals as mere names and collections of bones.

STEGOSAUR PALEOBIOLOGY AND PALEOECOLOGY:
Gaits, Pates, and Plates

As we have seen, the phylogeny of stegosaurs can be quite complicated and often-times very elusive. The same is true for their "wildlife biology." There is nothing like a stegosaur alive today; thus, we have much to surmise about even the most obvious of structures or ways of life.

Stance and Gait

The general body plan of any of the stegosaur taxa gives the impression that these animals were more on the plodding, than on the sprinting, end of the getting-around spectrum. In all but *Huayangosaurus*, the forelimbs were short and stocky and the hindlimbs, elongate as they were, had very short lower limb segments (the tibia, fibula, and foot) compared with the length of the femur. All of these features conspire to suggest that these animals were not built for great speed. Instead, they give a sense of great stability and strength, almost like elephants do today.

As we have already described, members of Stegosauridae have much longer hindlimbs than forelimbs. Indeed, this disparity in limb length is a shared, derived feature of that group of stegosaurs that excludes *Huayangosaurus*. Not surprisingly, it has some important biomechanical consequences for stegosaur locomotion, which have been comprehensively studied by R. A. Thulborn of the University of Queensland. As Thulborn pointed out, the maximal stride for the short forelimbs is limited to the arc available from all the short limb elements. In the same way, the long hindlimbs ensure that hindlimb stride was proportionately long. As a consequence, stegosaurs had a problem: At the same **cadence** (the rate of feet hitting the ground), the hindlimbs would have been able to outrun the forelimbs. At high speeds, therefore, the rear end of the animal would have overtaken its head: not a likely scenario.

Thulborn observed that the problem could be avoided in two ways: (1) Running stegosaurs could have drawn up the forelimbs up from the ground (that is, temporarily been bipedal while running), or (2) stegosaur locomotion may have been limited to a slow walking gait. The sheer bulk of a stegosaur makes the first option unlikely and the second option far more tenable. So far, we have only these anatomical observations to judge walking or running speeds in these animals; no footprints or trackways are yet known. Nevertheless, it appears likely that the pace of stegosaur life was leisurely, on the order of 6.5 to 7.0 km/hr maximal speed. Among contemporaries, stegosaurs may have been slower than the multi-ton sauropods or such large ornithopods as *Camptosaurus*. More critical to health and longevity, stegosaur running speeds were much lower than those of both the small and large theropods that must have hunted – and devoured – stegosaurs. Indeed, small theropods such as *Ornitholestes* could have run circles around stegosaurs (what *Ornitholestes* would have done when the stegosaur was caught is quite another matter; see Chapter 12: Theropoda).

Obviously, even these running speeds were not of much importance to a hungry stegosaur, for these animals were herbivorous. Their general body form (large abdominal region for a capacious gut), small head, toothless snout, and simple blunt teeth clearly speak strongly that these animals ate plants.

Dealing with Mealing

For a stegosaur – as for all gnathostomes – the business end of feeding begins at the front of the jaws. Here, a rhamphotheca covered the fronts of both the upper jaw (the premaxilla) and the predentary bone of the lower jaw, similar to rhamphothecae seen in modern turtles and birds. The rhamphothecae of the upper and lower jaws were probably sharp edged, but not hooked like on a bird of prey or a snapping turtle. By bringing these sharp edges together, the rhamphotheca became quite an effect apparatus for cropping or stripping foliage from plants. Only in *Huayangosaurus* – where the upper rhamphotheca was relatively small – was this cropping ability aided by the premaxillary teeth (remember, in those other stegosaurs where we have evidence, these teeth are absent).

Once enough food was in the mouth, chewing began. But how this was accomplished is hard to tell. The chewing (i.e., cheek) teeth of stegosaurs were relatively small, simple, and triangular (Figure 6.6) and apparently did very little grinding, because they lack regularly placed, well-developed wear surfaces on their

---

BOX 6.1

19TH-CENTURY DINOSAUR WARS:
Boxer vs. Puncher

ONE OF THE STRANGEST EPISODES in the history of paleontology was the extraordinarily nasty and personal competition between late-19th-century paleontologists Edward Drinker Cope and Othniel Charles Marsh (Figure Box 6.1). In many respects, it was a boxer vs. puncher confrontation: the mercurial, brilliant, high-strung Cope vs. the steady, plodding, bureaucratic Marsh. Their rivalry resulted in what has been called the "Golden Age of Paleontology," a time when the richness of the dinosaur faunas from western North America first became apparent; when the likes of *Allosaurus*, *Apatosaurus*, *Stegosaurus* were first uncovered and brought to the world's attention. But the controversy had its down side, too. Who were these men, and why were they at each other's throat?

Cope was a prodigy – one of the very few in the

Figure Box 6.1 **The two paleontologists responsible for the Great North American Dinosaur Rush of the late 19th century.** (Left) Edward Drinker Cope of the Academy of Natural Sciences, Philadelphia; and (right) Othniel Charles Marsh of the Yale Peabody Museum of Natural History. *Photograph courtesy of the American Museum of Natural History.*

history of paleontology. By 18, he had published a paper on salamander classification. By 24, he became Professor of Zoology at Haverford College. Blessed with independent means, within four years he had moved into "retirement" (at the grand old age of 28) to be near Cretaceous fossil quarries in New Jersey. He quickly became closely associated with the Academy of Natural Sciences, Philadelphia, where he amassed a tremendous collection of fossil bones that he named and rushed into print at a phenomenal rate (during his life he published over 1,400 works). He was capable of tremendous insight, made his share of mistakes, and was girded with the kind of pride that did not admit to errors.

Marsh, nine years older than Cope, was rather the opposite, with the exception that he, too,

crowns. Furthermore, the jaw musculature doesn't seem to have been particularly powerful: The coronoid process on the lower jaw was low, lending little mechanical advantage to the jaw musculature.

In view of the small teeth and weak jaw musculature, one might look elsewhere for the means by which stegosaurs broke up their food. Where to look would be in the stomach, for it is known that some animals (birds and crocodilians, as well as sauropodomorphs and psittacosaurs) use stones within the muscular part of the stomach (gastroliths) to fragment food (see Chapter 9: Ceratopsia, and Chapter 11: Sauropodomorpha). Although it might seem logical that stegosaurs, with their seemingly inadequate chewing apparatus, might

rushed his discoveries into print almost as fast as he made them (some thought faster!) and that he, too, did not dwell upon his mistakes. Marsh's own career started off inauspiciously: He reasoned that if he performed well at school he could obtain financial support from a rich uncle, George Peabody. This may have been Marsh's most important insight: Marsh persuaded Peabody to underwrite a natural history museum at Yale (which to this day exists as the Yale Peabody Museum of Natural History), and while he (Peabody) was at it, an endowed chair for Marsh at the Museum.

The careers of the two men moved in parallel. At first, Marsh slowly published but acquired prestige and rank, while Cope frenetically published paper after paper. When they first met, there was no obvious acrimony, but this changed when Marsh apparently hired one of Cope's New Jersey collectors right out from under him. Suddenly, the fossils started going to Marsh instead of to Cope! Then, in 1870, Cope showed Marsh a reconstruction of a plesiosaur, a long-necked, flippered marine reptile. The fossil was unusual to say the least, and Cope proclaimed his findings in the Transactions of the American Philosophical Society. Marsh detected at least part of the reason why the fossil was so unusual: The head was on the wrong end (the vertebrae were reversed). Moreover, he had the bad manners to point this out. Cope, admitting no error, attempted to buy up all the copies of the journal. Marsh kept his.

Cope sought revenge in the form of correcting something that Marsh had done. The rivalry ignited, and the battle between the two spilled out into the great western fossil deposits of the Morrison Formation (Chapter 15). Both hired collectors to obtain fossils; the collectors ran armed camps (for protection against each other's poaching!), and between about 1870 and 1890, eastbound trains continually ran plaster jackets back to New Haven and Philadelphia. Marsh and Cope rushed their discoveries into print, usually with new names. Discoveries (and replies) were published in newspapers as well as scholarly journals, lending a carnival atmosphere to the debate. Because Philadelphia and New Haven were not that far apart by rail, it was possible for one of the men to hear the other lecture on a new discovery and then rush home that night, describe it, and claim it for himself. Because many fossils in their collections were similar, this was easy to do and each accused the other of it.

Both Cope and Marsh eventually aged, and in Cope's case, his private finances dwindled. Moreover, a new generation of paleontologists arose that rejected the Cope–Marsh approach to paleontology, believing, not unreasonably, that it had caused more harm than good. Both men ended their lives with somewhat tarnished reputations. History has viewed the thing a bit more dispassionately, and it is fair to state that the result ultimately was an extraordinary number of spectacular finds and a nomenclatorial nightmare that has taken much of the past one hundred years to disentangle.*

---

* A superlative account of this rivalry can be found in A. J. Desmond's book, *The Hot-Blooded Dinosaurs*.

have had such gastroliths, skeletons of these animals have never been found with associated gastroliths.

Stegosaurs were animals with relatively weak chewing adaptations for handling food at the mouth, but they probably had cheeks, an adaptation normally associated with sophisticated oral food processing. The coexistence in stegosaurs of simple, irregularly worn teeth, cropping rhamphothecae, cheeks, and a large gut capacity all conspire to make stegosaurs unique, but still poorly understood herbivores.

At least we can get a glimmer of some more obvious aspects of feeding styles in these plant eaters. First, recall that the snout is quite narrow in all stegosaurs, suggesting a fair degree of selectivity in the food these animals were seeking. Second, remember that all but *Huayangosaurus* have a great disparity in size between the forelimb and hindlimb. From these observations, it is obvious that the head must have been brought very close to the ground, most likely near the 1-m level. And if the head was naturally positioned low to the ground, it is highly likely that these stegosaurs were principally low-browsing animals, consuming in great quantities the leaves and perhaps even succulent fruits and seeds of such ground-level plants as ferns, cycads, and other herbaceous gymnosperms.

But were stegosaurs confined to such low browsing – behavior that could have put them into direct competition with other low browsers, such as the smaller and more agile hypsilophodontid ornithopods? Not necessarily so, suggests R. T. Bakker of Boulder, Colorado, who argues that some forms, particularly *Stegosaurus*, may have been able to rear upon their hindlimbs in order to forage at higher levels, perhaps even into the crowns of tree. This kind of posture is similar to that generally assumed by kangaroos and occasionally by elephants. But stegosaurs may have gone the elephant one better. First, with the center of gravity near the hips, the hindlimbs would already be supporting nearly 80% of the weight of the body. In addition, the strong, flexible tail would have been able to act as a third "leg" to form a tripod under the animal as it attempted to rear up. Should this kind of behavior have been utilized by stegosaurs, then these animals may have been able to feed on leaves quite high up in the trees (perhaps as much as 2 m for *Kentrosaurus*, and 4 m for *Stegosaurus*).

Brains

Turning back to stegosaur anatomy and what it can tell us about stegosaur biology, we can learn a great deal about these animals in terms of the structure of their nervous system. Here we enter the realm of brain size and the nature of the spinal cord, neither of which is preserved. Still, the spaces where these neural organs once lay have proven informative.

Whatever the pretensions of dinosaurs to deep thought, stegosaurs cannot be ranked among the crowning luminaries. With very small brains (an estimated 0.001% of body weight), it is certainly well justified to consider stegosaurs near the bottom of the gray-matter scale. Only some sauropods, and perhaps ankylosaurs, had proportionately smaller brains (see Chapters 7 and 11).

The means by which these measures are obtained and their meanings are interpreted, are exceedingly important and worth describing. How, after all, do we know the size and shape of a dinosaur brain? The brain is soft tissue, and commonly decomposes long before the process of fossilization can begin. As many workers have shown, however, casts can be obtained of the interior of the brain-cases of fossil vertebrates. Latex is painted onto the inside of a well-preserved, three-dimensional braincase. When the latex has dried (and is flexible), it can be peeled off the inside of the braincase and pulled through the **foramen magnum**, the opening through which the spinal cord enters the skull. The result is a three-dimensional cast of the region occupied by the brain (see Figure 14.8). Such casts give some inkling about the shapes and sizes of brains. Dinosaur brains can be no larger than the volume of these casts, but they were probably significantly smaller. Observations made of the brains of living lizards, snakes, and crocodilians show that these brains take up less room within the braincase than do those of mammals or birds. Researchers have long thought that the brains of non-avian dinosaurs should be similarly smaller than the entire volume of the braincase, and thus correction factors have been developed, which are used in calculations of dinosaurian gray matter.

On the basis of studies of fossil "brains" by H. J. Jerison of the University of California at Los Angeles and further, more detailed research by J. A. Hopson of the University of Chicago (see Box 14.3), it is now clear that brain size in vertebrates scales negatively allometrically with body size (technically, to the 0.67 power). What this means is that as animals get bigger, their brains also get larger, but at a rate not equal to their size. Even for large-brained mammals like ourselves, as we reach maturity and stop growing, our brains have grown less than have our bodies.

Hopson used this relationship between estimated dinosaurian brain size (calculated from the expected brain size of lizards, snakes, and crocodilians scaled up to dinosaur size) and dinosaurian body size to make comparisons with measured brain size in actual dinosaurs. Among stegosaurs, casts of the braincase are available only for *Kentrosaurus* and *Stegosaurus*, both of which are reasonably similar in shape and size. The brains of these two stegosaurs are relatively long, slightly flexed, and above all else *small*! The only aspect of the stegosaur brain that appears to have been somewhat large is its extraordinary olfactory bulbs, those portions of the brain that provide the animal its sense of smell. Otherwise, the mental and sensory faculties of stegosaurs suggest an assuredly unhurried life style and possibly an uncomplicated range of behaviors.

Brains between the Thighs

Yet the small-brained stegosaur defied even those who considered dinosaurs to be dullards because, in a quirk of anatomy, stegosaurs apparently looked elsewhere in order to embellish their anomalous nervous systems. It was O. C. Marsh who, over a century ago, observed the very much enlarged canal – upwards of 20 times the volume of the brain! – that accommodated the spinal cord in the hip region. Here began the legend of the dinosaur with two brains: a diminutive one in the head that monitored and controlled the front half of the animal, and another in the sacral region that did the same for the remainder. The subject of poetry (Box 6.2), this sacral enlargement has caused a great deal of controversy. Some suggested, as did Marsh, that the enlarged spinal cord "covered" for what the standard brain could not handle, whereas others hypothesized that it was the logical enlargement for the hindlimb and tail region in an animal that emphasized both for reasons of locomotion (hindlimb domination) and protection (the tail). Although this latter expla-

---

BOX 6.2

THE POETRY OF DINOSAURS

DINOSAURS HAVE BEEN THE SUBJECTS of numerous poems, limericks, and other bits of doggerel, virtually since the time of their earliest discovery. Most have centered around their enormity and putative lack of brain power and social graces, with few recent efforts to balance such dismal views.

The most famous dinosaurian poem celebrates the mental claims of *Stegosaurus*, in particular the cerebral gymnastics supplied by its double brains – the standard issue and the one in its keester. The piece, by Bert L. Taylor, a columnist in the 1930s and 1940s for the *Chicago Tribune*, goes like this:

> Behold the mighty dinosaur,
> Famous in prehistoric lore,
> Not only for his power and strength
> But for his intellectual length.
> You will observe by these remains
> The creature had two sets of brains –
> One in his head (the usual place),
> The other at his spinal base.
> Thus he could reason *a priori*
> As well as *a posteriori*.
> No problem bothered him a bit
> He made both head and tail of it.
> So wise was he, so wise and solemn,
> Each thought filled just a spinal column.
> If one brain found the pressure strong
> It passed a few ideas along.
> If something slipped his forward mind
> 'Twas rescued by the one behind.
> And if in error he was caught
> He had a saving afterthought.
> As he thought twice before he spoke
> He had no judgement to revoke.
> Thus he could think without congestion
> Upon both sides of every question.
> Oh, gaze upon this model beast,
> Defunct ten million years at least.

Another effort, by William S. Mills, treats dinosaurs – as well as other extinct life forms – in an even more lyrical fashion. This issue of brain power yet again makes an appearance as the principal dinosaurian Achilles' heel:

> A family name was this Dino,
> A tribe unaccountably queer,
> Yet, purpose all-wise, you and I know,
> Directed and planned their career.
> To name every "sauros" comprises
> A task one would deeply deplore;
> The list would include all the sizes
> From one foot to eighty and more.
> There is no wild extravaganza
> Displayed in this effort to say
> How creatures here pictured in stanza
> Once lived where we now have our day.
> Beneath our farms, blooming in tillage,
> No history tells where or when,
> They swashed in cantankerous pillage
> And warfare, in bayou and fen.
> They tore on another asunder,
> And floundered and splashed in the slough;
> The noise must have sounded like thunder,
> So savage the strenuous row.
> Elasmo, and Mosasaur, Allo,
> Atlanto, and Stego, were kin,
> All sauruses, small-brained, and shallow,
> Deplorable plight to be in.
> Take Brontosaur, just for example,
> His weight forty thousand, immense;
> His brain but two pounds; could such sample
> Of hulk be possessed of much sense?
> When Stego, out foraging, tarried
> To serve up the herbs near his track,
> He felt well equipped, for he carried
> Two rows of bone plates on his back . . .

The poem goes on in this vein for many more pages, but we all get the general idea.

nation is more in keeping with comparative neuroanatomy, neither has garnered much acceptance in the paleontological literature.

In 1990, E. B. Giffin of Wellesley College reexamined the issue of enlargements of the neural canals in sacral vertebrae in dinosaurs. She did so with an eye for how large the neural canals should be, given the size of the spinal cord, in much the same way that Hopson and, earlier, Jerison had done with brain size. First, she found that the sacral enlargements in stegosaurs (also in sauropods; see Chapter 11) were more than large enough to accommodate the normal expansion of the nerve cord that passes through this region, giving off nerves to the hindlimbs and continuing backward to the tail. That is, it was much too large just for accommodating nerves. Only in living birds is this enlarged anatomy known to occur, and in these forms there is a structure called a glycogen body that takes up the extra space. The function of glycogen bodies is not well understood, but it is thought to be to supply glycogen (one of the carbohydrate reserves of the body) to the nervous system, where it is used in the synthesis of specialized kinds of nervous tissue. By comparison, then, it is possible that the inordinately large expansion of the neural canal of the sacrum in stegosaurs also housed a glycogen body, much like the one all modern birds have. It did not function as a second brain and, in all likelihood, did not provide space for more than the usual volume of nerves to the hindlimbs and tail.

Behavior

Simply put, we don't have much of an idea about the social behavior of stegosaurs, nor much of an idea about their life histories. For example, no nests, isolated eggs, eggshell fragments, nor hatchling material is yet known for any stegosaur. In fact, only a few juvenile and adolescent stegosaur specimens are known thus far for *Dacentrurus*, *Kentrosaurus*, *Lexovisaurus*, and one of the *Stegosaurus* species. Among fully adult individuals, it appears that there is some **sexual dimorphism** – that is, differences between the sexes. This shows up in, of all places, the number of ribs that contribute to the formation of the sacrum. Whether it is the male or the female that has the greater number of ribs is anybody's guess! Other ways in which the differentiation of the sexes may have been manifested are unknown, but it might be that sexual dimorphism would be found in the size and shape of the spines and/or plates if only we had better samples. Within species, little is known about the degree of sociality among stegosaurs. The mass accumulation of disarticulated, yet associated, *Kentrosaurus* material from Tendaguru (Box 11.1) provides us with a hint that – in this stegosaur, at least – there was some degree of herding behavior, either seasonal or perennial. In other species, however, we have no such information – they may have been solitary creatures or gregarious. The fossil record is silent on this issue – so far.

Plates and Spines

Whether or not stegosaurs engaged in herding, there are some features of these animals that assuredly reflect their paleobiology: the spines and plates. The majority of stegosaurs have at least one row of osteoderms along the dorsal margin of each side of the body, and these osteoderms generally have the form of spines, either long and drawn-out spikes (as in *Kentrosaurus*) or as blunt conelike affairs (as in *Tuojiangosaurus*). Only in *Stegosaurus* do the majority of osteoderms appear as large, leaf-shaped plates. In all cases, at the end of the tail were pairs of long spines. All of these plates and/or spines, regardless of their position on the body, did not articulate directly with the underlying neural spines of the vertebrae, but instead were embedded in the skin (Figure 6.12).

Once thought to have solely protective and defense importance, the spines and plates of stegosaurs have begun to take on complex behavioral significance, relating not only to these aforementioned functions, but also to display and thermoregulation. On the one hand, enlarged spines and, particularly, plates would have provided these animals with a much larger, more formidable appearance. Based on studies of *Stegosaurus*, Bakker argues strongly that the osteoderms were mobile at their bases and hence able to rotate from a folded-down to an erect position, giving these animals a greater degree of protection from, and deterrence to, predators. Others, including V. de Buffrénil and his colleagues at the University of Paris VII, disagree. Working on the microscopic structure of the plates of *Stegosaurus*, they instead suggest that these plates were not particularly mobile, but rose nearly vertically on the back. In this position, the plates would have been quite useful in display, as well as in thermoregulation.

As noted by Russian evolutionary biologist L. S. Davitashvili, as well as by N. B. Spassov of the National Natural History Museum of Bulgaria and, most recently, by de Buffrénil and his colleagues, the shapes and patterns of plates and spines in stegosaurs are nearly always species specific. That is, more often than not, we use the shape and size of the plates and spines as derived features for stegosaur species. In all cases, osteoderms were arranged for maximal visual effect and thus have made their greatest impact during lateral display. If this interpretation is correct – and wouldn't it be great if we could identify sexual dimorphism in their size, shape, or placement pattern? – then it is possible that stegosaurs could have used

Figure 6.12 **Diagram of one of the best skeletons of *Stegosaurus*** as it was found in the field. Note that the plates do not articulate directly with the vertebrae and are thus believed to have been embedded in soft tissue. Scale = 50cm.

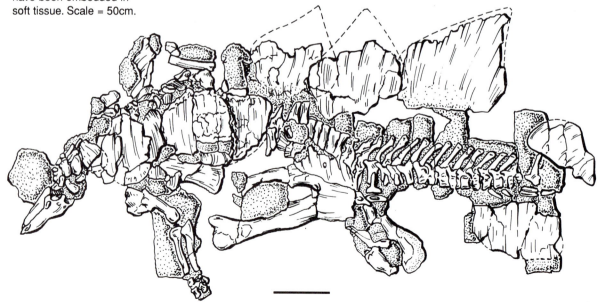

spines and plates not only to tell each other apart, but also to gain dominance in territorial disputes and/or as libido enhancers during the breeding season.

Yet our picture of the functional significance of these osteoderms, at least for plates, is incomplete. Also arguing that plates would have offered poor protection from predators, J. O. Farlow (then at Yale University) and colleagues analyzed these structures as heat radiators and/or solar panels for regulating body temperature. The plates of *Stegosaurus* are covered with an extensive pattern of grooves, and the insides are filled with a honeycomb of channels (Figure 6.13). These external grooves and internal channels most likely formed the bony walls for an elaborate network of blood vessels.

With such a rich supply of blood from adjacent regions of the body, Farlow and colleagues argued, these plates were well designed to cool the body by dissipating heat in a breeze or to warm the body by absorbing solar energy. The "fine control" for cooling and heating is the regulation of blood flow to plates.

As a test of these ideas, a crude model of a stegosaur with plates along its back was placed in a wind tunnel, and temperature changes were monitored by a thermocouple placed inside the "animal." How these plates were arranged on the model becomes critical to their thermoregulatory performance. In the case of this experiment, two patterns were tested.

First, plates were positioned as symmetrical pairs. This arrangement proved to provide more than adequate heat dissipation. When the plates were placed in alternating positions, however, they functioned much better to dissipate internal heat loads. Thus on biophysical grounds, the plates of *Stegosaurus* could have been arranged in alternating pairs. Yet is there any evidence as to their actual arrangement, regardless of their relative ability to absorb or dump heat?

Suggestions of how stegosaur plates were positioned date back almost to the inception of studies of the group. It was Marsh, in 1891, who first advocated not pairs of plates, but a single row down the back. This hypothesis lasted only a

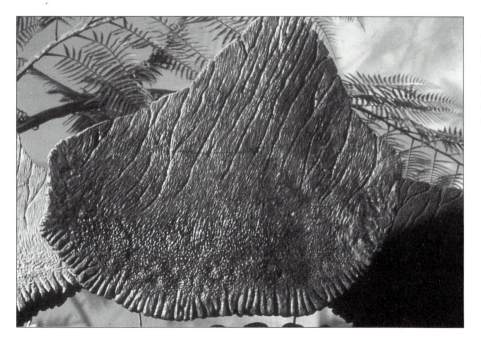

Figure 6.13 **Lateral view of one of the dermal plates of *Stegosaurus*.** Note the great number of parallel grooves, presumably conveying blood vessels across the outer surface of the plate. *Photograph courtesy of the Royal Ontario Museum.*

decade, when Lucas replaced Marsh's idea with the suggestion that the back of *Stegosaurus* supported paired, but staggered, rows of plates. Not to be outdone, R. S. Lull of Yale argued that this was only half right: There were indeed two rows of plates, but they were symmetrically placed (he argued that the staggered interpretation was in error, because the plates of a specimen had slipped after death).

As it turns out, it is the staggered arrangement of plates in paired rows that looks like the best interpretation. Based upon newly discovered *Stegosaurus* material, K. Carpenter and B. Small of the Denver Museum of Natural History suggest that the beast does indeed have the familiar staggered double-plate arrangement.

One final point on the plates of *Stegosaurus*. Juveniles may not have developed osteoderms on their backs. If true, and if we are right in presuming that plates functioned in display and/or thermoregulation, then their supposed absence in small, sexually immature individuals may reflect already adequate ability in dumping or absorbing heat and the irrelevance of looking big and sexy.

We are left with the functional significance of the long, pointed parascapular spines and the pairs of terminal tail spikes. In both cases, these appear to be the main means of defense for stegosaurs. Although the parascapular spines were likely immobile, the tail spikes could have been swished from side to side on the powerful tail in order to injure, or to deter rear assaults by, predators. How the flanks of the animal were protected – even with the parascapular spines – is far from clear, and perhaps stegosaurs were particularly susceptible to attack on this account.

It is this kind of information – plate and spine function – that can, and should, be integrated with stegosaur phylogeny to help understand the evolutionary history of the group. For example, we might imagine that, basally in stegosaur history, portions of the neck and all of the back and tail of these animals were covered with backwardly projecting pairs of spines. This primitive condition may have been tied to a protective function, but it is equally likely that spines across the back and tail may have been part of the visual display complex.

Whatever the principal function(s) of this primitive stegosaur condition might have been, these large, spiny projections could not have avoided collecting some solar radiation and/or providing an avenue for dumping heat, even if they were initially horn covered. Display and thermoregulatory functions became more important during stegosaur phylogeny as osteoderms became more plate-like down the back and, at the same time, more vascular. Finally, with the acquisition of a full complement of plates in *Stegosaurus*, there is a commitment of all osteoderms but the terminal spines to both display and thermoregulation.

Seen in this way, the story of stegosaur evolution is one of changing osteodermal patterns and functions. Moving from spiny osteoderms, whose primary functions are defensive and possibly display, to a condition where display and thermoregulation are inextricably linked via plate-like osteoderms, stegosaurs appear to have mastered the business of looking and feeling hot.

Important Readings

Alexander, R. McN. 1989. Dynamics of Dinosaurs and Other Extinct Giants. Columbia University Press, New York, 166 pp.

Bakker, R. T. 1986. Dinosaur Heresies. William Morrow, New York, 481 pp.

Bakker, R. T. 1987. Return of the dancing dinosaur; pp. 38–69 *in* S. J. Czerkas and

E. C. Olsen (eds.), Dinosaurs Past and Present, Vol I. Natural History Museum of Los Angeles County, Los Angeles.

Buffrénil, V. de, J. O. Farlow, and A. de Riqlès. 1986. Growth and function of *Stegosaurus* plates: evidence from bone histology. Paleobiology 12:459–473.

Carpenter, K. 1993. New evidence for plate arrangement in *Stegosaurus stenops* (Dinosauria). Journal of Vertebrate Paleontology (Supplement) 13:28A–29A.

Czerkas, S. 1987. A reevaluation of the plate arrangement of *Stegosaurus stenops;* pp. 82–99 *in* S. J. Czerkas and E. C. Olsen (eds.), Dinosaurs Past and Present, Vol II. Natural History Museum of Los Angeles County, Los Angeles.

Davitashvili, L. S. 1961. The Theory of Sexual Selection. Izdatel'stov Akademia Nauk SSSR, Moscow (in Russian).

Galton, P. M. 1990. Stegosauria; pp. 435–455 *in* D. B. Weishampel, P. Dodson, and H. Osmólska (eds.), The Dinosauria. University of. California Press, Berkeley.

Giffin, E. B. 1991. Endosacral enlargements in dinosaurs. Modern Geology 16: 101–112.

Gilmore, C. W. 1914. Osteology of the armored Dinosauria in the United States National Museum, with special reference to the genus *Stegosaurus*. Bulletin of the U.S. National Museum 89:1–136.

Hopson, J. A. 1977. Relative brains size and behavior in archosaurian reptiles. Annual Reviews of Ecology and Systematics 8:429–448.

Hopson, J. A. 1980. Relative brain size in dinosaurs: implications for dinosaurian endothermy; pp. 278–310 *in* R. D. K. Thomas and E. C. Olson (eds.), A Cold Look at the Warm-Blooded Dinosaurs. American Association for the Advancement of Science, Selected Symposium no. 28.

Jerison, H. J. 1973. Evolution of the Brain and Intelligence. Academic Press, New York, 482 pp.

Sereno, P. C. 1986. Phylogeny of the bird-hipped dinosaurs (Order Ornithischia). National Geographic Research 2:234–256.

Sereno, P. C., and Dong Z.-M. 1992. The skull of the basal stegosaur *Huayangosaurus taibaii* and a cladistic analysis of Stegosauria. Journal of Vertebrate Paleontology 12:318–343.

Spassov, N. B. 1982. The "bizarre" dorsal plates of *Stegosaurus*: ethological approach. Compte Rendu de l'Académie Bulgare des Sciences 35:367–370.

Thulborn, R. A. 1982. Speeds and gaits of dinosaurs. Palaeogeography, Palaeoclimatology, Palaeoecology 38:227–256.

CHAPTER 7

ANKYLOSAURIA:
Mass and Gas

NATURE HAS A PENCHANT for inventing suits of armor. Today, we have armadillos, pangolins, turtles, and pill bugs. And in the past, there were glyptodonts (large, squat, armored relatives of sloths and armadillos that died out about 10,000 years ago) and glyptosaurs (armored lizards now extinct 40 million years), and among the arthropods, the long-gone trilobites. But nobody did it like ankylosaurs, the Mesozoic armor-plated, spiked-shouldered dinosaurs who made self-defense-by-hunkering-down an art.

As their name implies, ankylosaurs (*ankylo* – fused; *sauros* – lizard) were encased in shell-like dermal armor (Figure 7.1). Not only did this pavement of bony plates and spines – each embedded in skin and interlocked with adjacent plates – form a continuous shield across the neck, throat, back, and tail, it also covered the top of the head and often the cheeks. This armor varied and is used to identify particular ankylosaurs even from scrappy material.

Under the armor covering, the ankylosaur body was round and broad, an appropriate design for housing a large gut to digest food. Likewise, the head was low and broad, and equipped with simple, leaf-shaped teeth for pulverizing whichever plants an ankylosaur chose to feed on.

Few specimens are preserved from head to tail; however, our best estimates suggest that these stocky quadrupeds were rarely over 5 m in length, although some (such as *Ankylosaurus*) may have reached 9 m. Like stegosaurs, the limbs were short, with the hindlimb exceeding the length of the forelimb by 50%. These proportions are in keeping with the general construction of the body and the presumed slow pace of the animals.

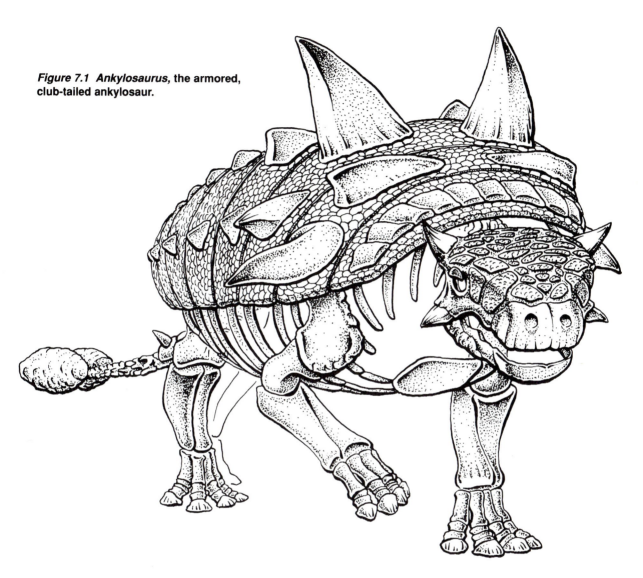

Figure 7.1 *Ankylosaurus,* the armored, club-tailed ankylosaur.

HISTORY OF ANKYLOSAUR DISCOVERIES

19th Century

Ankylosaurs were there at the very beginning of the history of dinosaur studies. Along with *Iguanodon* and *Megalosaurus* (an ornithopod and a theropod respectively), *Hylaeosaurus* (*hylaios* – forest-dwelling) was a charter member of Sir Richard Owen's Dinosauria. This ankylosaur from England stands today almost as completely known as it did in the mid-1800s: mostly the front half of the skeleton and parts of the armor shield.

It took another 34 years to find a rear end! This creature, named *Polacanthus* (*poly* – many; *akantha* – spine), was closely followed by *Acanthopholis* (*pholis* – scale). Both were named by T. H. Huxley, most famous for placing his debating abilities in the service of evolution.

For the next 30 or so years, ankylosaurs – big, squat herbivores – were somehow confused by their discoverers for slim carnivores. First there was E. Bunzel of the University of Vienna, who in 1871 identified some Upper Cretaceous bones as *Struthiosaurus* (*strouthion* – ostrich). Then R. Lydekker of the British Museum (Natural History) in London named *Sarcolestes* (*sarkos* – flesh; for flesh-robber) for an unusual lower jaw from central England. It took another 90 years before the true identities of both of these were resolved.

Things got off on a better footing in North America, where discoveries began in the late 1880s. Yale paleontologist O. C. Marsh announced a new armored dinosaur, *Nodosaurus* (*nodus* – knob) in 1889. American ankylosaurs had to await the early 20th century, however, for their renaissance.

20th Century

Shortly after *Nodosaurus*, *Hoplitosaurus* (*hoplites* – Greek, shield-carrying foot soldier) was reported from Lower Cretaceous rocks of South Dakota. Named in 1902 by F. A. Lucas of the National Museum of Natural History (Smithsonian Institution), this dinosaur is so far based on no more than fragmentary forelimb and hindlimb material and some spines and plates. At the same time,

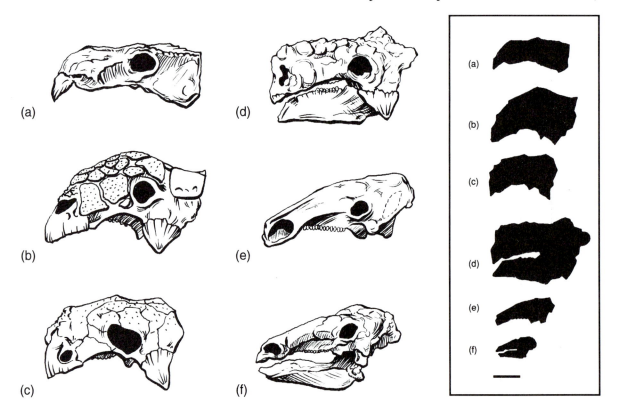

Figure 7.2 Left lateral view of the skulls of (a) *Shamosaurus*, (b) *Ankylosaurus*, (c) *Pinacosaurus*, (d) *Tarchia*, (e) *Silvisaurus*, and (f) *Panoplosaurus*.
Scale for silhouettes = 10 cm.

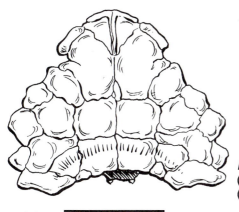

(a)

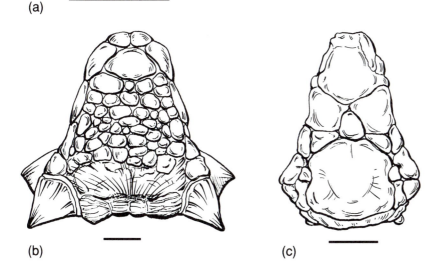

Figure 7.3 Dorsal view of the skulls of (a) *Pinacosaurus,* **(b)** *Ankylosaurus,* **and (c)** *Panoplosaurus.* Scale = 10 cm.

(b)   (c)

Euoplocephalus (*eu* – true; *hoplon* – shield; *kephale* – head), one of the all-time best known of these armored dinosaurs, was discovered in the great badlands of the Red Deer River of Alberta and given its first name, *Stereocephalus,* in 1902 by L. M. Lambe of the Geological Survey of Canada. Unfortunately, this name had already been given to an insect,[1] so in 1910 Lambe gave it its presently recognized name – *Euoplocephalus.*

Thereafter, on a nearly regular basis (a new ankylosaur every 10 years!), new ankylosaurs began to be discovered and named. *Ankylosaurus* (Figures 7.2b and 7.3b) was discovered in Montana by B. Brown of the American Museum of Natural History in 1908. *Panoplosaurus* (*pan* – all; Figures 7.2f and 7.3c), like *Euoplocephalus,* was found in Alberta and named by Lambe. C. M. Sternberg of the National Museums of Canada nabbed *Edmontonia* (for the Edmonton Formation of Alberta) in 1928.

Then came the booty from Mongolia. In 1933, C. W. Gilmore named

[1] Insects continue to plague dinosaur nomenclature. As recently as 1993, the Mongolian Cretaceous bird "*Mononychus*" was described. Unfortunately, the name was already occupied by a beetle. Consequently, "*Mononychus*" had to be renamed *Mononykus,* the name this creature carries to this day (see Chapter 13).

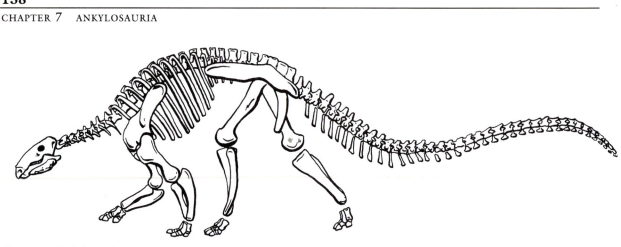

Figure 7.4 Left lateral view of the skeleton of *Sauropelta* without its armor shield.
Scale = 50 cm.

Pinacosaurus (*pinak* – plank; Figures 7.2c; and 7.3a), recovered by an American expedition led by R. C. Andrews (see Box 7.1). Next came *Talarurus* (*talaros* – basket; *oura* – tail), another ankylosaur from Mongolia, this time collected by the Joint Soviet-Mongolian Palaeontological Expeditions and named by E. A. Maleev of the Palaeontological Institute in Moscow. Back in North America, T. H. Eaton, Jr., of the University of Kansas discovered and named *Silvisaurus* (*silva* – forest; Figure 7.2e) in Kansas in 1960, and in 1970 J. H. Ostrom of the Yale Peabody Museum of Natural History named *Sauropelta* (*pelte* – shield; Figure 7.4) from Wyoming and Montana.

The end of the 1970s and the 1980s were the bonanza years for ankylosaur discovery. In 1977, two ankylosaurs – *Tarchia* (Mongolian: *tarchi* – brain; Figure 7.2d) and *Saichania* (Mongolian: *saichan* – beautiful) – were collected by the

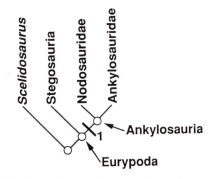

Figure 7.5 Cladogram of "higher" Thyreophora, emphasizing the monophyly of Ankylosauria. Derived characters at 1: low skull; closure of the antorbital and upper temporal openings; obliteration of sutures between cranial bones of the skull roof of adults; ossification of accessory antorbital elements; walling up of the back side of the eye socket; closure of the palate by the pterygoid bones; subdivision of the nasal cavity by fusion of the vomer bones; ossification and fusion of a keeled plate onto the side of the lower jaw; fusion of some of the ribs to their vertebrae and possibly to the ilium; fusion of some of the first tail vertebrae to the sacral vertebrae and ilium; rotation of the ilium to form flaring blades; closure of the hip joint; development of a dorsal shield of symmetrically placed bony plates (both large and small) and spines; reduction of the pubis and its removal from contributing to the hip joint; and loss of the prepubic process.

Polish-Mongolian Palaeontological Expeditions of the late 1960s and 1970s. *Minmi* (named for Minmi Crossing, Queensland) surprised the world when R. E. Molnar of the Queensland Museum described this first ankylosaur from Australia, indeed the first from the Southern Hemisphere! That same year, P. M. Galton described a new European ankylosaur, which he named *Dracopelta* (*drakon* – dragon). The 1980s closed out with two new ankylosaurs from Mongolia: *Shamosaurus* (Chinese: *shamo* – desert, a reference to the Gobi Desert; Figure 7.2a) and *Maleevus* (named in honor of Maleev), both named by T. A. Tumanova of the Palaeontological Institute in Moscow (in 1983 and 1987, respectively). And as recently as 1993, Tumanova named *Tsagantegia* from Bayn Shireh, Mongolia. Taken together, Ankylosauria presently contains over 20 genera and as many as 25 species. With this flood of activity, who knows what new ankylosaurs await our attention in the rocks or in the museum drawers?

ANKYLOSAURIA DEFINED AND DIAGNOSED

Ankylosauria is defined as the common ancestor of the two great clades of anky-losaurs – Ankylosauridae and Nodosauridae – and all its descendants (Figure 7.5). As we learned earlier, Stegosauria comprises the closest relative to Ankylosauria, together forming Eurypoda.

It is not surprising that armor and/or its support comprise the majority of derived features uniting Ankylosauria. In apparent pursuit of full protection these animals all share the features listed in Figure 7.5.

ANKYLOSAUR DIVERSITY AND PHYLOGENY: Actors in the Ankylosaur Armory

We begin this section on Ankylosauria with our standard moan: "If we only had more information . . ." We know very little about the pattern of armor and spines along the neck, back, and tail in all but three or four species. Two that we do

Figure 7.6 **Dorsal view of the body armor of *Euoplocephalus*.**

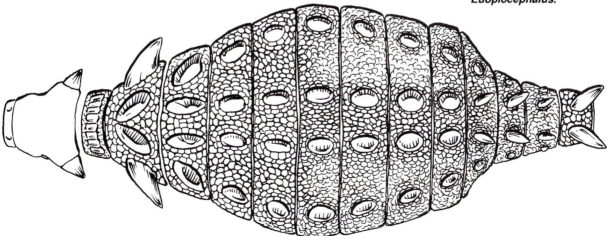

BOX 7.1

INDIANA JONES AND THE CENTRAL ASIATIC EXPEDITIONS OF THE AMERICAN MUSEUM OF NATURAL HISTORY

HE STANDS IN THE MIDDLE of the remote, rugged Mongolian desert: high leather riding boots, riding pants, broad-brimmed felt hat, leather-holstered sidearm hanging from a glittering ammunition belt. He carries a rifle and knows how to use it. Nobody else dresses like him, but then nobody else is the leader of the American Museum's Central Asiatic Expeditions to Mongolia (which in the 1920s could have been the moon). He is Roy Chapman Andrews, who 50 years later will be the inspiration, it is most plausibly rumored, for Indiana Jones (Figure Box 7.1).

Andrews always knew that he was a man with a destiny. Although he began his career at the American Museum of Natural History (AMNH) rather modestly (he scrubbed floors), his training in mammalogy (an M.A.), sheer will, and a *very* good idea carried him the distance. He had traveled extensively, spoke several Asian languages more or less fluently (at a time when very few Westerners did), and had fabulous contacts in Beijing (then called Peking).

His idea was simple: to run an expedition to

Outer Mongolia. Andrews's timing was superb: The Director of the AMNH, the august H. F. Osborn, had concluded that the cradle of humanity was located in Outer Mongolia, and so Andrews was effectively offering Osborn the opportunity to prove his thesis right (the possibility that Osborn could be *wrong* was not an option). The logistics of the expedition were extravagant: Dodge cars – resupplied by a caravan of camels – would bear the brunt of the expedition. The expedition itself would consist of paleontologists, geologists, and geographers to explore the Gobi Desert, the huge desert that forms the vast southern section of Outer Mongolia.

It was not without its risks. The Gobi Desert is a place of temperature extremes, beset by relentless strong winds. Politically, the region was in an uproar. China, the base of operations, was torn by civil strife. And in 1922, the year of the first of three expeditions, a revolution shook Mongolia. Moreover, only one fossil, a rhinoceros tooth, had ever been found in Mongolia up until that time.

As it turned out, the Central Asiatic

know well are *Euoplocephalus* (Figure 7.6) and *Sauropelta* (Figure 7.7). Still, many of the important ankylosaurs are known from different parts of the skeleton. Is *Hylaeosaurus* the same as *Polacanthus*, or are they two distinct animals? And what did *Acanthopholis* and *Struthiosaurus* really look like?

ANKYLOSAURIDAE. Ankylosauridae is a robust clade united on the basis of more than 26 derived features (Figure 7.8). Some of the most distinctive characters are found in the last tail vertebrae, which are partially or completely fused together to form a stiffened support for a terminal tail club, itself constructed from large lateral plates and two smaller plates (Figure 7.9).

Relationships of taxa within this clade are beginning to be recognized, primarily through the efforts of P. C. Sereno and of W. P. Coombs and T. Maryańska (Figure 7.8). Of those ankylosaurids that have been described, the most primitive is *Shamosaurus*, which comes from Mongolia in rocks that are Early Cretaceous in age. All that we know of this 5- to 6-m-long ankylosaur so far is a complete skull, a partial skeleton, and assorted armor. The beak is narrow and devoid of armor covering, and the eyes, although forwardly placed on the sides of the skull, faced

Expeditions were an unqualified success. Although Osborn's theory was not supported, Andrews brought back a wealth of fossils, including dinosaur material, that made Osborn's error easy to forget. Among the most famous dinosaur finds of his expedition, for example, were *Protoceratops* (the species name of this famous dinosaur is *andrewsi*) and eggs. Other incredible finds included *Velociraptor* and a group of tiny Mesozoic mammals (still the rarest of the rare). Andrews and his field parties also found the largest land mammal and the largest carnivorous land mammal of all time (both Cenozoic in age). Other fossils were obtained whose significance was not completely understood. For example, it was only in 1992 that a specimen of *Mononykus*, collected by Andrews's scientists in the 1920s, was finally correctly identified. All in all, it was quite a haul.

Andrews and his parties survived the Mongolian revolution of 1922, but eventually the expeditions came to an end when the political situation in China became too unstable and travel too dangerous. Andrews, himself, eventually went on to get the job held earlier by Osborn: Director of the AMNH. He assured his place in history, however, by leading the Central Asiatic Expeditions.

Figure Box 7.1 Roy Chapman Andrews, explorer, adventurer, and leader of what he called the "New Conquest of Central Asia." *Photograph courtesy of the American Museum of Natural History.*

laterally. As with other ankylosaurids, there are squamosal horns at the corners of the skull roof, but in *Shamosaurus* these are quite small.

The next more inclusive clade of ankylosaurids consists of a number of forms, including *Ankylosaurus, Euoplocephalus, Pinacosaurus, Saichania*, and others. This "higher" ankylosaurid clade appears to be united by four derived features (Figure 7.8). This clade, so far unnamed, whose origin dates to sometime during or before the middle of the Late Cretaceous (about 85 million years ago), has Mongolian and North American branches. The Mongolian clade, composed of *Pinacosaurus, Saichania*, and *Tarchia*, is united by the derived features in Figure 7.8. Of these three ankylosaurids, *Pinacosaurus* probably has the best record. Known from relatively abundant material, including a growth series of juveniles and adults, this dinosaur gave scientists a first inkling about what the skull looked like under all that dermal armor. *Pinacosaurus* is a medium-sized (5-m-long) and slenderly built ankylosaur. Like *Shamosaurus*, it too lacks armor on its snout. On its back and tail are bony spines and at the end of the tail is a bony club. The forefoot ends in four strong and hoof-covered fingers, and the hindfoot in five, equally massive toes. *Saichania* is larger than *Pinacosaurus*, estimated at upward of 7 m long. *Tarchia*,

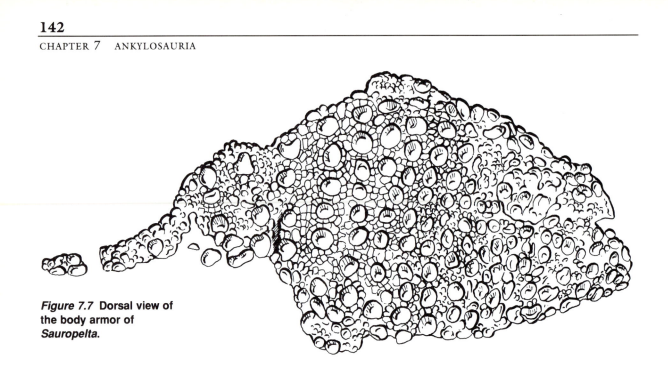

Figure 7.7 Dorsal view of
the body armor of
Sauropelta.

however, claims the honor of being Mongolia's largest ankylosaur at an estimated 8 m. Unique to *Saichania*, there is armor not only on its back, but on its belly as well. Perhaps as a consequence of all of this weight, the postcranial skeleton is extremely strong and massive. Unlike both *Saichania* and *Pinacosaurus*, *Tarchia* has a wide beak and a remarkably large tail club, suggesting, perhaps, a more aggressive personality.

Remaining ankylosaurids – *Ankylosaurus*, *Euoplocephalus*, and *Talarurus* – are a mostly North American clade (Figure 7.8). Of these, *Ankylosaurus* is an 8- to 9-m-long animal. Its beak is narrow and there are large squamosal horns that project bull-like from the back of the head. Not nearly so large as *Ankylosaurus* (an estimated 6–7 m long), *Euoplocephalus* has a wide premaxillary beak for grazing in a voracious way through the undergrowth, and short, yet very powerful legs. With both squamosal horns and massive tail clubs, *Ankylosaurus* and *Euoplocephalus* were walking "don't mess with me" advertisements. Finally, *Talarurus* is a medium-sized (4–5-m-long), close relative of *Ankylosaurus* and *Euoplocephalus*, with a relatively long and narrow skull (for an ankylosaurid) and a small tail club.

NODOSAURIDAE. Turning to the other great clade of ankylosaurs, Nodosauridae consists of *Hylaeosaurus* and the clade comprising all remaining forms, all of which share a host of derived features of the skeleton (Figure 7.10). We know very little of *Hylaeosaurus*, because it still resides partially hidden in the rocks that originally entombed it. Still, what can be seen in its partially prepared state is the front half of a skeleton. In all, it appears to be a 4-m-long animal. Rows of large, curved spines line the neck and shoulders, and the shoulder blade has a strikingly angled scapular spine, a sine qua non of nodosaurids. As already noted, Huxley described the spinose rear end of an ankylosaur, *Polacanthus*, known from the same strata as *Hylaeosaurus*. Unfortunately, there is very little overlap between what is known of *Polacanthus* and *Hylaeosaurus*, making it impossible to establish whether the two

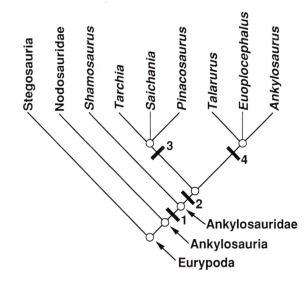

Figure 7.8 Cladogram of Ankylosauridae, with its two closest relatives, Nodosauridae and Stegosauria. Derived characters at 1: wide skull; arched snout; complex palate with highly altered respiratory passage; paired sinuses in the premaxilla, maxilla, and nasal bones; loss of premaxillary teeth; prominent wedge-shaped plates on the cheeks; large, triangular squamosal horns at the back of the skull; numerous small scutes covering the skull roof, snout, and above the eyes; last of the tail vertebrae fused to form a stiffened support for a terminal tail club (itself constructed from large lateral plates and two smaller plates); tail stiffening augmented by ossified tendons surrounding the caudal vertebrae; fusion of the sternal plates; shortening of the rear end of the ilium; and great reduction of the pubis; at 2: backward shift in the position of the eye sockets, and enlargement of the squamosal horns; at 3: as many as 18 large plates covering the skull roof, narrowing of the squamosal horns, flaring of the nostril openings of the skull, and relatively short premaxillary beak; at 4: skull roof covered by over 30 small plates and broad-based squamosal horns.

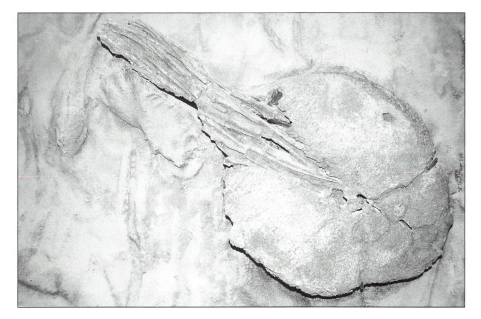

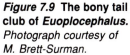

Figure 7.9 **The bony tail club of** *Euoplocephalus.* *Photograph courtesy of M. Brett-Surman.*

ankylosaurs are one and the same. Nevertheless, some paleontologists are hedging their bets, treating *Hylaeosaurus* and *Polacanthus* as the same animal and applying the name *Hylaeosaurus* to this beast now with front and back halves (remember, this dinosaur was recognized first).

The basal phylogenetic split between *Hylaeosaurus* and the remaining clade of "higher" nodosaurids – including *Acanthopholis, Struthiosaurus, Sauropelta, Silvisaurus, Panoplosaurus,* and *Edmontonia* –appears to have taken place no later than the Early Cretaceous, about 130 million years ago. Early on in the history of this clade, we encounter *Acanthopholis* and *Struthiosaurus*, both poorly known nodosaurids, the former from the mid-Cretaceous of England and the latter from the Late Cretaceous of Romania, Austria, and possibly France. Based on some teeth, vertebrae, foot bones, postcranial armor plates, and tall, conical spines, *Acanthopholis* is a small animal, probably no more than 3 or 4 m long. Likewise, *Struthiosaurus* is small. At 2 or 3 m body length, this nodosaurid, it has been argued, was an island-inhabiting dwarf. *Sauropelta* itself is a fairly large animal (7 m long). As the oldest well-known nodosaurid from North America (Early Cretaceous of Wyoming and Montana), *Sauropelta* possesses the primitive condition of having premaxillary teeth in its small mouth. Still, the narrow beak was probably horn-covered in life.

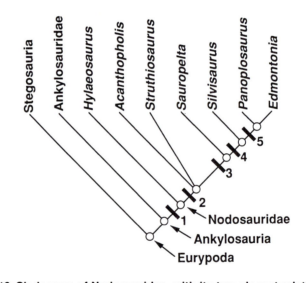

Figure 7.10 Cladogram of Nodosauridae, with its two closest relatives, Ankylosauridae and Stegosauria. Derived characters at 1: palate hourglass-shaped; pterygoid processes (projections extending from the undersurface of the braincase) roughened, stubby, and round; occipital condyle (that part of the braincase that articulates with first neck vertebra) very round and set off from the rest of the braincase on a short, angled neck and composed entirely of a single bone (the basioccipital); a large, long, and narrow coracoid (the bone that makes up the lower part of the shoulder blade); ischium bent downward; large osteoderm on the top of the skull; narrow osteoderm along the rear edge of the skull; unique scute pattern around, and in front of, the eye socket; and scapular spine bent toward shoulder joint; at 2: knob-like scapular spine with a space anterior to it; at 3: increase in body size, and fusion between the scapular and coracoid; at 4: lengthening of the secondary palate; at 5: loss of premaxillary teeth, and fusion of the atlas and axis.

Silvisaurus, *Panoplosaurus*, and *Edmontonia* are united by only a single derived character of the palate. The mid-Cretaceous *Silvisaurus* (from Kansas) grew to about 4 m long. It has a relatively wide skull for a nodosaurid, and at the front of this skull is still a set of premaxillary teeth. With *Panoplosaurus* and *Edmontonia*, we reach the culmination of the nodosaurid clade (Figure 7.10). Both are about the same size (6 m long) and come from Upper Cretaceous strata, *Edmontonia* from Montana and Alberta, and also now reported from Texas, and *Panoplosaurus* from Alberta, Texas, and New Mexico. Yet there are some important differences. *Panoplosaurus* lacks tall spines as part of its armor assembly but has a relatively wide skull. In contrast, *Edmontonia* is decked out in relatively long spines, including large, forwardly directed spines over the shoulders; it also has a long, narrow skull. In all, these two contemporaries must have had strikingly different habits.

But what of *Nodosaurus* itself? As it turns out, this ankylosaur, from the mid-Cretaceous of Wyoming, is hard to place phylogenetically. And that may not be surprising, considering that all we know of it are some fragmentary skeletal remains and armor. According to Coombs and Maryańska, it hovers around a basal relationship with *Silvisaurus*, *Panoplosaurus*, and *Edmontonia*. And *Sarcolestes*, *Dracopelta*, *Hoplitosaurus*, *Maleevus*, and *Minmi* are so poorly understood that their locations within Ankylosauridae or Nodosauridae are not known; in fact, we do not even know if they *are* ankylosaurids or nodosaurids!

HETERODOXY. We have spoken of ankylosaur relationships with an air of certainty, and we have done so based on the prevailing view that ankylosaurids and nodosaurids are their own closest relatives, with stegosaurs as first cousins. Never one to leave an orthodox position alone, R. T. Bakker has identified stegosaurs and nodosaurids as a monophyletic clade, with ankylosaurids as a more distantly related taxon. Under these relationships, an Ankylosauria consisting of Nodosauridae and Ankylosauridae cannot be monophyletic. Bakker identified five characters that he viewed as supporting the existence of a nodosaurid–stegosaur clade (Figure 7.11).

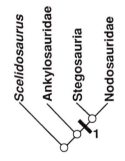

Figure 7.11 **Cladogram of "higher" Thyreophora, according to Bakker.** Note that Nodosauridae shares closest relationship with Stegosauria and only thereafter with Ankylosauridae. Derived characters at 1: greatly expanded area for muscle attachment (the supracoracoideus) on the shoulder blade; prominent muscle scar for the tendons of muscles (the latissimus dorsi and teres) on the humerus; expansion of the deltopectoral crest of the humerus; fusion across the midline of part of the pterygoids as they make up the back of the palate; and greatly enlarged scapular spines.

Which pattern of relationship is correct – Bakker's or that advocated here? To resolve this question, we ask which is more parsimonious, a monophyletic nodosaurid–stegosaur clade or a monophyletic nodosaurid–ankylosaurid clade? Using Bakker's cladogram, it takes 5 steps to evolve the five characters (1 step per character); however, at least 32 steps are required to evolve in duplicate all the features shared by ankylosaurids and nodosaurids.

In contrast, the cladogram that is preferred here takes these latter characters, interprets them as evolving once (i.e., all present in the common ancestor of nodosaurids and ankylosaurids), and then treats the five characters purporting to support a stegosaur–nodosaurid clade as either convergences (separately evolved in stegosaurs and nodosaurids) or reversals (evolved once in the common ancestor of stegosaurs, nodosaurids, and ankylosaurids, but secondarily lost in ankylosaurids). Either way, the phylogeny that portrays nodosaurids and ankylosaurids as closest relatives is more parsimonious than one that has stegosaurs and nodosaurids sharing closest relationships.

ANKYLOSAUR PALEOBIOLOGY AND PALEOECOLOGY:
Frisky, Running, Leaping Ankylosaurs – NOT!

Ankylosaurs have a worldwide distribution, predominantly from North America and Asia, but also from Europe and Australia (Figure 7.12). Remains questionably referable to ankylosaurs have even been found in Antarctica! The best-preserved ankylosaurs come from Asia, particularly Mongolia and China, where specimens are found as nearly complete, articulated skeletons, in some cases preserved in an upright pose or on their sides. In North America, only partial skeletons have yet been found; these are often found upside down, sometimes in rocks deposited along the seashore or even in the open seas. This suggests that ankylosaurs lived in terrestrial habitats sufficiently close to the sea that their bloated or partially dismembered carcasses might have been carried out with the tide. The flipping upside down presumably comes from the heavy nature of their armored backs: A floating ankylosaur carcass should turn over prior to settling out.

In addition to these general post-mortem considerations, most ankylosaur finds consist of individual skeletons or isolated partial remains. An exception is an accumulation in China of 12 *Pinacosaurus* specimens. Perhaps the rarity of bonebeds indicates that these animals had solitary habits or lived in very small groups. Even from our incomplete window on the past, it appears reasonably certain that ankylosaurs did not enjoy the company of huge herds.

Food and Mouths to Feed

In whatever size groups these armored animals lived, each animal clearly had a very low browsing range, foraging no more than a meter or so above the ground. But what were these ankylosaurs foraging for? Bugs? F. B. Nopcsa once suggested that ankylosaurs were insectivorous, but this idea is no longer given much credence. (How many beetles would it take to keep a 3,500-kg ankylosaur going? And for how long?) Instead, it is now thought that these animals fed

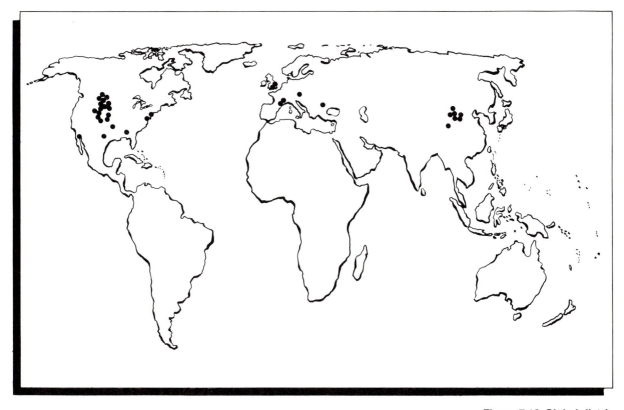

Figure 7.12 **Global distribution of Ankylosauria.**

exclusively on plants. Why this is so comes from a look at the digestive system; from the beak, mouth, and teeth at the front end; as well as from the far extreme of the gut.

Up front, it is the cutting edge of the beak that "makes its first impression" on plant food. Primitively in ankylosaurs, the beak is scoop shaped but relatively narrow (although slightly broader than in stegosaurs) and remains so in the nodosaurid clade and in *Shamosaurus*. In all other ankylosaurids, by contrast, the beak becomes very broad, and in a couple of these forms, the premaxillary teeth are lost. As noted by K. Carpenter of the Denver Museum of Natural History, this difference in beak shape may indicate a degree of feeding differentiation among ankylosaurs. A narrow beak in nodosaurids suggests that these ankylosaurs fed in a somewhat selective manner, plucking or biting at particular kinds of foliage and fruits with the sharp edge of the rhamphotheca. In contrast, the very broad beak of ankylosaurids (Figure 7.13a) may imply less-selective feeding, in which plant parts were indiscriminately bitten off from the bush or pulled from the ground.

How this food was then prepared for swallowing is a bit of a mystery. As in stegosaurs (and pachycephalosaurs; see Chapters 6 and 8), the triangular teeth of both nodosaurids and ankylosaurids are small, not particularly elaborate, and less tightly packed than those of other ornithischian dinosaurs (Figure 7.13b). Tooth wear is present on the crowns, indicating that chewing involved no more than simple slicing-and-dicing. Nevertheless, it is likely that ankylosaurs had a long, flexible tongue (they had large hyoid bones – which supported the tongue – in

Figure 7.13 **Palatal view of the skull of (a)** *Euoplocephalus* **and (b) a tooth of** *Edmontonia.*
Scale = (a) 10 cm; (b) 1 cm.

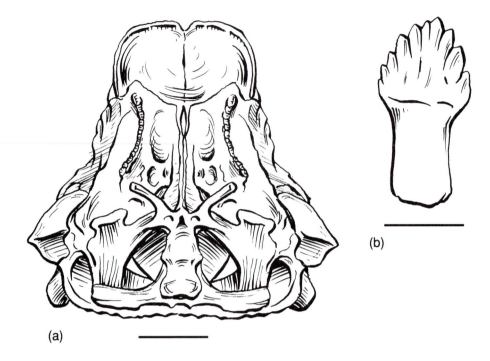

(b)

(a)

their throats) and, commonly, an extensive secondary palate that allowed them to chew and breathe at the same time. Moreover, deeply inset tooth rows suggest well-developed, deep cheek pouches to keep whatever food was being chewed from falling out of the mouth. The jaw bones themselves were relatively large and strong (although lacking enlarged areas for muscle attachment). All jaw features – except, ironically, the teeth – speak loudly that ankylosaurs were excellent chewers.

Perhaps the paradox of simple teeth placed in strong, cheek-bound jaws can be understood by looking not at how much chewing was done prior to swallowing (there obviously was some), but at the rear end of the animal, where the majority of plant digestion must have been accomplished by gut fermentation. Here, bacteria living within the gut chamber would decompose tough, woody plant material. Among modern mammals, this method of breaking down tough plant material is well known in horses, rhinos, and cattle.

In ankylosaurs, to digest the great amount of food that is needed to sustain these large animals, the gut must have been huge. Huge guts mean equally huge abdomens. In ankylosaurs, the very deep rib cage circumscribed an enormously expanded abdominal region and this can only mean that digestion took place in a very large, perhaps highly differentiated, fermentation compartment in these armored dinosaurs. Bakker dubbed this fermentation system "after-burner" digestion, analogizing, we fear, the exhaust of jet fighters with that of fermenting ankylosaurs!

Defensive Moves

Ankylosaurs may not have aspired to great thoughts, because their brain power was close to the bottom of the dinosaur range (only sauropods had smaller brains for their size; see Chapters 6 and 11, and Box 14.3). Similarly, ankylosaurs were among the slowest moving of all dinosaurs for their body weight. According to Australian paleontologist R. A. Thulborn's calculations from limb size and body mass, these animals were able to run no faster than 10 km/hr and walked at a considerably more leisurely pace (about 3 km/hr). The few known ankylosaur footprints bear this suggestion out. One series of prints, from the Lower Cretaceous Peace River locality (now under water) in British Columbia, Canada, has been referred by Carpenter to *Sauropelta*. This trackway indicates a rather slow pace of about 3 km/hr, the same as had been calculated from limb and body proportions (see Box 14.2).

Although ankylosaurs were slow of mind and foot, other aspects of the limb skeleton suggest that these Mesozoic plodders could actively defend themselves against predators. The ankylosaur shoulder region – particularly in nodosaurids – was exceedingly strongly muscled. And relatively longer hindlimbs and the wide pelvis may have had the effect of dropping the center of gravity more forward. Could these features mean that the fore- and hindlimbs were used in conjunction with body armor? The answer is a resounding Yes.

Consider that armor. First, in all ankylosaurs the entire upper surface of the body – the head, neck, torso, and tail – was covered by a pavement of bony plates. Nodosaurids, but not ankylosaurids, also retained the formidably sharp parascapular spines that were also seen in stegosaurs. Ankylosaurids were even more modified, producing armor literally everywhere, even over the eyes. And of course there was the famous ankylosaurid tail club (Figure 7.9), a mass of bone set at the end of a rigid tail (itself about half the length of the body).

How this tail and its terminal club worked was a masterful mechanical feat, fully as skillful and deadly as any mace-wielding knight. Although the front part of the tail was relatively free to swing side to side, the rear half was stiffened by modified vertebrae, as well as by a series of longitudinally running tendons. These tendons provided firm attachment for the powerful muscles, but more especially they added considerable stiffening to the end of the tail. Using the forward flexibility of the tail vertebrae and the stiffened posterior end, the club could have been forcefully swung side to side.

It is now easy to imagine how an ankylosaur might have behaved while under attack from one of its formidable, contemporary predators, among whom were *Deinonychus*, *Tarbosaurus*, and *Tyrannosaurus*. It was tail clubbing or hunkering down (or both). Who knows what neural pathways or synapse links might determine one or the other of these options for an ankylosaur, but ultimately the game must have been to persuade the predator to GO AWAY. The only offense available to ankylosaurids was a club bash to the shins; for nodosaurids it was a stab with the shoulder spines. In order to effectively wield their weaponry, these animals must have first planted their hindlimbs and then rotated their forequarters with the strong forelimb muscles, keeping watch on the threatening predator. For nodosaurids, this was a head-first (or shoulder-first) affair, but in ankylosaurids the tail must have been central to defense.

If such an offensive strategy failed to dissuade a predator, then the ankylosaur fail-safe was assuredly to hunker down. With its legs folded under its body, a 3,500-kg ankylosaur would have been very difficult to flip over. Safe under protective armor, both nodosaurids and ankylosaurids were, in short, the best-defended fortresses of the Mesozoic.

Important Readings

Bakker, R. T. 1986. Dinosaur Heresies. William Morrow, New York, 481 pp.

Bakker, R. T. 1988. Review of the Late Cretaceous nodosauroid Dinosauria, *Denversaurus schlessmani*, a new armor-plated dinosaur from the latest Cretaceous of South Dakota, the last survivor of the nodosaurians, with comments on stegosaur-nodosaur relationships. Hunteria 1:1–23.

Carpenter, K. 1982. Skeletal and dermal armor reconstruction of *Euoplocephalus tutus* (Ornithischia: Ankylosauridae) from the Late Cretaceous Oldman Formation of Alberta. Canadian Journal of Earth Sciences 19:689–697.

Carpenter, K. 1984. Skeletal reconstruction and life restoration of *Sauropelta* (Ankylosauria: Nodosauridae) from the Cretaceous of North America. Canadian Journal of Earth Sciences 21:1491–1498.

Carpenter, K., D. Dilkes, and D. B. Weishampel. 1995. The dinosaur fauna of the Niobrara Chalk Formation. Journal of Vertebrate Paleontology 15: 275–297.

Coombs, W. P., Jr. 1978. Theoretical aspects of cursorial adaptations in dinosaurs. Quarterly Review of Biology 53:393–418.

Coombs, W. P., Jr., and T. Maryańska. 1990. Ankylosauria; pp. 456–483 *in* D. B. Weishampel, P. Dodson, and H. Osmólska (eds.), The Dinosauria. University of California Press, Berkeley.

Farlow, J. O. 1978. Speculations about the diet and digestive physiology of herbivorous dinosaurs. Paleobiology 13:60–72.

Hopson, J. A. 1977. Relative brain size and behavior in archosaurian reptiles. Annual Review of Ecology and Systematics 8:429–448.

Hopson, J. A. 1980. Relative brain size in dinosaurs: implications for dinosaurian endothermy; pp. 287–310 *in* R. D. K. Thomas and E. C. Olson (eds.), A Cold Look at the Warm-Blooded Dinosaurs. Westview Press, Boulder, Colo.

Maryańska, T. 1977. Ankylosauridae (Dinosauria) from Mongolia. Palaeontologia Polonica 37:85–151.

Sereno, P. C. 1986. Phylogeny of the bird-hipped dinosaurs (Order Ornithischia). National Geographic Research 2:234–256.

FOR INFORMATION ON THE CENTRAL ASIATIC EXPEDITIONS

Andrews, R. C. 1929. Ends of the Earth. G. P. Putnam's Sons, New York, 293 pp.

Andrews, R. C. 1933. Explorations in the Gobi Desert. National Geographic Magazine 63:653–716.

Andrews, R. C. 1953. All About Dinosaurs. Random House, New York, 146 pp.

Preston, D. J. 1986. Dinosaurs in the Attic. St. Martin's Press, New York, 244 pp.

CHAPTER 8

PACHYCEPHALOSAURIA:
Head-to-Head, with Malice Aforethought

THROUGH A DENSE THICKET of shrubby angiosperms come the deep sounds of thuds, slaps, and scuffling. Beyond these shrubs in a large clearing are a dozen or more pachycephalosaurs – *Stygimoloch spinifer* – their broadly domed heads fringed with long knobs and horns. Many are foraging for succulent leaves and fruits in the undergrowth, but two of the largest individuals, some 2.5 m long, are kicking up a storm of dust in the center of the group. At regular intervals, they turn toward each other, then rapidly lunge forward, charging, head-to-sides, head-to-thighs. They go at it for more than an hour before one tires and is forced from the group by the victor.

Fact or fiction? How can we possibly tell? What *was* the lifestyle of these animals, whose profiles look like the Three Stooges' Curly and whose behavior may have rivaled that of Brahma bulls (Figure 8.1)?

HISTORY OF PACHYCEPHALOSAUR DISCOVERIES
19th Century

Human acquaintance with pachycephalosaurs started out in a sorry state of confusion. The story begins in 1856, when J. Leidy of the Academy of Natural Sciences, Philadelphia, studied a curious-looking, triangular, "cuspy" tooth from Montana, no larger than half a centimeter in height. Leidy thought that this tooth came from a fossil lizard, calling it *Troodon* (*troo* – wound; *odon* – tooth).

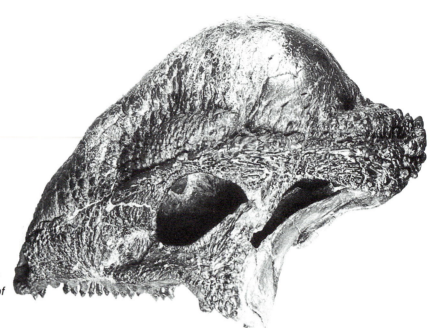

Figure 8.1 **Left view of the skull of Stegoceras.** *Photograph courtesy of H.-D. Sues.*

20th Century

Similar kinds of teeth were recovered as the century turned, and F. Nopcsa, the eccentric Hungarian paleontologist, suggested that the teeth belonged to a carnivorous dinosaur (theropod) instead of a lizard. L. M. Lambe's 1902 discovery of some rather peculiar, fragmentary, thick skull roof fragments[1] with big tubercles in the same beds where he had been finding *Troodon*-like teeth led him to the conclusion that he had found a new animal. He called it *Stegoceras* (*stego* – cover; *keras* – horn).

Nearly a quarter of a century later, a skull and partial skeleton were finally found and named *Troodon validus* by C. W. Gilmore of the National Museum of Natural History (Smithsonian Institution). No longer than 2 m, this animal had a 20-cm-long skull with an exceedingly thick, knobby-looking frontal-parietal region. On the back of this thickened skull cap was a short platform (called a parietal-squamosal shelf) ornamented with knobs and bumps. Beyond the head, the skeleton was much like many other bipedal ornithischians, having relatively short forelimbs, long four-toed hindlimbs, a stiff vertebral column, and a long tail. He believed that troodontids, as he now called them, had close relationships with ornithopods.

It took another 20 years for a new dome-headed dinosaur to be described. In 1943, as part of their larger study of existing material of these dinosaurs, the legendary dinosaur hunter B. Brown and his colleague E. Schlaikjer of the American Museum of Natural History described a new form – *Pachycephalosaurus* (*pachy* – thick; *kephale* – head; *sauros* – lizard) – which consisted of only a skull (Figure 8.2e). But what a skull! Nearly complete, it approached 65 cm in length, with a dome of solid bone 20 cm thick, and a snout studded with large and densely packed bumps. A skull this size would ride at the end of an estimated 8-m-long body.

---

[1]Technically, this part of the skull is formed from two bones, the frontal and parietal, which together form the bony vault that covers the top of braincase.

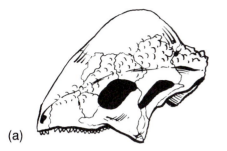

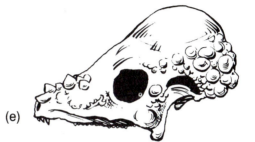

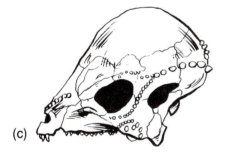

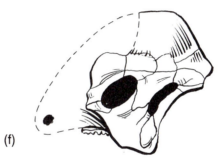

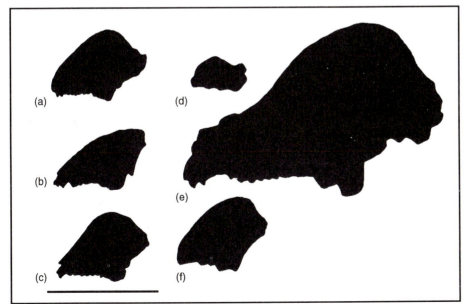

Figure 8.2 Left lateral view of the skulls of (a) *Homalocephale,* (b) *Prenocephale,* (c) *Stegoceras,* (d) *Majungatholus,* (e) *Pachycephalosaurus,* and *(f) Tylocephale.* Scale for silhouettes = 10 cm.

It took further studies to sort out *Troodon*. In 1945, C. M. Sternberg of the Geological Survey of Canada argued that Leidy's original *Troodon* tooth came from a small theropod dinosaur (see Chapter 12). Gilmore's *"Troodon"* was correctly renamed *Stegoceras* (Figures 8.1, 8.2c, and 8.3d), and *Pachycephalosaurus* became the namesake of the entire group of dome-heads, which was christened Pachycephalosauridae. Pachycephalosaurs had finally come into their own.

The next discovery, and the first and only pachycephalosaur from England, was named by P. M. Galton in 1971. Called *Yaverlandia* (from Yaverland Point; Figure 8.3a), this animal is known only from a small skull fragment, collected from the famous Wealden beds of the Isle of Wight, off the southern coast of England.

Thereafter, beginning in the mid-1970s, came a virtual flood of new pachycephalosaurs, this time from Mongolia and China. Two of these came from the Nemegt Formation of Mongolia: *Prenocephale* (*prenes* – sloping; Figure 8.2b) and *Homalocephale* (*homalos* – even; Figures 8.2a, 8.3b–c). *Prenocephale* – consisting of a beautiful skull and a partial postcranial skeleton – is spectacular. All that was needed after it had been found was to blow the sand grains off its outside and out of its nasal cavity! The very high dome, with its slightly roughened outer surface, and row of prominent bumps on lateral and rear margins of the skull dominate the profile of this animal.

In contrast to the dome-headed *Prenocephale*, *Homalocephale* is a flat-headed pachycephalosaur. Known from a nearly complete skull and, more important, a virtually complete, articulated skeleton, *Homalocephale* is perhaps best known

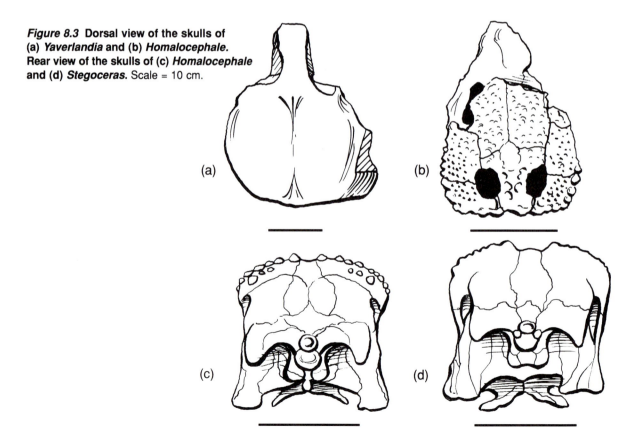

Figure 8.3 Dorsal view of the skulls of (a) *Yaverlandia* and (b) *Homalocephale*. Rear view of the skulls of (c) *Homalocephale* and (d) *Stegoceras*. Scale = 10 cm.

(a)

(b)

(c)

(d)

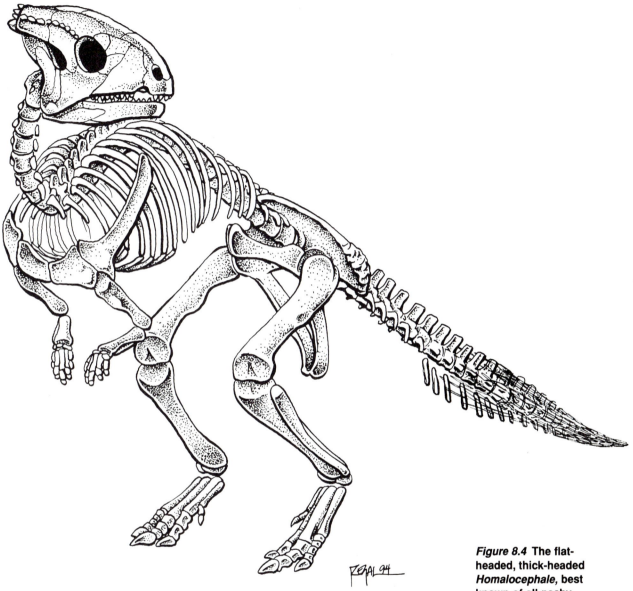

Figure 8.4 **The flat-
headed, thick-headed**
Homalocephale, **best
known of all pachy-
cephalosaurs.**

among all pachycephalosaurs (Figure 8.4). It still lacks a hand: To this day, none
has been found for any pachycephalosaur.

Tylocephale (*tyle* – swelling; Figure 8.2f), from Mongolia, is a full-domed
pachycephalosaur, known so far from only a partial frontal-parietal (skull roof)
specimen that appears to be somewhat similar to that of *Stegoceras*. Nobody yet
knows what the rest of the skeleton of this animal is like.

On the other side of the border, Chinese pachycephalosaurs began showing up
in the late 1970s. *Wannanosaurus* (from the Chinese name for a southern part of
Anhui Province; Figure 8.5), so far known only from a partial skull and associated
skeletal fragments, was described in 1977 by Hou L.-H. (Institute of Vertebrate

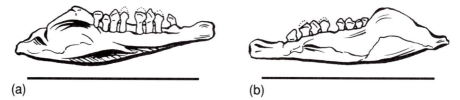

(a)   (b)

Figure 8.5 (a) Internal and (b) external views of the lower jaw of *Wannanosaurus*.
Scale = 5 cm.

Paleontology and Paleoanthropology in Beijing, China). This was followed shortly
thereafter by the discovery of *Micropachycephalosaurus* (*micro* – small). Named
in 1978 by the dean of Chinese dinosaur paleontology, Dong Z.-M.,
Micropachycephalosaurus is presently known only from a lower jaw and an associated,
but fragmentary, postcranial skeleton from Shandong Province, China. In the same
paper in which he identified *Micropachycephalosaurus*, Dong coined the name
Homalocephalidae for those pachycephalosaurs with thick, yet flattened, skull roof
and kept Gilmore's Pachycephalosauridae for those with a rounded skull roof.

Perhaps the most surprising of all pachycephalosaurs, not so much for the distinctiveness of its dome but for its place of discovery, is *Majungatholus* (Figure
8.2d). This form, whose name means "dome from Majunga," hails from the
Majunga region of Madagascar. At the same time, another new North American
pachycephalosaur reared its unimpeachably ugly head: *Gravitholus* (*gravis* – heavy;
tholos – dome) from Alberta.

The last three pachycephalosaurs to be discovered so far are *Goyocephale* (*goyo* –
decorated), *Stygimoloch* (*Stig [Styx]* – river of Hades [Hell Creek]; *moloch* – demon),
and *Ornatotholus* (*ornatus* – adorned). *Goyocephale*, a flat-headed pachycephalosaur,
was described in 1982 by A. Perle from the Geological Institute in Ulan Baator
(Mongolia), T. Maryańska, and H. Osmólska. This animal is known from a fragmentary skull and a nearly complete postcranial skeleton from Mongolia.
Interestingly, in the front of both upper and lower jaws are short canine-like teeth;
the lower canine actually projects into a notch in the upper jaw! *Stygimoloch*[2] so far
consists only of a domed skull cap and postcranial fragments from Montana and
Wyoming. Finally, *Ornatotholus*, described by H.-D. Sues and Galton in 1983, consists of dome fragments found in Alberta and Montana. Today we recognize over a
dozen pachycephalosaur species, principally from the Northern Hemisphere and all
but one (*Yaverlandia*) from the Late Cretaceous.

PACHYCEPHALOSAURIA DEFINED AND DIAGNOSED

The most obvious thing about pachycephalosaurs is their thick-headedness. It is
not surprising that Pachycephalosauria, with such obvious modifications of the
skull, has long been considered a monophyletic group. More controversial is
which ornithischian group represents the next-closest relatives of pachy-

[2] Fragments of a pachycephalosaur were described by E. B. Giffin and colleagues as a new genus,
Stenotholus, in 1987. Although the opinion has never been published, many specialists believe that
Stenotholus is actually *Stygimoloch*. If *Stenotholus* proves to be *Stygimoloch*, *Stygimoloch* will be one of
the better-known pachycephalosaurs, because *Stenotholus* is a reasonably complete specimen, making
Stygimoloch, if this is true, a much better known animal.

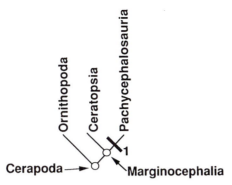

Figure 8.6 Cladogram of Cerapoda emphasizing the monophyly of Pachycephalosauria. Derived characters at 1: thickened skull roof, modified cheek region that shrouds the jaw joint, extensive ossification of the orbit (including additional bony elements fused to its upper margin), shortening of the floor of the braincase, expansion of the back of the skull, abundant and strongly developed ornamentation of the external surfaces of skull, the development of osteoderms on the rim of the skull roof, special ridge-and-groove articulations between articular processes on the back and tail vertebrae, "basketwork" of ossified tendons that cover the end of the tail, short forelimbs, and reduction of the pubic bone to the point that it does not contribute to the formation of the hip joint.

cephalosaurs. Our best call is that these animals are close to ceratopsians, within the clade Marginocephalia (see Part II: Ornithischia).

Pachycephalosauria can be defined as the common ancestor of the two great clades of pachycephalosaurs – Pachycephalosauridae and *Wannanosaurus* – and all its descendants (Figure 8.6). Members of the pachycephalosaur clade share a host of derived features, most of them cranial, but even a few features of the rest of the skeleton where we know it well enough.

PACHYCEPHALOSAUR DIVERSITY AND PHYLOGENY

As we have seen, pachycephalosaurs come in two varieties – flat-headed and full-domed – which formally have been designated Homalocephalidae and Pachycephalosauridae, respectively. Are they each monophyletic, descended from a single common ancestor that includes all its descendants? There are two views on this matter. On the one hand, Sues and Galton, and later Maryańska, regarded homalocephalids as a natural group (Figure 8.7). On the other hand, P. C. Sereno suggested these flat-headed pachycephalosaurs to be a sequence of increasingly more-derived forms with independent origins from the line leading to the very highly domed pachycephalosaurs (Figure 8.8). Who's right?

A monophyletic Homalocephalidae (Figure 8.7) would have as its members *Wannanosaurus*, *Homalocephale*, and *Goyocephale*, all united by the common possession of an evenly thickened, but completely flat, skull roof that bears a pitted outer surface. Of the three forms, *Goyocephale* and *Homalocephale* are most closely related, sharing the derived characters shown in Figure 8.7.

Both Sues and Galton, and Maryańska regarded *Wannanosaurus* as more distantly related to *Homalocephale* and *Goyocephale*. Known only from a very small, but adult, individual, *Wannanosaurus* appears to occupy this primitive position

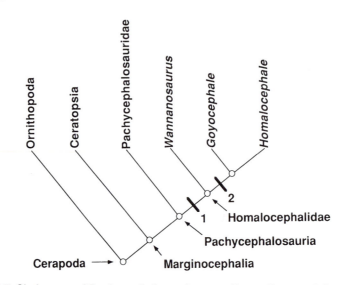

Figure 8.7 Cladogram of Pachycephalosauria according to Sues and Galton and according to Maryańska, with two more distantly related taxa: Ceratopsia and Ornithopoda. Derived characters at 1: evenly thickened, but completely flat, skull roof that bears a pitted outer surface; at 2: reduced upper temporal openings, and a row of large tubercles along the posterior margin of the parietal-squamosal shelf.

primarily because its relatively large upper temporal openings are bigger than those in *Homalocephale* and *Goyocephale*. Two additional characters that support this interpretation are the ornamentation – granulated rather than pitted – which is coarser on the front of the skull roof than on the back, and the forelimb, which is more strongly bowed than in other pachycephalosaurs.

How does this character-support for monophyly of Homalocephalidae stack up against the alternative pattern of pachycephalosaur phylogeny in which flat-headed forms are serial branches along the evolutionary way toward fully domed pachycephalosaurs? Sereno has argued that *Wannanosaurus* is the most primitive pachycephalosaur, with all remaining taxa united on the basis of at least four derived features; these include the pattern of ornamentation on the back of the skull and on the lower jaw, and the configuration of the bones of the skull roof (Figure 8.8). Within this clade of "higher" pachycephalosaurs (which Sereno called **Goyocephalia**), the most primitive member is (naturally) *Goyocephale*, with all remaining forms united by the characters on Sereno's cladogram (Figure 8.8). Because these successively more-derived pachycephalosaurs are the familiar, fully domed forms, Sereno has dismembered Homalocephalidae.

What meaning does this have for trying to decide the issue of homalocephalid monophyly? The first thing we notice is that there are more characters supporting serial relationships of the flat-heads leading successively to the dome-heads. In this case, only one feature – flat-headedness itself – has to reverse in the evolution of Pachycephalosauria. In contrast, the phylogenetic tree involving a monophyletic Homalocephalidae is much longer than the serial arrangement. This is because many more characters have to reverse or to arise twice with a monophyletic Homalocephalidae. Using parsimony to help us select the tree that uses our characters most economically, we naturally select the serial arrangement of flat-headed pachycephalosaurs: Homalocephalidae does not appear to be a true clade.

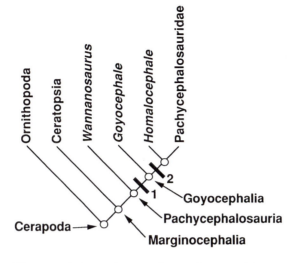

Figure 8.8 **Cladogram of Pachycephalosauria according to Sereno, with two more distantly related taxa, Ceratopsia and Ornithopoda.** Derived characters at 1: linear row of 5 to 7 tubercles on the rear margin of the squamosal, prominent ornamentation on the rearward bones of the lower jaw, and wide bar between the orbit and the lower temporal opening; at 2: widening of the parietal part of the skull roof, and broadening of the inside of the rear half of the ilium.

Once these flat-headed pachycephalosaurs have been left behind, we encounter the less ambiguous dome-headed dinosaurs, Pachycephalosauridae, which contain as many as 10 presently recognized species. All share as a derived feature the doming of the frontal and/or parietal bones – the beginnings of the development of the full domes seen in derived pachycephalosaurs. Within the pachycephalosaurid clade, individual forms appear to be arranged in a pattern of serial relationships. Overall, this group shows progressive enlargement of the frontal-parietal dome and the consequent diminishing of the parietal-squamosal shelf and upper temporal openings (Figure 8.9).

Three pachycephalosaurs have not been included in this discussion of phylogeny: *Micropachycephalosaurus*, *Gravitholus*, and *Stygimoloch*. The first – from Upper Cretaceous rocks of Shandong Province, China – is a relatively poorly known pachycephalosaur whose affinities within the clade cannot yet be properly assessed. For dinosaur trivia buffs, though, *Micropachycephalosaurus* is the dinosaur with the longest generic name. *Gravitholus* and *Stygimoloch*, on the other hand, both come from the Late Cretaceous of the northern realm of the Western Interior of North America but are as yet only very poorly known. *Gravitholus* is characterized by having a high and very wide dome with a large depression toward the back of the skull cap. On the basis of these features, it is clearly a pachycephalosaurid, but where within the clade it resides is not yet known. *Stygimoloch*, also from the Late Cretaceous of the Western Interior, exhibits completely closed upper temporal openings and a prominent parietal-squamosal shelf. Perhaps it is a member of the pachycephalosaurid clade consisting of *Tylocephale*, *Prenocephale*, and *Pachycephalosaurus*. Interestingly, *Stygimoloch* has massive, pointed horn-cores (over which cornified horns would have grown – much like a modern cow horn) that extend from the back of the parietal-squamosal shelf. In this feature, *Stygimoloch* is unlike any other pachycephalosaur.

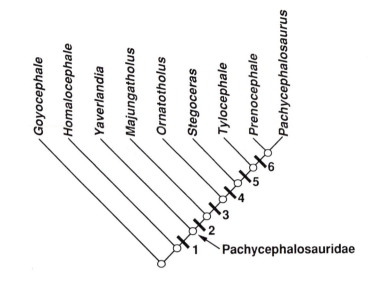

Figure 8.9 Cladogram of Pachycephalosauridae, with successively more distant relatives, *Homalocephale* and *Goyocephale*. Derived characters at 1: doming of the frontal and/or parietal bones; at 2: increase in size of the frontal dome; and enlargement of the nerves involved in the sense of smell; at 3: prominent doming of both the frontal and parietal bones; at 4: stronger doming of the entire frontal-parietal region, with the further incorporation of adjacent skull elements over the eye and onto the snout; fusion of the frontal and parietal bones to each other in adulthood; continued reduction of the upper temporal openings; and greater projection of the shelf over the back of the skull; at 5: very abbreviated shelf; at 6: fully closed upper temporal openings; pronounced doming of the skull cap, which incorporates nearly all of the top of the skull, including the region around the top of the eye socket, adjacent parts of the snout, and all of the parietal-squamosal shelf; and contact between the jugal and quadrate.

PACHYCEPHALOSAUR PALEOBIOLOGY AND PALEOECOLOGY: Food and Sex

DISTRIBUTION.   All pachycephalosaurs but *Majungatholus* (from Madagascar) are known from the Northern Hemisphere – in plate tectonics terms, Laurasia (see Chapter 2). In fact, all of the primitive members of Pachycephalosauria (e.g., *Wannanosaurus, Goyocephale,* and *Homalocephale*) are known from central or eastern Asia, suggesting that they arose in that part of the world. Thereafter, several dispersals are thought to have taken place between Asia, North America, Africa, and Europe (Figure 8.10).

Concentrating on the material from North America, most of what we know about pachycephalosaurs comes from isolated skull caps, many of which are highly waterworn. In fact, only single specimens of *Stegoceras* and *Pachycephalosaurus* are represented by more than just these skull caps, the former by a well-preserved skull and partial postcranial skeleton, and the latter by a nearly complete skull. Still, these skull caps are not uncommon in Upper Cretaceous rocks of North America, suggesting that some pachycephalosaurs – particularly *Stegoceras* – may have constituted up to 10% of the dinosaur fauna of the time. What is controlling the bias against the preservation of better specimens? Sternberg and, later,

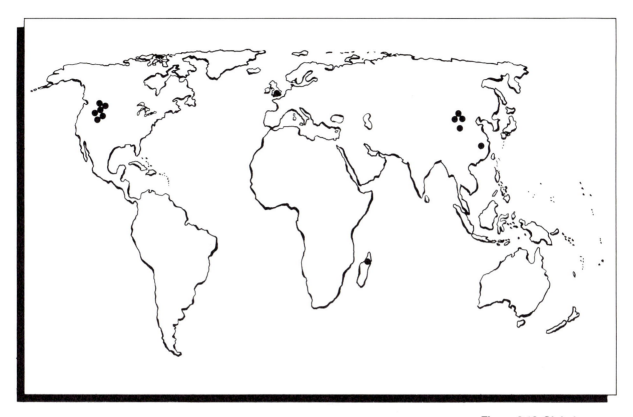

Figure 8.10 Global
distribution of
Pachycephalosauria.

P. Dodson have suggested that North American pachycephalosaurs may have lived some distance away from those rivers and lakes whose sediments now preserve their remains.

In contrast, when we turn to Asia, the situation is quite different. Here, the combination of desert conditions and fluvial environments of Mongolia and China must have created favorable conditions for pachycephalosaur preservation. The nearly complete skulls and associated skeletons clearly lack traces of the kinds of long-distance transport seen in North American forms, and it may be that these animals lived (and died) much closer to streams, dunes, and lakes.

FOOD CONSUMPTION. Whether close to, or far from, the rivers of the Late Cretaceous, pachycephalosaurs spent a great deal of their time foraging on many of the low-growing plants of the time (because of their stature, pachycephalosaurs probably browsed no more than a meter or so above the ground). The herbivorous habits of these animals are evident not only by their dentition, but also by the impressive volume of their abdominal region. At the front of the jaws are simple, peglike gripping teeth, the last of which is sometimes enlarged in a canine-like fashion. It is likely that these teeth were surrounded by a small, horny rhamphotheca. Farther back, the cheek teeth of pachycephalosaurs are uniformly shaped, with small, triangular crowns (Figure 8.11). The front and back margins of these crowns bear coarse serrations, the better for cutting or puncturing plant leaves or fruits. This kind of chewing produced a variety of kinds of tooth wear, from which we infer that different pachycephalosaurs fed on different kinds of vegetation.

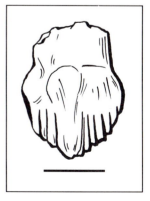

Figure 8.11 An upper
cheek tooth of
Pachycephalosaurus.
Scale = 5 mm.

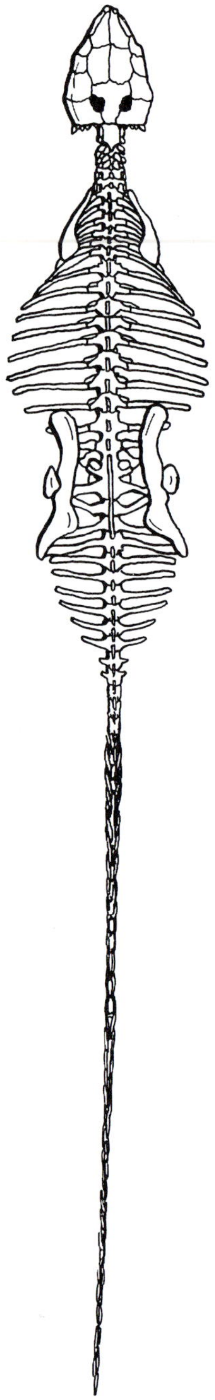

Figure 8.12 Dorsal view of the skeleton of *Homalocephale.* Scale = 50 cm.

At the other end of the feeding apparatus, pachycephalosaurs must have had a large vat for fermenting their food. The rib cage is very broad, a great girth that extends backward to the base of the tail (Figure 8.12). These anatomical modifications over the more primitive condition seen in ceratopsians (Chapter 9), ornithopods (Chapter 10), and other ornithischians suggest a backward migration, and enlargement, of the digestive tract to occupy a position between the legs and under the tail. Much as is seen in thyreophorans (see Part II: Ornithischia, and Chapter 9) and sauropods (Chapter 11), simple styles of chewing may combine with more extensive chemical digestion via the development of a huge gut to solve the problem of making a living as a plant-eating dinosaur.

THOUGHTS OF A PACHYCEPHALOSAUR.   Pachycephalosaur neuroanatomy suggests that, despite having only an average-sized brain for their body size, these animals may have had a quite acute sense of smell: The olfactory lobes of the brain were enlarged. The front half of the brain (the cerebrum) was highly flexed relative to the rest of the brain (to give it a horizontal orientation); the back half (the pontine region) was less flexed than in other dinosaurs. With a smaller degree of pontine flexure, this region of the brain is downwardly inclined. As reported by E. B. Giffin of Wellesley College, the reduction in pontine flexure appears to reflect the rotation of the back of the skull (the occiput) to face not only backward, but slightly downward. She has also shown that the most extreme degree of occipital inclination is associated with the most prominent doming of the frontal-parietal skull cap. And prominent doming, of course, is the feature that has led paleontologists to suggest that pachycephalosaurs head-butted.

Butting

The idea that pachycephalosaurs used their thickened skull roofs as battering rams comes from suggestions made in 1955 by E. H. Colbert (then at the American Museum of Natural History) and, quite independently, by the Russian evolutionary biologist L. S. Davitashvili in 1961. Butting among pachycephalosaurs was analyzed in considerable detail by Galton (University of Bridgeport, Connecticut) and Sues (Royal Ontario Museum) in the 1970s (Figure 8.13).

Galton and Sues noted that the outside of the very thick dome is often very smooth. Some specimens bear what look like scars on this external surface. Internally, the structure of the dome is very dense, consisting of fine bony columns that radiate so as to be approximately perpendicular to the external surface of the dome. Such an arrangement is ideal for resisting forces that come from strong and regular thumps to the top of the head and for transmitting such forces around the brain, much like a football player's helmet channels forces around the head. Adding strength to this functional interpretation of bone structure is Sues's important study that simulates how such forces would pass through the dome. Using special clear plastic cut to resemble a cross section of the dome of *Stegoceras* (Figure 8.14), Sues stressed this plastic model in ways that simulated head-butting and, *voilà*, the stress lines (which can be seen under ultraviolet light) had the same orientation as the columnar bone. The match between stress lines and bony columns strongly suggests that the latter have optimal orientation to resist the former.

If domes were used as pachycephalosaur battering rams, we might expect that

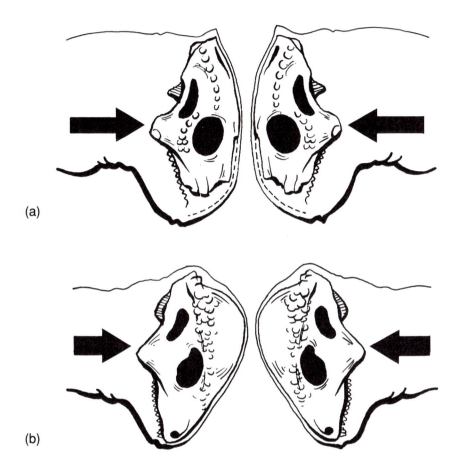

(a)

(b)

these structures should show some degree of sexual dimorphism. The first suggestion of such dimorphism in domes was made by Galton and his then student, W. Wall, but by far the most comprehensive treatment is an important study by Smithsonian paleontologist and morphometrician R. E. Chapman and several of his colleagues. This work assessed variation in the size and shape of the frontal-parietal dome using a large sample of a single pachycephalosaur species, *Stegoceras validum*. Many of these domes were from juveniles, but others – the largest specimens – obviously came from old individuals. On the whole, variation within the *Stegoceras validum* sample was fairly homogeneous – none of the specimens stood out from the main group.

   Some heterogeneity in size and shape occurs in the domes of *Stegoceras*. Two groups could be recognized among Chapman's sample of **sympatric** (i.e., living in the same place at the same time) *Stegoceras*, based upon the sizes of domes and braincases (Figure 8.15). The members of one group reveal a slight acceleration in the rate of growth of the dome relative to that of the braincase; this group inevitably ended up forming larger domes. These larger-domed individuals also had a slightly greater growth rate in dome thickness. The larger, more-convex-domed individuals were designated males. The other group, distinguished by having thinner, flatter domes (as a consequence of lower growth rates than those in

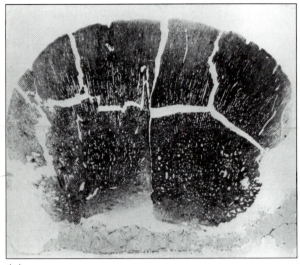

(a)

Figure 8.14 **The dome of *Stegoceras*. (a)** Vertical section through the dome of *Stegoceras*. Note the radiating organization of internal bone. **(b)** Plastic model of the dome of *Stegoceras*, in which forces were applied to several points along its outer edge and seen through polarized light. Note the close correspondence of the stress patterns produced in this model and the organization of bone indicated in **(a)**. *Photographs courtesy of H.-D. Sues.*

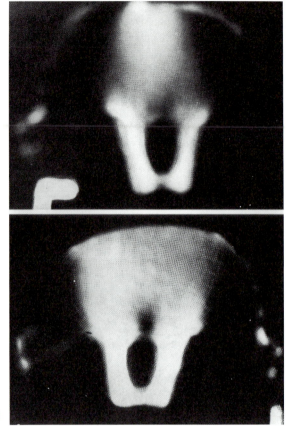

(b)

the other, more prominently domed group), were presumed to be females.[3] Interestingly enough, these assignments to male or female came out to be about a one-to-one ratio.

Chapman and his colleagues' study led them to a remarkable insight into pachycephalosaur behavior. The "males" with their large and more convex domes were better designed for initiating and receiving head-butting blows; the "females" had less-well-developed domes; in fact, the general profile of their domes looked very similar to juvenile or young adult "males." These, noted Chapman and colleagues, were exactly those individuals not yet likely to engage in head-butting. If Chapman and his colleagues are correct, then we have obtained a perspective on the relationship of dome size and shape to gender, and a subtle insight on behavioral differences between the sexes: One gender – designated "males" – head-butted, and the other did not. But what else can be said about pachycephalosaur butting in terms of other regions of the body? We turn next to the orientation of the head on the neck.

[3]Why is maleness assigned to the larger-domed forms, and femaleness assigned to the smaller-domed forms? Although we might not exonerate all of our colleagues from the charge of sexism, Chapman and his co-workers' sex designations were based upon the fact that in crocodiles and birds – the closest living relatives of pachycephalosaurs – it is predominantly the males who are larger and who are disposed toward sexual display, predominantly through coloration and/or behavior.

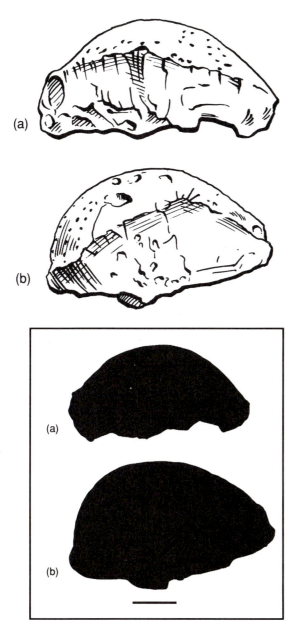

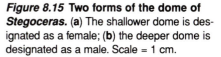

Figure 8.15 **Two forms of the dome of** **Stegoceras.** (**a**) The shallower dome is designated as a female; (**b**) the deeper dome is designated as a male. Scale = 1 cm.

As we have already noted, the back of the pachycephalosaur skull is progressively rotated forward beneath the skull roof, which is the principal reason for the high degree of pontine flexure of the brain in these animals. In this way, the line of action of any impact force to the top of the head passes close to the occipital condyle (the bony joint between the skull and the neck vertebrae). With the head in a downward position – the only position that makes sense for butting – rotation of the back of the skull minimizes the chance of violent rotation, or even dislocation, of the head on the neck. These sorts of injuries represent the Cretaceous equivalent to something like automobile whiplash: immediately debilitating and possibly life threatening. Therefore, we might expect some corrective measures to be seen in the neck region as well. Unfortunately, the neck itself is not known in any pachycephalosaur. It is nevertheless clear from the back of the skull of these animals that the extremely well developed and very strong neck musculature was used to position the head at the right attitude for head-butting and for resisting some potentially injurious rotations of the skull.

What about farther down the animal? Are there skeletal measures to ensure that the body did not torque around itself, injuring the spinal cord and the nerves that extended from it? The answer is a resounding Yes. Unique to pachycephalosaurs, the tongue-and-groove articulations between vertebrae along the back are very much like the carpentry joints of the same name, which give strength to, and make rigid, adjoining pieces of wood. In a similar fashion, the pachycephalosaur tongue-and-groove condition would have provided a great deal of rigidity to the back, preventing the kinds of violent lateral rotations of the body that would otherwise have been suffered at the time of impact.

The skull and the back vertebrae all speak strongly for butting in pachycephalosaurs. With the phylogeny of the group in mind, we can now develop some of the history of this kind of behavior in the clade. Most primitively, pachycephalosaurs all bore thickened, yet flat, heads. This feature alone strongly suggests that these animals were pushers. Pushing encounters were probably made with the vertebral column held horizontally – the primitive condition for these bipedal ornithischians and an obvious advantage when competing males approached one another. Analogous kinds of pushing matches are also found in modern marine iguanas, which use these types of confrontations to establish social hierarchies. Perhaps the social structuring described here for primitive pachycephalosaurs had an even wider phylogenetic distribution, as ceratopsians are also thought to have evolved a similar social order. If so, then perhaps social hierarchies and head-on confrontations constitute the ancestral behavioral pattern for at least Marginocephalia and are retained in all members of Pachycephalosauria.

It was not until the evolution of fully domed forms that battering-ram behavior appears to have developed. Again, approaches of competing males were probably done with the vertebral column held horizontally. Perhaps these males faced each other and made dome-to-dome blows or else charged at each other and lowered their heads just before impact. But head-to-head collisions would not have come cost free: Very severe injuries to the head and neck would have been expected simply because these animals lacked self-correcting mechanisms of the kind seen in modern head-to-head butters like goats and bighorn sheep, which would have kept the tremendous forces aligned with the rest of the skeleton. Glancing blows would have been the worst for these head-on battering rams. Debilitating or even lethal, injuries to the brain or spinal cord were to be avoided at all cost. So without precision head-to-head butting, these animals instead may have simply pushed head-to-head or butted each other along their flanks. It was much safer that way and the results were presumably the same: winners and losers.

Whatever region of the body pachycephalosaurids butted, many anatomical modifications took place to reduce the possibility of injury. As we have already outlined, the tremendous forces applied to the top of the dome would have been transmitted by the reoriented occiput through the long axis of the neck, thereby reducing the chances that the skull would be dislocated by the jolt of head collision. Any additional tendency toward head dislocation would have been prevented by the large and powerful muscles of the neck. Finally, whiplash to the back would have been prevented by the rigid construction of the vertebral column. All in all, these animals were very well designed for the rough-and-tumble lifestyle of the ramrods of the Late Cretaceous.

Integrated with butting is a suite of features related to visual display. First, there are the canine-like teeth. These were likely used in threat display or biting combat between rival individuals, much like pigs and primitive deer do today. If we only knew more about them, we might discover that these teeth are sexually dimorphic, large in one sex and less prominent, more normal-sized in the other. Alas, all this remains speculation. More informative are the knobby and spinous osteoderms that cover the snout and the side of the face and that occur most extensively on the back of the parietal-squamosal shelf. Assuredly these distinctive knobs and spines – especially in *Stygimoloch* – were used to show off, males displaying alternately to females and to threatening rival males.

SEXUAL SELECTION. The "glue" holding all of these phylogenetic modifications together is sex. The establishment of dominance hierarchy gave some males – those that sent the right signals about their qualifications for breeding – preferred access to females. And it was the females who chose not only their mates but, by their very actions, what those signals were. These same males must also have fended off competitors for access to females. To do so required some kind of threat display and/or offensive weaponry that was used to establish dominance between greater and lesser males. In modern bighorn sheep, it is the curl of horns; in modern anole lizards, it is the colorful dewlap that unfolds under the chin. In general, this practice of female choice and its effect on establishing dominance hierarchies constitute what is called **sexual selection**, selection not among all individuals within a species, but among males alone. Sexual selection emphasizes features related to display and combat, principally of males to females and males to males. In pachycephalosaurs, this emphasis was on the prominence of domes, knobs, and spikes – structures that acted in ritual display and, should the need have arisen, in actual, violent clashes. The winner, presumably the male with the best-fashioned head-thumper (either in terms of showiness or safest design), got to perpetuate his family line but nevertheless had to be ever vigilant for other males that wanted to literally knock his block off.

Important Readings

Chapman, R. E., P. M. Galton, J. J. Sepkoski, and W. P. Wall. 1981. A morphometric study of the cranium of the pachycephalosaurid dinosaur *Stegoceras*. Journal of Paleontology 55:608–616.

Colbert, E. H. 1955. Evolution of the Vertebrates. Wiley, New York, 479 pp.

Davitashvili, L. S. 1961. The Theory of Sexual Selection. Izdatel'stov Akademia Nauk SSSR, Moscow (in Russian).

Galton, P. M. 1971. A primitive dome-headed dinosaur (Ornithischia:

Pachycephalosauridae) from the Lower Cretaceous of England, and the function of the dome in pachycephalosaurids. Journal of Paleontology 45:40–47.

Giffin, E. B. 1989. Pachycephalosaur paleoneurology (Archosauria: Ornithischia). Journal of Vertebrate Paleontology 9:67–77.

Maryańska, T., and H. Osmólska. 1974. Pachycephalosauria, a new suborder of ornithischian dinosaurs. Palaeontologia Polonica 30:45–102.

Sereno, P. C. 1986. Phylogeny of the bird-hipped dinosaurs (Order Ornithischia). National Geographic Research 2:234–256.

Sues, H.-D. 1978. Functional morphology of the dome in pachycephalosaurid dinosaurs. Neues Jahrbuch für Geologie und Paläontologie Monatshefte 1978:459–472.

Sues, H.-D., and P. M Galton. 1987. Anatomy and classification of the North American Pachycephalosauria (Dinosauria: Ornithischia). Palaeontographica A 198:1–40.

Wall, W. P., and P. M. Galton. 1979. Notes on pachycephalosaurid dinosaurs (Reptilia: Ornithischia) from North America, with comments on their status as ornithopods. Canadian Journal of Earth Sciences 16:1176–1186.

CHAPTER 9

CERATOPSIA:
Horns, Frills,
and Slice-and-Dice

THE EARTH HAS NEVER SEEN the likes of ceratopsians. Diverse and abundant, herds of these hooved, horned, frilled herbivores must have dominated the Late Cretaceous landscape, foraging in the shrubs and thundering across the open terrain in places that now are called Alberta, Montana, and Wyoming.

Most of our knowledge about these animals comes from North America, a place when ceratopsian evolution reached its acme. But although the North American forms ranged upwards of 6 or 7 metric tons, a host of smaller, non-horned ceratopsians from slightly earlier in the Late Cretaceous hail from Asia. Even without the horns and size, they are easy to recognize as ceratopsians. The recognition factor is their sharp, parrotlike beak and a large sheet of bone called the **frill** (Figure 9.1). Extending from the back of the skull and made up of the parietal and squamosal bones, the frill varied considerably in size. In some cases, the ceratopsian frill could be quite large: In *Torosaurus*, the 2.7-m-long skull is nearly two-thirds frill! In some genera, the margin of the frill was ornamented by long spikes or by extra bones, known as **epoccipitals**, that formed a sort of scalloping.

Still, large, aggressive-looking horn cores are perhaps the most memorable feature of ceratopsians (Figures 9.1d–i). Some ceratopsians had two horns that were prominent – one over each eye – and a third horn over the nose; some had a nose horn and no eye horns; and several had a large block of very roughened bone over the nose and eyes. A few had no horns at all (Figures 9.1a–c).

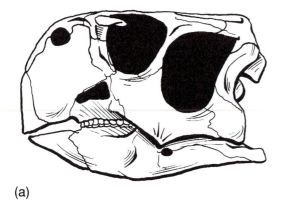

(a)

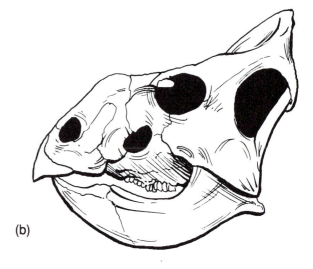

(b)

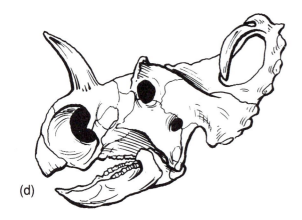

(d)

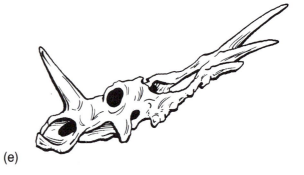

(e)

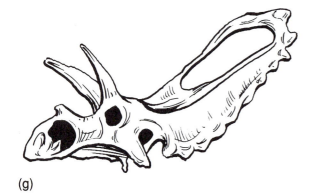

(g)

(h)

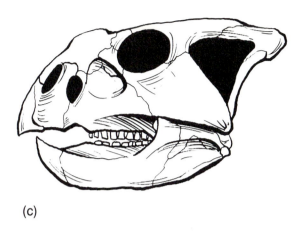

(c)

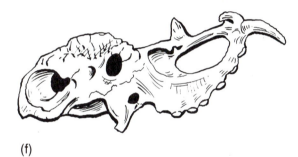

(f)

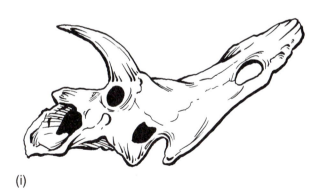

(i)

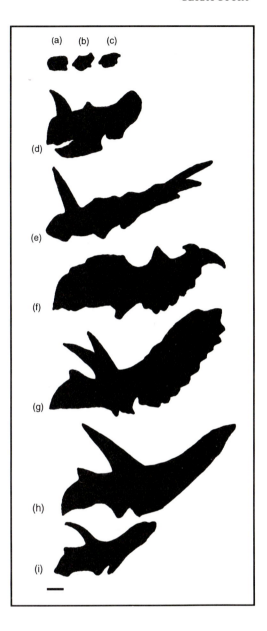

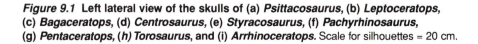

Figure 9.1 Left lateral view of the skulls of (a) *Psittacosaurus,* (b) *Leptoceratops,*
(c) *Bagaceratops,* (d) *Centrosaurus,* (e) *Styracosaurus,* (f) *Pachyrhinosaurus,*
(g) *Pentaceratops,* (h) *Torosaurus,* and (i) *Arrhinoceratops.* Scale for silhouettes = 20 cm.

HISTORY OF CERATOPSIAN DISCOVERIES

19th Century

The initial ceratopsian finds were not auspicious. The first was a partial pelvis discovered in Wyoming and named *Agathaumas* (*aga* – very; *thauma* – wonder) by E. D. Cope in 1872. It would prove to be of little value in understanding these great dinosaurs. *Monoclonius* (*mono* – single; *klon* – sprout, referring to the root of the tooth), recovered from Montana and named in 1876 by Cope on the basis of a fragmentary pelvis, some vertebrae, a few teeth, and a parietal bone, gave no real inkling of what was to follow. So little, in fact, that 1·1 years later Cope's rival, O. C. Marsh, announced the discovery of a pair of very large ceratopsian horn cores near Denver, Colorado, that he believed belonged to an extinct bison (*Bison alticornis*)!

It wasn't until the late 1880s that the true identity of the horn cores, pelvis, and collection of *Monoclonius* bones and teeth was finally revealed. The enlightenment was due to none other than *Triceratops* (*treis* – three; *keras* – horn; *tops* – face), originally uncovered in Wyoming. Named by Marsh in 1889 on the basis of an imperfect skull from a huge animal, *Triceratops* is now known from over 50

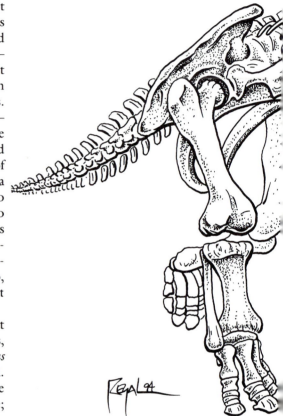

complete and partial skulls, as well as a number of well-preserved skeletons that have been collected throughout the northern Great Plains of North America (Figure 9.2).

In 1891, Marsh named another of these Late Cretaceous ceratopsians – *Torosaurus* (*toreo* – perforate, referring to the holes in the frill) – again from Wyoming. Today it is one of the most widespread of dinosaurs, having been documented across the entire Western Interior of North America. In spite of its broad distribution, however, *Torosaurus* is rare: Although it has been known for over one hundred years, a complete skull of it has never been collected!

20th Century

In the first quarter of this century, ceratopsians were discovered with a vengeance, particularly in Alberta during the years of prospecting up and down Canada's Red Deer River. In 1904, *Centrosaurus* (*kentron* – sharp point; now believed to be syn-

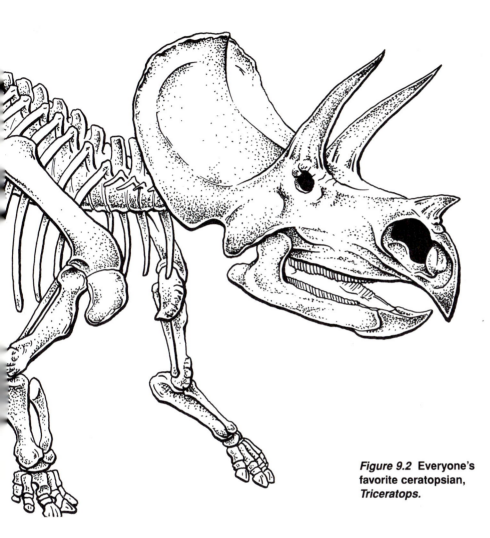

Figure 9.2 **Everyone's favorite ceratopsian, *Triceratops*.**

onymous with *Monoclonius*) gave us – and L. M. Lambe, who described it – a glimmer of the wealth of ceratopsians when it was first discovered (Figures 9.1d and 9.13c).[1] This was followed shortly thereafter by *Diceratops*, from Wyoming and named by J. B. Hatcher.

Yet it was not until the 1910s and 1920s that ceratopsian riches were to be fully realized. In 1913 and 1914 alone, *Styracosaurus* (*styrak* – spike; Figures 9.1e and 9.5c), *Chasmosaurus* (*chasma* – wide opening; Figures 9.3 and 9.5d), *Anchiceratops* (*anchi* – close), and *Leptoceratops* (*leptos* – slender; Figure 9.1b) –

[1]A somewhat morbid story accompanies the first major description of many of these early ceratopsian discoveries. They, as well as aspects of ceratopsian anatomy and paleoecology, were presented in comprehensive fashion in a large study begun by O. C. Marsh of Yale University. The work was later taken over by J. B. Hatcher, Marsh's principal collector and a researcher in his own right, after Marsh died. Hatcher, however, died tragically from typhoid fever at the age of 43, before *he* could complete the work. The work was then taken over by Yale's R. S. Lull, whose apparent lack of superstition was rewarded by the eventual publication of his monograph in 1907. The work stands today as one of the most important references on those ceratopsians known at the turn of the century.

Figure 9.3 The long-frilled ceratopsian, ***Chasmosaurus.*** *Courtesy of the Royal Ontario Museum.*

Canadian dinosaurs all – were discovered, described, and named by Lambe and by B. Brown. Across the border, C. W. Gilmore was providing Montana with another ceratopsian, this time *Brachyceratops* (*brachys* – short). And, from New Mexico, *Pentaceratops* (*penta* – five; Figure 9.1g) was described by H. F. Osborn in 1923.

On the other side of the globe, the Gobi Desert was beginning to yield its dinosaurian treasures (Box 7.1). The most famous of all was *Protoceratops* (*protos* – first; Figures 9.5b and 9.13b), announced to the world in 1923 by W. Granger and W. K. Gregory. Now known from both Mongolia and China, *Protoceratops* was the first dinosaur for which a reasonably complete growth series, and even sexual dimorphism, were documented.[2] Also from the Gobi Desert, and no less important, were the remains of *Psittacosaurus* (*psittakos* – parrot; Figures 9.1a, 9.5a, and 9.13a), described by Osborn in 1923. *Psittacosaurus* is small (about 2 m), making it one of the smallest of all ceratopsians. Juveniles of the *Psittacosaurus* clade are known; hatchlings were no more than 23 cm long, about the size of an adult pigeon! *Psittacosaurus* is now also known from a variety of localities in China and eastern Russia. A return to North America completes the research cycle of the 1920s. In 1925, W. A. Parks, vertebrate paleontologist from Toronto's Royal Ontario Museum, described yet another long-frilled ceratopsian, *Arrhinoceratops* (*a* – without; *rhinus* – nose; Figure 9.1i), from the great fossil beds of Alberta.

In the years that intervened between the 1920s and the present, new studies of ceratopsians came out in what might best be described as fits and starts. It was not until 1942 that C. M. Sternberg identified another new ceratopsian,

---

[2] An abundance of eggs and *Protoceratops* specimens led R. C. Andrews and his team of researchers to conclude that they had found eggs belonging to *Protoceratops*. For the next 70 years, the eggs were considered to belong to *Protoceratops*. An *Oviraptor* skeleton lying across one of the nests was presumed to be stealing the eggs, hence the name *Oviraptor* (see Chapter 12). In 1993, however, American researchers in Mongolia recovered for the first time "*Protoceratops* eggs" with embryos preserved within. Imagine their surprise when they discovered that the "*Protoceratops* eggs" contained *Oviraptor* babies! All this time, *Oviraptor* had a bum rap as an egg stealer, when actually the skeleton from which the name derived was most likely a parent. So to this day, the eggs of *Protoceratops* remain unknown.

Montanoceratops (from Montana). Meanwhile (1950), he was also was busy with the announcement of the Upper Cretaceous *Pachyrhinosaurus* (*pachy* – thick; Figure 9.1f) from Alberta, one of the ugliest (or most magnificent, depending upon one's perspective) of ceratopsians. *Pachyrhinosaurus* came replete with masses of exceedingly roughened bone extending the length of the snout from nose to eyes; the huge **boss**, as this mass of bone is sometimes called, is thought to have supported a sheath of fingernail-like material. *Pachyrhinosaurus* turns out to have an extraordinary geographic distribution: Montana, Alberta, and now on the north slope of Alaska, where 70 million years ago it was within 5° of the Late Cretaceous North Pole!

Three years after Sternberg's description of *Pachyrhinosaurus*, B. Bohlin – paleontologist with the joint Sino-Soviet Palaeontological Expedition to the Gobi Desert – published his account of *Microceratops* (*micro* – small). Much like *Leptoceratops* and *Protoceratops*, this small ceratopsian was without horns but sported a modest frill behind the head. *Microceratops* is now known from several localities in northern China and southern Mongolia. It was another 22 years before these Asian ceratopsians were again in the limelight. In 1975, T. Maryanska and H. Osmólska, the two principal dinosaur specialists in the Polish-Mongolian Palaeontological Expedition (1963–1971), described a new form that they called *Bagaceratops* (Mongolian: *baga* – small; Figure 9.1c).

During the last decade, all kinds of new ceratopsians have been described. The first, *Avaceratops* (to honor Ava Cole, a collector of fossils), comes from a 1986 study by P. Dodson and represents the first advanced horned dinosaur to have been discovered since Sternberg's discovery of *Pachyrhinosaurus*. Nearly ubiquitous *Protoceratops*-like ceratopsians continue to tumble out of the rocks. One, discovered in Kazakhstan by L. Nessov and co-workers from the Leningrad Natural History Museum, was named *Turanoceratops* (from Turan, Kazakhstan) in 1989. Another – *Breviceratops* (*brevis* – short) – was originally thought by Maryanska and Osmólska in 1975 to be a new species of *Protoceratops*. It was S. Kurzanov of the Palaeontological Institute in Moscow who rechristened it *Breviceratops* in 1990. The most recent Asian ceratopsian thus far discovered (1992) is *Udanoceratops* (from Udan Sair), again the product of ongoing field research in the Gobi Desert conducted by Kurzanov. And in North America, two strange and unexpected ceratopsians emerged from Montana: *Einiosaurus* (*einio* – from the Blackfoot word meaning "buffalo," a reference to its horns) and *Achelousaurus* (*Achelos*, the river god in Greek mythology who fought Hercules by changing shapes, a reference to the differences in shape between it and *Einiosaurus*). Both have two large, distinctive, rearward-directed spines on the frill, but *Achelousaurus* has large roughened bosses over the eyes and nose, and *Einiosaurus* bears a single, bizarre nose horn uniquely bent forward at a 90° angle.

CERATOPSIA DEFINED AND DIAGNOSED

Ceratopsia is a monophyletic taxon that consists of the common ancestor of members of Psittacosauridae and Neoceratopsia and all the descendants of this common ancestor (Figure 9.4). Among the rich array of unambiguously derived features shared by this ceratopsian clade, the most important is the rostral bone, a unique bone at the tip of the snout, whose parrot-beak shape gives ceratopsians

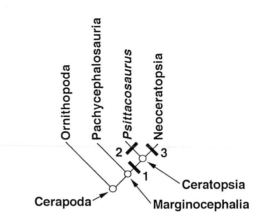

Figure 9.4 **Cladogram of Cerapoda, emphasizing the monophyly of Ceratopsia,**
Psittacosaurus, **and Neoceratopsia.** Derived characters at 1: rostral bone, skull that is
narrow at the beak end and flaring and deep in the cheek region, frill composed principally
of the paired parietal bones, and strongly vaulted palate beneath the beak; at 2: short pre-
orbital region of the skull, high external nares, broad lateral premaxillary process, loss of
antorbital opening, and loss of digit V in the hand; at 3: pointed, sharply keeled rostral bone
and predentary point, reduction or loss of premaxillary teeth, loss of the external mandibular fenestra, and very short retroarticular process.

their distinctive snout. In life, the rostral was covered with a beak. Other characters uniting Ceratopsia are listed in Figure 9.4.

Ceratopsians and pachycephalosaurs (Chapter 8) appear to share a unique
common ancestor, thus forming the larger clade that P. C. Sereno has called
Marginocephalia (see Part II: Ornithischia).

CERATOPSIAN DIVERSITY AND PHYLOGENY

Ceratopsia primitively consists of Psittacosauridae and the much more diverse
monophyletic Neoceratopsia (Figure 9.4). Psittacosauridae is presently the most
species-rich, genus-poor clade yet known among dinosaurs – consisting of one
genus (*Psittacosaurus*) and seven species. Thus far, only the bare rudiments of the
evolutionary relationships of these species are known. Still, all members of this
clade share as many as 12 derived features, listed in Figure 9.4.

Neoceratopsia, the remaining clade of ceratopsians, is clearly monophyletic,
based on at least 10 important shared, derived characters, also listed in Figure 9.4.
These include the loss of the external opening in the lower jaw (the **external
mandibular fenestra**), and a very short projection of the lower jaw beyond the
jaw joint (the **retroarticular process**).

The cladogram in Figure 9.4 reveals a fundamental fact about the evolution
of Ceratopsia. The basal member of the clade, *Psittacosaurus*, is a biped, as are all
members of the groups most closely related to Ceratopsia: Pachycephalosauria and
Ornithopoda. With the exception of *Leptoceratops* and *Microceratops*, however, all
other ceratopsians are quadrupeds. Because the ancestral condition for all tetrapods
is quadrupedality but the ancestral condition for Ceratopsia is bipedality, the

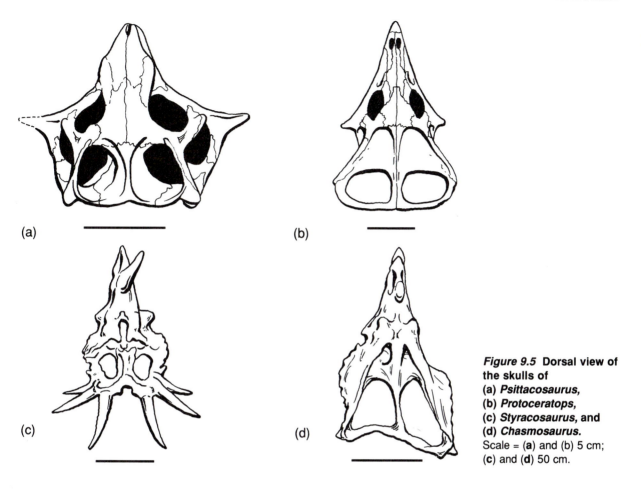

Figure 9.5 Dorsal view of the skulls of
(a) *Psittacosaurus,*
(b) *Protoceratops,*
(c) *Styracosaurus,* and
(d) *Chasmosaurus.*
Scale = (a) and (b) 5 cm;
(c) and (d) 50 cm.

cladogram implies that the evolution of Ceratopsia has involved the *return* to quadrupedality for most of its members. Like stegosaurs and ankylosaurs, quadrupedal ceratopsians are secondarily so.

Protoceratopsidae . . . or Not?

There is some debate about the relationships of the most primitive members within Neoceratopsia. Do *Leptoceratops, Bagaceratops, Microceratops, Protoceratops,* and *Montanoceratops* belong to a monophyletic Protoceratopsidae (Figure 9.6) or do they not (Figure 9.7)? Those who regard Protoceratopsidae as a clade – for instance, Dodson (University of Pennsylvania's School of Veterinary Medicine) and P. J. Currie (Royal Tyrrell Museum of Palaeontology in Drumheller, Alberta) – have argued that a number of cranial characters support this monophyly, including the inclination of a splint of bone on the **palatine** (one of the bones of the palate) called the **parasagittal process**, and the presence of a **sinus** (opening) within the **maxilla** (the upper jaw bone). Figure 9.6 shows this phylogenetic hypothesis.

Sereno has offered an alternative to a monophyletic Protoceratopsidae. He has suggested that each of the animals otherwise regarded as a conventional pro-

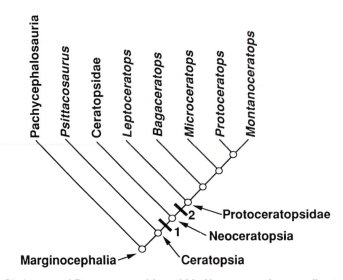

Figure 9.6 **Cladogram of Protoceratopsidae within Neoceratopsia according to Dodson and Currie, with the more distantly related *Psittacosaurus* and Pachycephalosauria.** Derived characters at 1: sharply keeled rostral bone and predentary both ending in a point, reduction or loss of premaxillary teeth, loss of the external mandibular fenestra, and very short retroarticular process; at 2: shallow and circular excavation of the antorbital opening, inclination of the parasagittal process on the palatine bone, and the presence of a sinus within the maxilla. No characters have been advanced as supporting the remaining relationships among protoceratopsids.

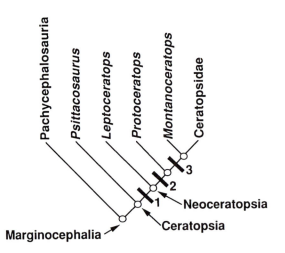

Figure 9.7 **Cladogram of Neoceratopsia according to Sereno, with the more distantly related *Psittacosaurus* and Pachycephalosauria.** Derived characters at 1: sharply keeled rostral bone and predentary both ending in a point, reduction or loss of premaxillary teeth, loss of the external mandibular fenestra, and very short retroarticular process; at 2: development of a frill through the expansion of the parietals and squamosals, and presence of a nasal horn and at least 10 other features; at 3: development of a prominent, wide-based nasal horn, modification of the neck vertebrae, and at least 6 additional features.

toceratopsid[3] is sequentially more closely related to Ceratopsidae (Figure 9.7).

Sereno's Neoceratopsia is still supported by the same features indicated in Figure 9.4. The clade of *Protoceratops* (and possibly *Bagaceratops*?) + all remaining neoceratopsians (called "Coronosauria" by Sereno) is supported by at least 10 derived features. The clade of *Montanoceratops* + Ceratopsidae (called "Ceratopsoidea" by Sereno) shares at least 6 important derived features (Figure 9.7).[4]

Which phylogenetic pattern is best supported – a monophyletic Protoceratopsidae or a monophyletic neoceratopsian clade that consists of forms sequentially more closely related to ceratopsids? Using Dodson's cladogram, it takes 3 steps to evolve the three characters (1 step per character) and at least 32 steps to account for the multiple evolution of all of the other characters (used by Sereno) necessary for the establishment of Protoceratopsidae. In contrast, the cladogram suggested by Sereno has these latter characters evolving once. We then apply the three characters (used by Dodson) purporting to support monophyly of protoceratopsids. In the context of Sereno's cladogram, these are either convergences (separately evolved in *Leptoceratops*, *Protoceratops*, and *Montanoceratops*) or reversals (evolved once in the common ancestor of Neoceratopsia, but secondarily lost in Ceratopsidae). Convergence takes 9 steps, whereas the primitive acquisition of these features in Neoceratopsia – with their loss in Ceratopsidae – takes 6 steps to account for the evolution of these characters. Regardless, on the basis of the most parsimonious use of characters distributed among all of these primitive neoceratopsians, Protoceratopsidae disappears, to be replaced by a sequential unfolding of the history of all of these primitive neoceratopsians (Figure 9.7).

Ceratopsidae

Now that we have dealt with the question of primitive neoceratopsians, we can finally turn to Ceratopsidae. This monophyletic clade, which includes Centrosaurinae (those ceratopsids with short squamosals) and Chasmosaurinae (those with long squamosals), is supported by upwards of 50 important diagnostic features, a few of which are listed in Figure 9.8.

All members of the chasmosaurine clade uniquely share a suite of modifications of the snout, nasal horn, frill, and external nares. Within this clade, the most primitive members are *Chasmosaurus* and *Pentaceratops*, which together appear to be each other's closest relatives. They share a number of derived features (Figure 9.9, node 2).

All remaining, more-derived members of Chasmosaurinae consist of the clade containing *Anchiceratops*, *Arrhinoceratops*, *Torosaurus*, *Diceratops*, and *Triceratops*. Sharing as many as seven derived features (Figure 9.9, node 3), this group has as its primitive members the small clade of *Anchiceratops* + *Arrhinoceratops* (Figure 9.9, node 4).

That leaves us with *Torosaurus*, *Diceratops*, and *Triceratops*. This clade, which shares a number of new modifications of the snout and nasal horn, has *Torosaurus* as its most primitive member (Figure 9.9, node 5). *Diceratops* and *Triceratops* are

[3] Note that Sereno did not include *Microceratops* or *Bagaceratops* in his analysis, although more recent research by C. Forster and Sereno suggests that *Bagaceratops* and *Protoceratops* might be each other's closest relative.

[4] A special note: The newly discovered *Turanoceratops* may be – according to new work by Forster and Sereno – a neoceratopsian closer still to ceratopsids than *Montanoceratops*.

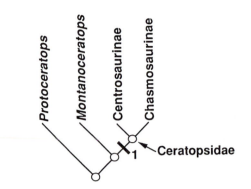

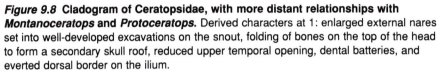

Figure 9.8 Cladogram of Ceratopsidae, with more distant relationships with *Montanoceratops* and *Protoceratops*. Derived characters at 1: enlarged external nares set into well-developed excavations on the snout, folding of bones on the top of the head to form a secondary skull roof, reduced upper temporal opening, dental batteries, and everted dorsal border on the ilium.

the only chasmosaurines to have reduced the size of their frills (Figure 9.9, node 6).

The other great ceratopsian clade is Centrosaurinae. This group ancestrally acquired a number of important features (Figure 9.10, node 1). Centrosaurinae has at least five members: *Avaceratops, Brachyceratops, Centrosaurus, Pachyrhinosaurus,* and *Styracosaurus.*

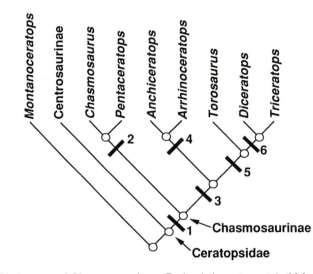

Figure 9.9 Cladogram of Chasmosaurinae. Derived characters at 1: thickening of the bones of the snout, additional ossification for the nasal horn, enlargement of the rostral bone, keyhole-shaped opening between the two frontal bones, thickening of the squamosal bone where it joins the frill, and epoccipitals broad-based and triangular; at 2: parietal bar that is square or rectangular in cross section, laterally convex deltopectoral crest on the humerus, and huge openings in the parietal part of the frill; at 3: sinuses in the horns that are continuous with the frontal sinus, and ischial shaft dorsally expanded; at 4: square frill that has very numerous traces of blood vessels on its undersurface; at 5: shallow pockets in the narial fossa, two pairs of openings in the premaxillary palate, increase in body size, and modifications of the snout and nasal horn; at 6: reduced frill.

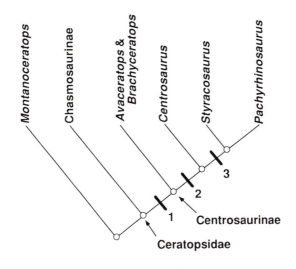

Figure 9.10 Cladogram of Centrosaurinae. Derived characters at 1: large nasal horn, long and narrow opening in the top of the skull roof, short brow horns, short squamosals, and broadly rounded epoccipitals decorating the rim of the frill; at 2: hooks on the back of the frill; at 3: long spikes extending from the frill margin.

Figure 9.11 Global distribution of Ceratopsia.

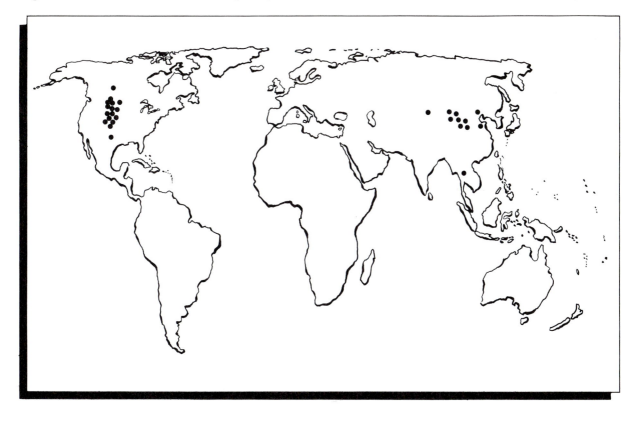

Avaceratops and *Brachyceratops* appear to be the most primitive members of Centrosaurinae. These two dinosaurs lack the development of hooks that ornament the back of the frill. *Centrosaurus*, *Pachyrhinosaurus*, and *Styracosaurus* are among the best known of all ceratopsids, and there is abundant material from both adults and juveniles. Within this more restricted clade of centrosaurines, *Centrosaurus* itself appears to be the most primitive member, lacking the development of long spikes that extend from the frill margin, a feature that unites *Pachyrhinosaurus* and *Styracosaurus* (Figure 9.10, node 2). Compared to *Pachyrhinosaurus*, *Styracosaurus* is much more conventional in its cranial adornment. Nonetheless, sprouting from the margin of the frill like the eagle feathers on a Native American war bonnet, the series of spikes along the rim of the squamosals and parietals complements the enlarged nasal horn and makes this centrosaurine one of the most spiny of all dinosaurs.

Ceratopsians are now known from material found in eastern Asia and western North America (Figure 9.11). It is not coincidence that the Asian forms tend to be slightly older, smaller, and more (phylogenetically) basal, nor that the North American forms tend to be younger, larger, and more derived. The clear implication from the geographic, temporal, and phylogenetic distributions of ceratopsians is that the group, sometime in the mid-Cretaceous to early Late Cretaceous, migrated from Asia to North America, and there radiated into a spectacular variety of large, horned dinosaurs. As we shall see, the evolution of horns may have come hand in hand with changes in social behavior.

CERATOPSIAN PALEOBIOLOGY AND PALEOECOLOGY: Feeding, Locomotion, Horns, and All the Frills

Feeding

Ceratopsians were herbivores and, to all appearances, were mighty good at it. A hooked rhamphotheca, a **dental battery** (a dense cluster of cheek teeth), a sturdy coronoid process, and evidence for the existence of fleshy cheeks – all reflect sophisticated manipulation of food in the mouth.[5]

Recall that chewing is a derived behavior among tetrapods. None of the groups that we have so far discussed in this book has had the wherewithal for extensive chewing; hence, we have inferred the evolution of fermentation systems for breaking down plant matter. Ceratopsians, however, chewed well, and in ways that no other vertebrates appear to have tried. Indeed, a likely consequence of ceratopsian masticatory prowess[6] is the fact that the remainder of the digestive tract does not appear to have been disproportionately large.

The driving force behind the high-efficiency mastication came from a great mass of jaw-closing musculature, which in the frilled forms crept through the upper temporal openings and onto the base of the frill. The other side of this mus-

---

[5] This anatomical arrangement – a cropping beak followed by a diastema, dental batteries toward the rear of the mouth, strong muscles to manipulate these, and muscular cheeks to keep things in the mouth – is similar to that seen in hadrosaurids (Chapter 10), as well as in many groups of herbivorous mammals (such as cows, sheep, and horses, to name three of the most familiar).

[6] If the kind of chewing we've described here was not enough for *Psittacosaurus*, then a packet of gastroliths lodged in the gizzard doubly pulverized its meal. Only in *Psittacosaurus* among ceratopsians (in fact, among all ornithischians) are gastroliths known.

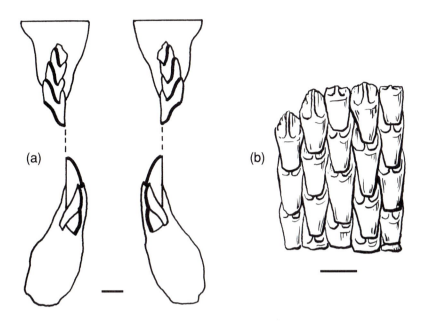

Figure 9.12 **(a) Cross section through the upper and lower jaws of** *Triceratops,* **illustrating high-angle slicing-and-dicing motion of the teeth; (b) internal view of the dental battery in the lower jaw of** *Triceratops.* Scale = 20 cm.

cle attached to a sturdy, large coronoid process on the mandible. Plant matter was undoubtedly ground down between the occlusal surfaces of the rugged dentition. Worn teeth were constantly replaced, so that the active surface of the dental battery was continually refurbished. The grinding action was bequeathed from the common ancestor of all ceratopsians; however, through the phylogenetic history of the group, the angle of **occlusion** – that is, the angle at which the teeth in the upper jaw intersect the teeth in the lower jaw – became more and more vertical, until in the latest forms, the near-vertical, highly efficient slicing and dicing characteristic of the later large ceratopsids was achieved (Figure 9.12).

Ceratopsians never browsed particularly high above the ground. Even for the largest (*Triceratops* and *Torosaurus*), browse height was probably no more than 2 m, and no one has seriously entertained the possibility that ceratopsids reared up on their hind legs to forage at higher levels (as has been suggested for sauropods and stegosaurs). Nevertheless, they may have been able to knock over trees of modest size in order to gain access to choice leaves and fruits. The sharp, hooked beaks suggest the capability for careful selection of the food.

DIET. *Which* food remains a mystery. Once thought to be feeders on the fibrous fronds of cycads and palms, the majority of ceratopsians are rarely found in the same areas as these kinds of plants. The principal plants whose statures match browsing heights of ceratopsians were a variety of shrubby angiosperms, ferns, and perhaps small conifers. In fact, it has been argued by two paleobotanists, S. Wing of the National Museum of Natural History (Smithsonian Institution), and B. Tiffney of the University of California at Santa Barbara, that ceratopsians, along with other large yet low-browsing, generalist-feeding herbivorous dinosaurs, were engaged in a mutually advantageous evolutionary waltz with early flowering plants. Suffice it to say for the present that herbivorous dinosaur feeding habits may have contributed to the extraordinary rise of flowering plants during the Late Cretaceous. This hypothesis and a related one are discussed in Chapter 15.

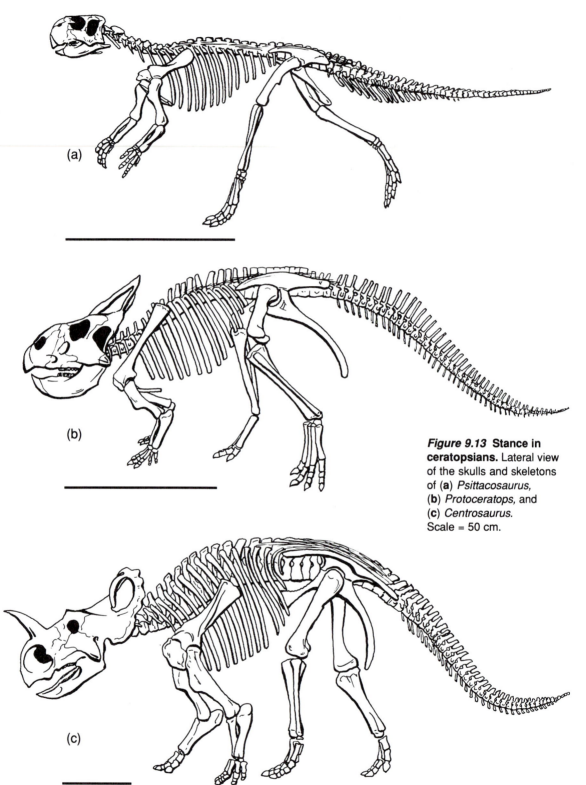

(a)

(b)

(c)

Figure 9.13 **Stance in ceratopsians.** Lateral view of the skulls and skeletons of (**a**) *Psittacosaurus,* (**b**) *Protoceratops,* and (**c**) *Centrosaurus.* Scale = 50 cm.

Locomotion

The legs of horned dinosaurs are unlike those of any mammal – living or extinct – and thus exactly how ceratopsians cruised over Late Cretaceous landscapes remains a matter of conjecture. Did they thunder along like enraged rhinos, or were their legs (and perhaps metabolism, as well) built for a kinder, gentler locomotion?

How fast bipedal ceratopsians may have traveled is not known, because there are no known trackways for any of these animals. However, those estimates that have been made for quadrupedal forms – based principally on biomechanical analyses of the limbs – range from 20 to 50 km/hr. A problem with this kind of analysis, however, is that the orientations of the front limbs in quadrupedal ceratopsians are not fully understood. R. T. Bakker has argued for what is essentially a mammal-like posture – that is, fully erect front limbs that mimic the well-known posture of the back limbs. Others have argued that this cannot be so and that a more sprawling posture in the front legs is indicated (Figure 9.13; see also Chapter 14).

For the present, R. A. Thulborn has presented the most detailed speed estimates. His work indicates that average walking speed for both large and small ceratopsians was somewhere between 2 and 4 km/hr, and maximum running speeds ranged from 30 to 35 km/hr (see also Box 14.2).

Socializing the Dinosaur

WHAT IS THE FUNCTION OF HORNS? At virtually *any* speed, a *Triceratops* fully equipped with horns rumbling across the Late Cretaceous countryside was apt to pack a serious wallop if something got in the way. Long supposed to function for individual defense against predators, horns may have had an altogether different function as well.

E. H. Colbert (then at the American Museum of Natural History) examined the relationship between horns, frills, and defense among ceratopsids in 1951. He noted that in living African bovids (for example, impalas, antelopes, and gnus), horns, whatever their shape, were used for resisting predators during **interspecific** (*inter* – between; between different species) combat. By analogy, ceratopsian horns would also have functioned to ward off predators at close quarters – for example, in the classic Late Cretaceous face-off between *Tyrannosaurus* and *Triceratops*.

Subsequent interpretations have also drawn on analogy with these same bovids, but have instead centered on the **intraspecific** (*intra* – within; among members of the same species) functioning of horns; that is, their role in display, ritualized combat, establishment of dominance, and defense of territories.

The link between dominance, defense, and horns comes from important studies of mammals in their natural habitats, research that has gained increasing prominence ever since humans began taking stock of how much they have disturbed virtually all terrestrial (and aquatic) ecosystems. For example, we know that in many horned mammals, larger males tend to breed more often than do small individuals, because females apparently choose them more often than not. Dominance in these mammals (and in other tetrapods) is accentuated by the development of structures, including horns and antlers, that "advertise" the size of the animal. On the basis of sexual selection (Chapter 8), these features come to be highly linked with reproductive success.

Can the intraspecific behavior and sexual-selection explanations observed in mammals be used to interpret the horns and frills in ceratopsians? Several dinosaur paleontologists, among them L. S. Davitashvili, J. O. Farlow and P. Dodson, R. E. Molnar, N. B. Spassov, and J. H. Ostrom and P. Wellnhofer, think so.

No one has ever doubted that ceratopsian horns were used for combat, they only wondered who they were aimed at. Puncture wounds on faces, frills, and bodies of ceratopsians, preserved in at least five specimens, provide strong evidence of the bloodletting that probably came from head-on engagements between competing ceratopsians. Using the mammal analogue makes it clear that the large nasal and brow horns of ceratopsians may have functioned during intraspecific combat (for example, territorial defense and establishing dominance).

WHAT IS THE FUNCTION OF FRILLS? Can the frills be interpreted as the horns were? The spikes and scallops of the ceratopsian frill are quite distinctive, visually separating one species from another. However, the fact that *humans* find scalloped and spiked frills distinctive doesn't guarantee that *ceratopsians* sized each other up that way. Can we provide any more evidence that frills were used in intraspecific recognition?

In 1966, Ostrom (Yale University) suggested that the frill provided a platform for the attachment of the major mass of jaw-closing muscles; these were thought to have extended through the upper temporal fenestra and onto the upper surface of the frill. Thus, in long-frilled ceratopsians, the jaw musculature must have been very extensive and, consequently, exceedingly powerful. The phylogenetic changes in the size and shape of the frill among ceratopsians, Ostrom suggested, reflected changes in the attachment and action of muscles running between the frill and the lower jaw, ultimately increasing the power of the bite.

Dodson and others have remained suspicious of such an explanation. They have noted that there is a marked sexual dimorphism in the size and shape of adult frills, suggesting that jaw mechanics are unlikely to be the *sole* factor governing frill morphology. Possibly, ceratopsian frills answered to other important imperatives of ceratopsian biology.

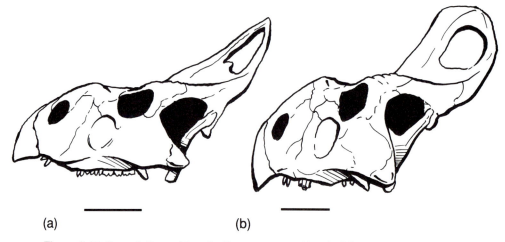

(a)   (b)

Figure 9.14 **Sexual dimorphism in *Protoceratops*.** Note in **(a)**, a presumed female, the frill is less showy and the nasal ridge is less prominent, quite the opposite of **(b)**, a presumed male. Scale = 10 cm.

As noted previously, the development of *Protoceratops* from juvenile to adult is well documented. Because of this, we can investigate when during development the frill begins to grow and when it becomes most expansive. Thanks to statistical studies of growth in *Protoceratops* by Dodson, it is now clear that frill development takes off when individuals are approaching fully adult body sizes. In the group arbitrarily designated "males" by Dodson, the frill becomes inordinately larger and showier than in the "females" (Figure 9.14). This pattern suggests that the onset of frill growth occurs with sexual maturity and therefore that there is a reproductive connection to frill size and shape. Does this sound like sexual selection? Similar patterns are now thought to occur also in other ceratopsians for which juveniles and adults are well known, among them *Centrosaurus* and *Chasmosaurus*. In these forms, the development of scallops and spikes on the frill margin might also reflect gender.

SOCIAL BEHAVIOR. It can be claimed that many, if not all, ceratopsians lived in large herds, at least during part, if not all, of the year. The justification for this comes from our ever-increasing catalog of ceratopsian bonebeds. These mass accumulations of single species of ceratopsians are known for at least nine separate species, including several in which the minimum number of individuals may exceed well over one hundred! It is not hard to imagine that such an abundance of bones can come only from populations of ceratopsians that themselves were

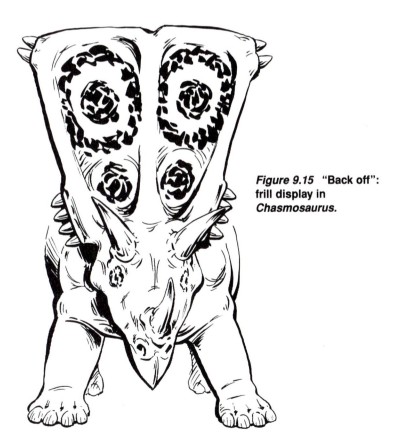

Figure 9.15 "Back off": frill display in *Chasmosaurus.*

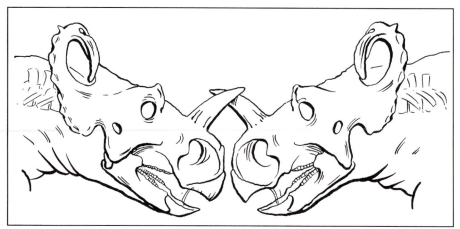

Figure 9.16 "Crossing of the nasal horns": combat between male *Centrosaurus.*

locally abundant. Moreover, such gregariousness also makes sense when putting frills and horns into their behavioral context. Territoriality, ritualized combat and display, and the establishment of dominance hierarchies are to be expected in animals that are thrown together in such highly social circumstances as herds. Perhaps in ceratopsians, we are seeing an example of the most complex of dinosaur intraspecific behavior.

Despite the evidence for complex social behavior, ceratopsian brain sizes were not all that large (see Box 14.3). For example, although being near opposites in terms of body size and display-related anatomy, both *Protoceratops* and *Triceratops* had brains less than the size expected of a similarly sized crocodilian or lizard. Cerebrally, they were above average compared to sauropods, ankylosaurs, and stegosaurs, but commanded considerably less gray matter than either ornithopods or theropods. As J. A. Hopson has suggested, perhaps these brain size measures indicate that ceratopsians had relatively unhurried and uncomplicated lifestyles.

We noted earlier that behavior may have evolved along with morphology. Farlow and Dodson outlined the evolution of intraspecific behavior among ceratopsians. Display- and combat-related behavior could have been present primitively among neoceratopsians and perhaps even among all ceratopsians. With their frill serving as a visual dominance symbol and their small nasal horns acting as weaponry, display in *Leptoceratops*, *Protoceratops*, and *Montanoceratops* could have initially involved nodding and showing a broad frill. Should this have failed to impress, these animals may have rammed their blunt nasal horns full tilt into the flanks of their opponent.

The more-derived ceratopsids share more elaborate frills and either nasal or brow horns. Among the long-frilled chasmosaurines (for example, *Chasmosaurus*, *Pentaceratops*, and *Torosaurus*), the display function of the frill may have been emphasized. The very long frills of these dinosaurs would have provided a very prominent frontal threat display, exhibited not only by inclining the head forward (Figure 9.15) but also by nodding or shaking the head side-to-side. Should such display have failed to send the message to one or the other of the opponents, combat may have involved frontal engagement of the nasal and brow horns, with shoving and charging determining the winner.

In contrast, the short-frilled centrosaurines (such as *Centrosaurus* and

Avaceratops) may have used their horns more readily (Figure 9.16). We imagine that opponents crossed horns, thus reducing to a degree the amount of damage inflicted to the eyes, ears, and snout. Nevertheless, the possibility of injury assuredly may have been very much greater in short-frilled ceratopsians than in other taxa.

In this context of frills, horns, and behavior, two interesting anomalies stand out. First, *Styracosaurus* was a short-frilled centrosaurine whose frill appears inordinately large because of long spikes along the frill margin. *Styracosaurus* may have relied more heavily than other centrosaurines upon the display qualities of the frill. In contrast, *Triceratops* is a chasmosaurine ceratopsid whose frill is secondarily shortened, suggesting that threat display may have been less utilized than in other chasmosaurines. Instead, combat may have taken place without extensive frill display. In *Triceratops*, the solid frill may have served as a shield against the parry and thrust of the opponents' horns.

Weaponry and display almost seem to drive the evolution of ceratopsian dinosaurs. In this diverse group, we witness a world where display and competition were apparently of great significance and where – when push came to shove – it may have been better to nod vigorously than to lock horns.

Important Readings

Dodson, P. 1976. Quantitative aspects of relative growth and sexual dimorphism in *Protoceratops*. Journal of Paleontology 50:929–940.

Dodson, P. 1993. Comparative craniology of the Ceratopsia. American Journal of Science 293-A:200–234.

Dodson, P., and P. J. Currie. 1990. Neoceratopsia; pp. 593–618 *in* D. B. Weishampel, P. Dodson, and H. Osmólska (eds.), The Dinosauria. University of California Press, Berkeley.

Farlow, J. O., and P. Dodson. 1975. The behavioral significance of frill and horn morphology in ceratopsian dinosaurs. Evolution 29:353–361.

Hatcher, J. B., O. C. Marsh, and R. S. Lull. 1907. The Ceratopsia. U.S. Geological Survey Monograph 49:1–300.

Lull, R. S. 1933. A revision of the Ceratopsia or horned dinosaurs. Peabody Museum of Natural History Memoirs 3:1–175.

Maryańska, T., and H. Osmólska. 1975. Protoceratopsidae (Dinosauria) from Mongolia. Palaeontologia Polonica 33:133–182.

Norell, M. A., J. M. Clark, D. Demberelyin, R. Barsbold, L. M. Chiappe, A. R. Davidson, M. C. McKenna, A. Perle, and M. J. Novacek. 1994. A theropod dinosaur embryo and the affinities of the Flaming Cliffs dinosaur eggs. Science 266:779–782.

Ostrom, J. H. 1964. A functional analysis of the jaw mechanics in the dinosaur *Triceratops*. Postilla 88:1–35.

Ostrom, J. H. 1966. Functional morphology and evolution of ceratopsian dinosaurs. Evolution 20:290–308.

Ostrom, J. H., and P. Wellnhofer. 1986. The Munich specimen of *Triceratops* with a revision of the genus. Zitteliana 14:111–158.

Sereno, P. C. 1986. Phylogeny of the bird-hipped dinosaurs (Order Ornithischia). National Geographic Research 2:234–256.

Sereno, P. C. 1990. Psittacosauridae; pp 579–592 *in* D. B. Weishampel, P. Dodson, and H. Osmólska (eds.), The Dinosauria. University of California Press, Berkeley.

CHAPTER 10

ORNITHOPODA:
The Tuskers, Antelopes, and the Mighty Ducks of the Mesozoic

ORNITHOPODS HAD IT ALL. Some had tusks projecting from the corners of their mouths, some had spikes on their thumbs, some had more teeth than just about any other kind of animal, some sported hollow crests atop their heads, and many had long tails that projected straight back!

The ornithopods had one of the longest reigns of all dinosaur groups, lasting for most of the Mesozoic. From the Early Jurassic, when they first show up in the fossil record, until the end of the Cretaceous, when all went extinct, ornithopods evolved into an extremely diverse clade, boasting more than 80 species at present count. In so doing, they also managed to get all over the globe.

In size, ornithopods run the gamut, from an adult length of about 1–2 m (*Heterodontosaurus*) to more than 12 m (*Shantungosaurus*). Of course, younger individuals were smaller still: A near-term embryo of *Orodromeus* measures approximately 30 cm in length (Figure 10.1)!

And some of them are well known. *Iguanodon*, for example is a charter member of Sir Richard Owen's original 1842 Dinosauria, an Early Cretaceous ornithopod known from everything from isolated skeletal parts to huge quantities of skeletons, particularly from England, Belgium, and Germany. But even more famous than *Iguanodon* are the "duckbills," whose fossil record extends back some 25 million years prior to the close of the Cretaceous. These are known not only from abundant skeletal remains (including delicate bones such as sclerotic rings, stapes, and hyoid bones) but also from skin impressions and ossified tendons. Paleontologists also have found eggs and nests, and growth series that range from hatchlings through juveniles and "teenagers" to adults. Duck-billed dinosaurs, as well as many other kinds of ornithopods, offer a smorgasbord of information about their anatomy, biology, and evolution (Figure 10.2).

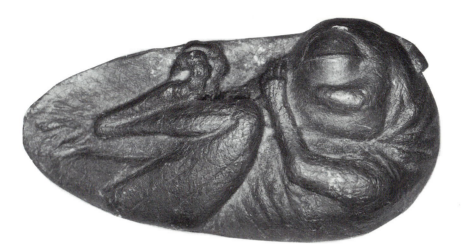

Figure 10.1 A reconstruction of a full-term embryo of the Late Cretaceous hypsilophodontid, *Orodromeus.* *Model courtesy of M. Smith.*

HISTORY OF ORNITHOPODA DISCOVERIES

19th Century

In 1822, Gideon and Mary Ann Mantell discovered peculiar-looking teeth from the Wealden beds of the Tilgate Forest in what is now West Sussex, England. These teeth, later given the name *Iguanodon* (*don* – teeth; *Iguana*, a lizard; Figures 10.3 and 10.9h) by Gideon Mantell in 1825, are generally reckoned to be the start of dinosaur discoveries in the Western world. From that time forward,

Figure 10.2
***Corythosaurus*, a hollow-crested hadrosaurid from the Late Cretaceous of western Canada.**
Photograph courtesy of the Royal Ontario Museum.

Figure 10.3 iguanodon, the great beast of Bernissart, Belgium.

Iguanodon fossils began to accumulate in Gideon Mantell's collection and in other collections by early English paleontologists. Out of the initial studies came a picture of dinosaurs as gigantic (estimated length in excess of 35 m), pachydermous reptiles that lumbered on all fours. In keeping with this picture, a spikelike bone found with *Iguanodon* remains was fitted to its nose, much like the modern rhinoceros. Who could have guessed that within 30 years, it would become clear that the spike on the nose actually belonged on the thumb of the hand?!

Nearly a quarter-century later and some 6,000 km across the Atlantic Ocean, J. Leidy of the Academy of Natural Sciences, Philadelphia, obtained some large bones that he named *Hadrosaurus* (*hadro* – stout) in 1858. *Hadrosaurus* was the first dinosaur to be exhibited to the public, and museum attendance skyrocketed: The thing apparently walked on two legs! In 1869, T. H. Huxley and P. Matheron named two relatively small (2–3 m long) ornithopods: *Hypsilophodon* (*hypsi* – high; *lophos* – ridge; Figures 10.4b and 10.9b) from the same beds as *Iguanodon*, but from the Isle of Wight off the southern coast of England; and *Rhabdodon* (*rhabdo* – rod) from southern France.

All this was eclipsed in April 1878 by the recovery of the Bernissart bonanza: 31 complete *Iguanodon* skeletons from a coal seam some 300 m beneath the small mining town of Bernissart in southern Belgium. Over the next seven years, L. Dollo, curator of paleontology at the Institut Royal de Science Naturelle de Belgique in Brussels, described and interpreted this extraordinary collection. Later, Dollo described another, more enigmatic ornithopod from Belgium. This animal – known only from isolated teeth – he called *Craspedodon* (*craspedo* – border or edge).

The Great Dinosaur Rush (Box 6.1) provided its share of ornithopods: In short order, discovery in 1872 of new hadrosaurid material from Kansas (subsequently dubbed *Claosaurus* [*klao* – break] by Yale's O. C. Marsh in 1890) was followed by the 1885 announcement of *Camptosaurus* (*kamptos* – flexible), a

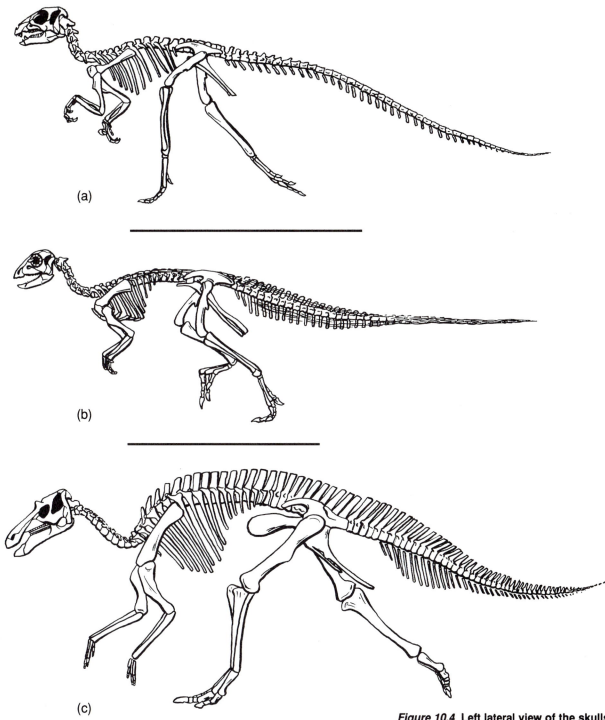

Figure 10.4 Left lateral view of the skulls
and skeletons of (a) *Heterodontosaurus,*
(b) *Hypsilophodon,* and (c) *Maiasaura.*
Scale = 50 cm.

Figure 10.5 **The Late Jurassic iguanodontian,** *Camptosaurus,* **from the Western Interior of the United States.** *Photograph courtesy of the Royal Ontario Museum.*

new, medium-sized (5–7 m long) ornithopod dinosaur from Wyoming (Figures 10.5 and 10.9g). Nine years later, in 1894, Marsh again described a new ornithopod, this time a smaller contemporary of *Camptosaurus*, which he named *Dryosaurus* (*dryos* – oak tree; Figure 10.9f). A relatively small iguanodontian, *Dryosaurus* is one of the few dinosaurs with an intercontinental distribution; it is known not only from the Western Interior of the United States, where it was first discovered, but also from southern Tanzania, where it is known from abundant adult and juvenile material.

20th Century

The first discoveries of the 1900s, however, were made not in the United States, but in Transylvania. Here, dinosaur fossils found on the estate of the Nopcsa family were described by F. Nopcsa as *Telmatosaurus* (*telmat* – swamp; Figure 10.14a) in 1900. This hadrosaurid ornithopod is now known from southern France and northern Spain.

Thereafter, the first quarter-century of the 1900s saw 11 new kinds of ornithopod dinosaurs, all but 3 collected in Canada. Floating on log rafts and dragging horse-drawn wagons, the American Museum of Natural History's ace collector, B. Brown, excavated *Saurolophus* (*lophos* – crest; Figure 10.14f) in 1912, *Hypacrosaurus* (*hypakros* – highest, referring to spines; Figure 10.17b) in 1913, *Corythosaurus* (*koryth* – crown; Figure 10.17c) in 1914, and *Prosaurolophus* (*pro* – before; Figure 10.14e) in 1916. Close behind were two Canadian dinosaur paleontologists, L. M. Lambe (Geological Survey of Canada) and W. A. Parks (Royal Ontario Museum). Lambe's discoveries include *Gryposaurus* (*grypos* – hooked; Figure 10.14c), named in 1914, and *Edmontosaurus* (from Edmonton; Figure 10.14g), named in 1920; Parks christened *Parasaurolophus* (*para* – near; Figure 10.17a) in 1922 and *Lambeosaurus* (after Lambe; Figure 10.17d) in 1923. He also described a new hypsilophodontid, which C. M. Sternberg was to call *Parksosaurus* (after Parks) in 1937.

Meanwhile, from New Mexico, Brown described a new hadrosaurid, *Kritosaurus* (*kritos* – separate), in 1910. *Thescelosaurus* (*theskelos* – astonishing), a hypsilophodontid ornithopod, was discovered in Wyoming by C. W. Gilmore (National Museum of Natural History [Smithsonian Institution]) in 1913, and *Lycorhinus* (*lykos* – wolf; *rhinus* – snout) – originally thought to be a synapsid but now known to be a heterodontosaurid ornithopod – was collected from South Africa and described by S. H. Haughton (South African Museum) in 1924.

The second quarter-century produced a host of discoveries from elsewhere in the world, principally Asia and Australasia. *Tanius* (named for H. C. Tan, the Chinese geologist who discovered its remains), a flat-headed hadrosaurid from China, was described by the Swedish paleontologist C. Wiman in 1929. F. von Huene, renowned dinosaur paleontologist from Tübingen, Germany, announced the hypsilophodontid *Fulgurotherium* (*fulgur* – lightning; *therion* – beast, referring to Lightning Ridge) from New South Wales, Australia, in 1932; unfortunately, it was regarded as a theropod dinosaur until the 1980s, when it was reexamined by R. E. Molnar of the Queensland Museum.

With the 1930s came the first descriptions of ornithopod dinosaurs from the Gobi Desert. *Bactrosaurus* (*baktron* – club, referring to vertebrae) and *Mandschurosaurus* (from Manchuria) – two hadrosaurids from the Late Cretaceous of the Inner Mongolian region of China – were described by Gilmore in 1933. *Mandschurosaurus* was renamed *Gilmoreosaurus* in 1979 by M. K. Brett-Surman (National Museum of Natural History [Smithsonian Institution]) in honor of Gilmore's efforts to understand these first hadrosaurids from the Gobi Desert.

Although not so well known as the Gobi discoveries, there were dinosaurs from elsewhere in Asia. *Nipponosaurus* (from Japan), the first and most famous hadrosaurid from Japan, was named by T. Nagao (Hokkaido Imperial University, Sapporo, Japan) in 1936, and *Jaxartosaurus* (from the Jaxartes River, Kazakhstan) was described by A. N. Riabinin in 1939.

The years since 1950 have been marked by a healthy mixture of discovery, reflection, and revision on a worldwide scale. Beginning in Asia, studies published in the early 1950s by A. K. Rozhdestvensky (Palaeontological Institute in Moscow) introduced new species of *Saurolophus* and *Iguanodon*, both from Mongolia. Also from Asia came *Tsintaosaurus* (from Tsintao), described by C.-C. Young (Institute of Vertebrate Paleontology and Paleoanthropology in Beijing) in 1958. This hadrosaurid from Shandong Province, China, sports a unicorn-like horn.

Elsewhere, work continued on the dinosaurs collected at the beginning of the century in Tendaguru, Tanzania (see Box 11.1). In 1955, W. Janensch began publishing on his new ornithopod, *Dysalotosaurus* (*dysalotos* – uncatchable, later to

Figure 10.6
Brachylophosaurus, a
solid-crested hadrosaurid
from western North
America.

be referred to *Dryosaurus*).[1] And back in North America, a new hadrosaurid from the Late Cretaceous of Alberta was described as *Brachylophosaurus* (*brachys* – short; Figures 10.6 and 10.14d) by Sternberg in 1953; farther south and to the east, W. Langston, Jr. (University of Texas at Austin), christened *Lophorhothon* (*rhothon* – nose), a new hadrosaurid from Alabama.

In 1962, A. W. Crompton (then at the South African Museum, now at Harvard University) and A. J. Charig (British Museum [Natural History]) announced an ornithopod, *Heterodontosaurus* (*heteros* – different; *odont* – tooth) with canine-like teeth. *Heterodontosaurus* is now known from virtually complete skulls and an exquisite skeleton (Figures 10.4a and 10.9a).

Later that decade, results of the Sino-Soviet Palaeontological Expedition in China became available. The first ornithopod to come from these efforts, *Probactrosaurus* was described by Rozhdestvensky in 1966. Hailing from the Early Cretaceous of Inner Mongolia, China, *Probactrosaurus* has featured widely in discussions of the ancestry of Hadrosauridae. Hadrosaurids also received a new member, in the form of *Aralosaurus* (from the Upper Cretaceous Aral region of Kazakhstan), described by Rozhdestvensky in 1968.

Finally, from the close of the 1960s for about 15 years, University of Bridgeport's P. M. Galton began what can only be called a tour de force on ornithopod osteology, taxonomy, and phylogeny. He named *Othnielia* (for Othniel Charles Marsh), from the Late Jurassic of Colorado, Utah, and Wyoming; and *Valdosaurus* (*valdus* – Weald, from the Wealden deposits) from the Early Cretaceous of England and Niger.

[1]There is a double meaning in this fossil's name. The name *Dysalotosaurus* – "uncatchable lizard" – is sometimes thought to be a reference to its gracile, sleek morphology, but in fact the species name, *lettowvorbecki*, suggests the real intent behind the name. A crafty World War I German general, Lettow-Vorbeck, stationed in Tanzania, proved uncatchable to pursuing British and South African armies. In 1919 (just after the end of Word War I), the unrepentant German paleontologist H. Virchow, in naming this Tanzanian dinosaur, celebrated this fact with its name!

The 1970s were to be prolific. J. H. Ostrom started the ball rolling in 1970 with *Tenontosaurus* (*tenon* – tendon; Figure 10.9e), an ornithopod from Montana that is now regarded as a basal iguanodontian. Halfway around the world was discovered *Shantungosaurus*, a gigantic hadrosaurid described in 1973 by Hu S. from Shandong Province, China.

Two new heterodontosaurids were announced in 1975. The first, by C. E. Gow (Bernard Price Institute for Palaeontological Research, Johannesburg, South Africa), was named *Lanasaurus* (*lana* – wooly; named to honor Harvard University's A. W. Crompton, whose nickname is "Fuzz"!). Like *Heterodontosaurus*, *Lanasaurus* also comes from South Africa. The second heterodontosaurid, named *Abrictosaurus* (*abriktos* – awake) by J. A. Hopson in 1975, hails from South Africa and neighboring Lesotho. Unlike *Heterodontosaurus*, these two new forms are known from only fragmentary skull material.

Farther to the north on the African continent, we shift from heterodontosaurids to iguanodontians, for in 1976, P. Taquet (Muséum National d'Histoire Naturelle, Paris) announced the peculiar, high-spined *Ouranosaurus* (Nigerian: *ourane* – brave; Figures 10.9i and 10.21).

The new ornithopods of 1979 came shotgun-style from three of the four corners of the globe: from China, Argentina, and the United States. From the Middle Jurassic of Sichuan, China, the new hypsilophodontid *Yandusaurus* (Chinese: *yan* – salt; *du* – capital; Figure 10.9c) was described by He X.-L. of the Chengdu College of Geology. *Secernosaurus* (*secerno* – divide), the first hadrosaurid from the Southern Hemisphere, was announced by Brett-Surman (National Museum of Natural History [Smithsonian Institution]) from Rio Negro, Argentina. And J. R. Horner and R. Makela (Museum of the Rockies) stunned the world with *Maiasaura* (*maia* – good mother; Figure 10.14b), a hadrosaurid from Montana, whose remains included adult and hatchling specimens, thus providing the first inkling of parental care in dinosaurs (Figure 10.4c)!

The unleashing of new ornithopods slowed down only slightly during the 1980s. In Montana, *Zephyrosaurus* (Zephyros, Greek god of the west wind; Figure 10.9d) was an Early Cretaceous hypsilophodontid described by H.-D. Sues in 1980, and *Orodromeus* (*oros* – mountain; *dromeus* – runner; Figure 10.1), another hypsilophodontid, was named by Horner and D. B. Weishampel in 1988. In Asia, we were introduced to *Barsboldia* (named for Mongolian paleontologist R. Barsbold by T. Maryańska and H. Osmólska in 1981), a hadrosaurid from the Late Cretaceous of Mongolia; and *Gongbusaurus* (Chinese: *gong* – worker; *bu* – board; referring to Board of Works), a Late Jurassic hypsilophodontid from Sichuan and Xinjiang provinces, China, described by Dong Z.-M. and colleagues in 1983. Australia had a boom decade for ornithopods during the 1980s, with *Muttaburrasaurus* (from Muttaburra, described by A. Bartholomai and Molnar in 1981), an iguanodontian from the Early Cretaceous of Queensland. This was followed by *Atlascopcosaurus* (for Atlas Copco Co., which supplied excavation equipment) and *Leaellynasaura* (for Leaellyn Rich, who helped in the discovery), two hypsilophodontids from the Lower Cretaceous of Australia that were described in 1989 by T. H. Rich and P. Vickers Rich of the Museum of Victoria and Monash University, respectively.

Thus far in the 1990s, five new ornithopods have been discovered and named. The first, *Anatotitan* (*anat* – duck) is a rechristening of material known since Cope's

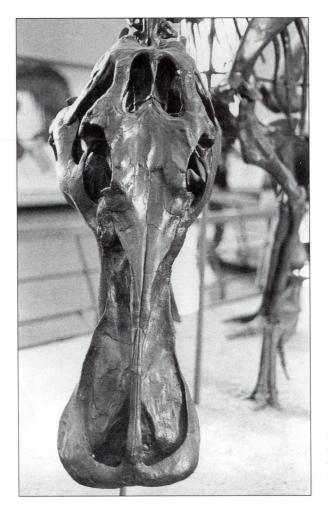

Figure 10.7 Anatotitan, a flat-headed hadrosaurid from the western United States.

time! This newly named hadrosaurid – recognized for the first time by Brett-Surman in 1990 – is known from Montana and South Dakota (Figure 10.7). Peng G.-Z. gave us *Agilisaurus* (*agili* – agile), a hypsilophodontid from Sichuan, China, in 1990, and R. T. Bakker and co-workers in Colorado in 1990 described *Drinker* (named in honor of Edward Drinker Cope), yet another hypsilophodontid from Wyoming. Then, in 1993, S. Lucas and A. P. Hunt, from the New Mexico Museum of Natural History and the University of New Mexico, respectively, described two more genera of hadrosaurids: *Anasazisaurus* (the Anasazi are ancient Native Americans of the U.S. southwest) and *Naashoibitosaurus* (Naashoibito – the name of the rock unit from which this dinosaur was recovered). Both of these hail from New Mexico. Who knows what will come next?

ORNITHOPODA DEFINED AND DIAGNOSED

The clade Ornithopoda is defined as all the descendants of the common ancestor of Heterodontosauridae and Euornithopoda (Figure 10.8). Heterodontosauridae and Euornithopoda are themselves monophyletic. Also containing

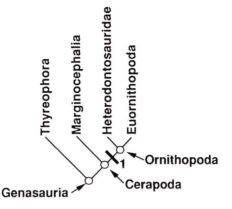

Figure 10.8 **Cladogram of Genasauria, emphasizing the monophyly of Ornithopoda.**
Derived characters at 1: pronounced ventral offset of the premaxillary tooth row relative to
the maxillary tooth row, crescentic paroccipital processes, strong depression of the
mandibular condyle beneath the level of the upper and lower tooth rows, and elongation of
the lateral process of the premaxilla to contact the lacrimal and/or prefrontal.

Hypsilophodontidae, Dryosauridae, and sequentially more-closely-related forms
to Hadrosauridae (the "iguanodonts"), Ornithopoda is diagnosed on the basis of
a number of derived features (Figure 10.8).

Where do ornithopods reside among the dinosaurs? They and marginocephalians form a monophyletic group that P. C. Sereno called Cerapoda (see
Part II: Ornithischia).

ORNITHOPOD DIVERSITY AND PHYLOGENY

Here we discuss the major divisions and the features of the ornithopod cladogram.
Much of what follows is based on studies by Sereno, D. B. Norman, and A. Milner
and Norman, as well as more recent work by Sereno, Norman, Weishampel, and
colleagues.

Heterodontosauridae

The Early Jurassic Heterodontosauridae is defined as all the descendants of the common ancestor of *Heterodontosaurus* (Figure 10.9a) and *Lanasaurus*. Primitively in
the history of this clade, which also contains *Lycorhinus* and *Abrictosaurus*, heterodontosaurids evolved high-crowned teeth, each bearing a chisel-shaped crown
ornamented with denticles. In addition, and the principal basis for the name "heterodontosaurid," a large canine-like tooth is present in both upper and lower jaws.
These "canines" are not true canine teeth like those found in mammals.

Euornithopoda

Euornithopoda constitutes the remaining ornithopod clade. Defined as all the
descendants of the common ancestor of Hypsilophodontidae and Iguanodontia,

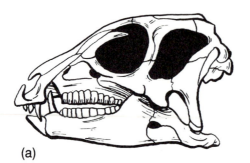

(a)

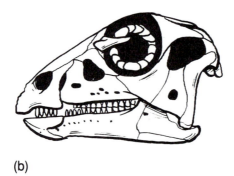

(b)

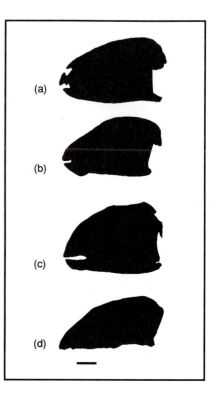

(a)

(b)

(c)

(d)

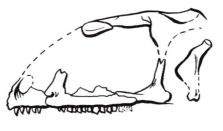

(c)

(d)

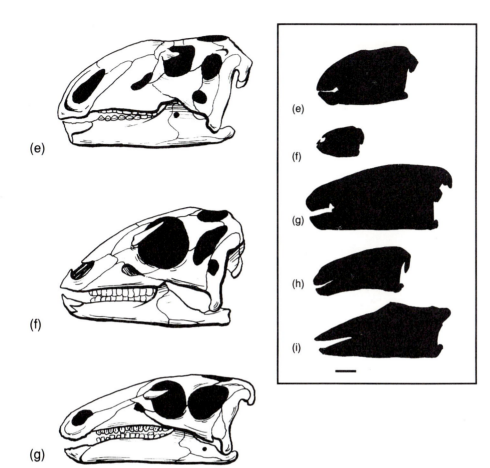

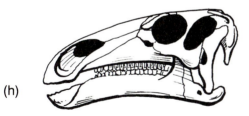

Figure 10.9 Left lateral view of the
skulls of (a) *Heterodontosaurus*,
(b) *Hypsilophodon*, (c) *Yandusaurus*,
(d) *Zephyrosaurus*, (e) *Tenontosaurus*,
(f) *Dryosaurus*, (g) *Camptosaurus*,
(h) *Iguanodon*, and (i) *Ouranosaurus*.
Scale for silhouettes a–d = 10 cm.
Scale for silhouettes e–i = 5 cm.

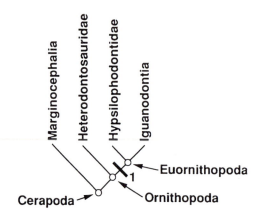

Figure 10.10 Cladogram of Euornithopoda, with more distant relationships with Heterodontosauridae and Marginocephalia. Derived characters at 1: absence of a prominent boss in the cheek region, and high angle between the prepubic process and body of the pubis.

euornithopodans are characterized by only two shared, derived features (Figure 10.10). This large clade is further split between Hypsilophodontidae – the clade containing a host of relatively small, agile ornithopods – and Iguanodontia, residence of such dinosaurian luminaries as *Camptosaurus*, *Iguanodon*, and hadrosaurids.

Hypsilophodontidae

The origin of Hypsilophodontidae primitively entails several character transformations, outlined in Figure 10.11. The most primitive member of the group is *Thescelosaurus*, paradoxically a latecomer to the fossil record of hypsilophodontids. Known from the Late Cretaceous of the Western Interior of the United States and Canada, *Thescelosaurus* stands as the sister-taxon to remaining hypsilophodontids, the better represented of which are *Othnielia*, *Yandusaurus*, *Orodromeus*, *Zephyrosaurus*, *Parksosaurus*, and *Hypsilophodon*. This clade of hypsilophodontids above *Thescelosaurus* is diagnosed in Figure 10.11, node 2.

Yandusaurus from the Middle Jurassic of Sichuan, China, and *Othnielia* from the Late Jurassic of the Western Interior in the United States together constitute the most primitive clade within this more restricted clade of "higher" hypsilophodontids (Figure 10.11, node 3).

The "crown" of hypsilophodontid evolutionary history involves a clade of four taxa (*Parksosaurus*, *Hypsilophodon*, *Orodromeus*, and *Zephyrosaurus*; Figure 10.11, node 4). *Parksosaurus* from the Late Cretaceous of Alberta and *Hypsilophodon* from the Early Cretaceous of England form a small clade unto themselves, uniquely identified by particular modifications of the side of the snout (Figure 10.11, node 6). Sibling to this group is the clade *Zephyrosaurus* + *Orodromeus*, both of which are known from Montana. *Zephyrosaurus*, however, is Early Cretaceous in age, whereas *Orodromeus* is Late Cretaceous (Figure 10.11, node 5).

How the many remaining hypsilophodontids will fit into this phylogeny remains to be seen. They may "shoehorn" in nicely, or they may radically reorga-

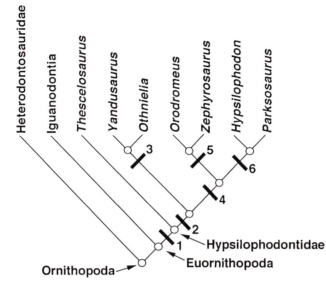

***Figure 10.11* Cladogram of Hypsilophodontidae.** Derived characters at 1: loss of ridges on the crowns of the cheek teeth, ossification of the more forwardly placed ribs where they join the sternum, and development of a rod-shaped prepubic process; at 2: relatively narrow skull roof, and modestly flexed braincase; at 3: reduction of the quadratojugal, and dorsal curvature of the pubis and ischium within the pelvis; at 4: reduction of the back end of the jugal bone, and rearward placement of the obturator process on the ischium; at 5: prominent boss on the side of the cheek region; at 6: exclusion of the jugal from the margin of the antorbital opening.

nize the cladogram in Figure 10.11. For now, however, we must regard *Atlascopcosaurus*, *Leaellynasaura*, *Fulgurotherium*, *Agilisaurus*, and *Gongbusaurus* as hypsilophodontids without a more exact phylogenetic home.

Iguanodontia

By far, Iguanodontia is the bushiest of ornithopod clades, claiming not only the very diverse Hadrosauridae, but a variety of more primitive forms. As many as 11 shared, derived characters unite the group (Figure 10.12).

Tenontosaurus is the most basal iguanodontian ornithopod. A large animal with a very long tail bundled up in a basketwork of ossified tendons, *Tenontosaurus* is one of the more recently discovered iguanodontians and hails from the Early Cretaceous of Montana and Wyoming. Remaining iguanodontians – called **Dryomorpha** after the most primitive member, *Dryosaurus* – form a robust clade (Figure 10.12, node 1). Climbing slightly higher, we encounter the more restrictive clade called **Ankylopollexia** – the "fused thumbs" (Figure 10.12, node 2). Within this clade, *Camptosaurus*, from the Late Jurassic of the Western Interior of the United States and England, is the most primitive. Yet what we do with the remainder of the ankylopollexian clade up to what is known as Hadrosauridae (remember, the duckbills) – a group that includes *Iguanodon*, *Ouranosaurus*, and possibly *Probactrosaurus* – is a bit of a quandary. Sereno's 1986 study places *Ouranosaurus* from the Early Cretaceous of Niger as the closest taxon to

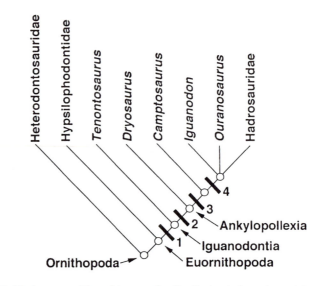

***Figure 10.12* Cladogram of basal Iguanodontia.** Derived characters at 1: eversion of the oral margin of the premaxilla, enlargement of the external nares, reduction of the antorbital opening, denticulate margin of the predentary, loss of premaxillary teeth, and reduction of the bones of digit II; at 2: second maxillary process that articulates with the premaxilla, modifications of the cheek region (including reduction of the quadratojugal), and upper teeth with relatively high crowns; at 3: relatively close-packed teeth, upper teeth with a prominent ridge on their outer side, reduction in the size of the antorbital opening, fusion of the wrist bones, and beginnings of the formation of the spiked thumb; at 4: enlargement of the external nares, narrowing of the back of the skull, reduction of the bones of the first four digits of the hand, loss of the fifth digit of the foot, and trelliswork of ossified tendons that extend from the base of the neck to the middle of the tail.

Hadrosauridae, with the Early Cretaceous *Probactrosaurus* from Kazakhstan and *Iguanodon* from the Early Cretaceous of England, Germany, Belgium, Spain, and the United States as progressively more distant relatives of both *Ouranosaurus* and Hadrosauridae. This relationship may well be true; however, in 1986 and 1990, Norman regarded *Ouranosaurus* and *Iguanodon* as sister-taxa (yet leaving out *Probactrosaurus*), and a more recent study (1993) by Weishampel, Norman, and D. Grigorescu places these two (again leaving out *Probactrosaurus*) in an unresolved trident with Hadrosauridae (Figure 10.12, node 3). Not surprisingly, we go with Weishampel, Norman, and Grigorescu!

Hadrosauridae

Phylogeny reconstruction is slightly easier once we ascend into Hadrosauridae (Fig 10.13, node 1). For one thing, this clade has been considered monophyletic by nearly everyone in the field, the only recent dissenter being Horner. He regards the "duck-billed" ornithopods as having separate origins deeper in iguanodontian evolutionary history; he does not believe duck-billed dinosaurs are monophyletic.

Is he right? To decide this issue, we use parsimony as a means of choosing between competing hypotheses, as we did with similar problems encountered in Ankylosauria and Pachycephalosauria. According to Horner's cladogram, it takes 33 characters to account for *Iguanodon*, *Ouranosaurus*, and the two separate

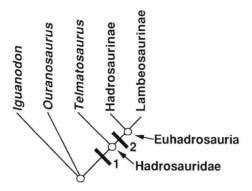

Figure 10.13 Cladogram of "higher" Iguanodontia, emphasizing the monophyly of Hadrosauridae and Euhadrosauria. Derived characters at 1: three or more teeth per tooth position, repositioning of the antorbital fenestra, loss of the opening between quadrate and quadratojugal bones, loss of the first digit in the hand, and an increase in the number of vertebrae in the sacrum; at 2: narrow quadrate part of the jaw joint, and reduction of the coarse denticulation of the oral margin of the premaxillary bones.

groups of duckbills. Adding other features used in two other studies advocating duckbill monophyly (one by Sereno documenting ornithischian relationships and one by Weishampel, Norman, and Grigorescu on hadrosaurid relationships), a relatively complex tree of 73 steps is obtained.

What about the other hypotheses (the ones by Sereno and by Weishampel and colleagues), that is, those advocating a monophyletic Hadrosauridae? Adding Horner's characters to theirs, a tree of 52 steps is obtained.[2] The phylogeny that portrays hadrosaurids as monophyletic (Figure 10.13) is more parsimonious than the one that has these duckbills with separate origins in Ornithopoda.

Within Hadrosauridae, the latest Cretaceous *Telmatosaurus* from Europe is the most primitive. All other hadrosaurids (recently dubbed Euhadrosauria by Weishampel and colleagues) form a monophyletic group (Figure 10.13, node 2); we know these forms best as the combination of Hadrosaurinae and Lambeosaurinae.

The relationships of taxa both within and between Hadrosaurinae and Lambeosaurinae are known only in a very general sense, and there is considerable work being conducted to build a meaningful phylogeny for these animals. As a result, what follows may ultimately prove to be a poor reflection of the state of hadrosaurid phylogeny in years to come. Nevertheless, it appears that there is a monophyletic group of hadrosaurids that includes nearly all the flat-heads and all of those forms with a crest made of solid bone (Figure 10.14, and Figure 10.15, node 1). This group includes *Gryposaurus, Maiasaura, Brachylophosaurus, Saurolophus, Prosaurolophus, Lophorhothon, Anatotitan, Edmontosaurus,* and

[2] Here is a paradox! How come if we add Sereno's and Weishampel & Company's characters to Horner's characters, we get a cladogram with 73 steps, but if we add Horner's characters to Sereno's and Weishampel & Company's characters, we get a cladogram with only 52 steps? What's the difference? Why aren't they equal? The answer comes from the characters themselves. Sereno and Weishampel & Company have 20 characters not used by Horner. When these are applied on the Horner cladogram, they must go on the tree twice to account for their separate origins in the two hadrosaurid groups that Horner hypothesizes. That makes 40 extra steps, which, added to Horner's original 33-step cladogram, equals a 73-step cladogram. Because the cladograms of Sereno and of Weishampel et al. call for a monophyletic Hadrosauridae, Horner's characters are added to it only once, producing a far simpler cladogram than that produced by Horner.

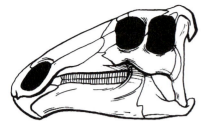
(a)

(b)

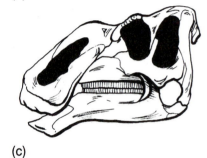
(c)

(d)

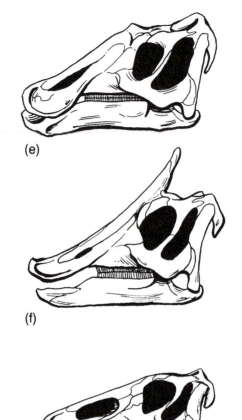
(e)

(f)

(g)

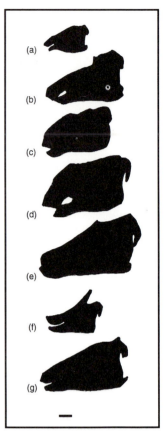

Figure 10.14 Left lateral view of the skulls of (a) *Telmatosaurus,*
(b) *Maiasaura,* (c) *Gryposaurus,* (d) *Brachylophosaurus,*
(e) *Prosaurolophus,* (f) *Saurolophus,* and (g) *Edmontosaurus.*
Scale for silhouettes = 10 cm.

Shantungosaurus. Within this large group are the nasal-arched gryposaurs
(*Gryposaurus* and *Aralosaurus*; Figure 10.15, node 3) and maiasaurs (*Maiasaura*
and *Brachylophosaurus*; Figure 10.15, node 4). Together, gryposaurs and
maiasaurs may form a monophyletic group (Figure 10.15, node 2).

The rest of the hadrosaurines include the solid-crested saurolophs
(*Saurolophus, Prosaurolophus,* and *Lophorhothon*) and the flat-headed edmon-
tosaurs (*Anatotitan, Edmontosaurus,* and *Shantungosaurus*).

Remaining hadrosaurids – lambeosaurines – are undoubtedly monophyletic
(Figure 10.16). These hollow-crested forms include a variety of taxa, among them
Bactrosaurus, Parasaurolophus, Corythosaurus, Hypacrosaurus, and *Lambeosaurus*
(Figure 10.17). It appears that the best known of these forms are successively

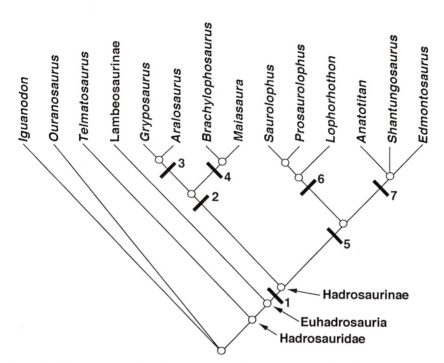

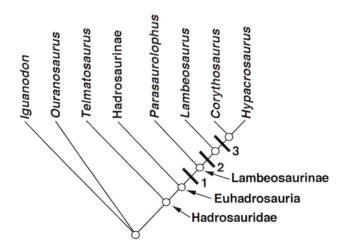

Figure 10.15 **Cladogram of Hadrosaurinae.** Derived characters at 1: highly flared snout, and circumnarial depression; at 2: prominent ventral flange of the jugal bone; at 3: arched nasals; at 4: solid and transversely broad crest positioned above the eyes; at 5: extremely long diastema between the front and lower jaws and the lower tooth row, and corresponding extension of the snout; at 6: backwardly pointed, narrow, and solid crest above the eye socket formed exclusively by the nasal bones; at 7: complex excavation of the circumnarial fossa, massive jugal bone, and pocket on the underside of the postorbital bone.

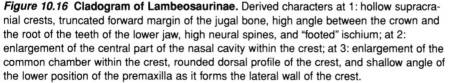

Figure 10.16 **Cladogram of Lambeosaurinae.** Derived characters at 1: hollow supracranial crests, truncated forward margin of the jugal bone, high angle between the crown and the root of the teeth of the lower jaw, high neural spines, and "footed" ischium; at 2: enlargement of the central part of the nasal cavity within the crest; at 3: enlargement of the common chamber within the crest, rounded dorsal profile of the crest, and shallow angle of the lower position of the premaxilla as it forms the lateral wall of the crest.

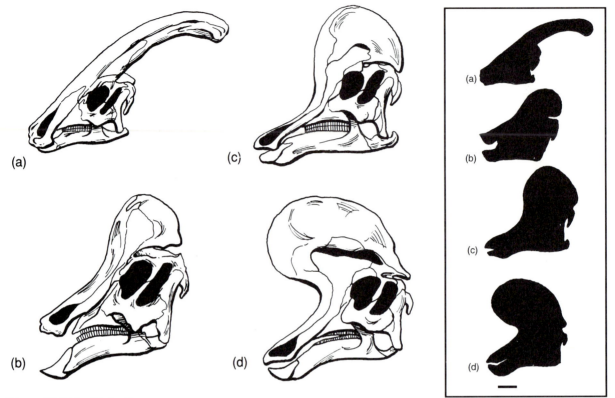

Figure 10.17 Left lateral view of the skulls of (a) *Parasaurolophus,* (b) *Hypacrosaurus,* (c) *Corythosaurus,* and (d) *Lambeosaurus.* Scale for silhouettes = 10 cm.

more closely related to *Hypacrosaurus,* based upon the features in the cladogram in Figure 10.16, including the distinctive "footed" ischium (Figure 10.18). Thus, we have – in stepwise fashion – *Parasaurolophus, Lambeosaurus,* and the crown-clade of *Corythosaurus* and *Hypacrosaurus.* These hierarchical transitions appear to involve first an enlargement of the central part of the nasal cavity within the crest, then elaborations of the vestibule (connecting the common chamber with the nostrils) and the bony profile of the crest.

As befits a group as prolific as Iguanodontia, there are many taxa that do not yet have a resolved position in the phylogeny of the group. Both *Valdosaurus* and *Rhabdodon* have been described as having a very basal relationship with other iguanodontians, but this suggestion has not yet been documented.

It is likely that the Early Cretaceous *Probactrosaurus* from China and *Muttaburrasaurus* from Australia will find resolution, although higher in the iguanodontian cladogram. For lack of material, we are not so optimistic about *Craspedodon.*

In addition, there is a host of hadrosaurids that – for one reason or another, mostly having to do with their incomplete preservation – are yet unresolved at the highest reaches of the iguanodontian tree. These include *Gilmoreosaurus, Tanius, Jaxartosaurus, Aralosaurus, Barsboldia, Tsintaosaurus,* and *Nipponosaurus* from Asia; *Secernosaurus* from South America; and *Hadrosaurus, Kritosaurus,* and *Claosaurus* from North America. As always, further discoveries and careful research will most likely provide us with that important bit of information to place these wayward ornithopods in their phylogenetic context.

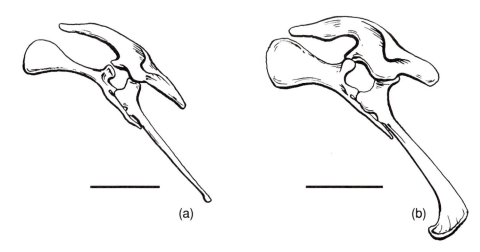

Figure 10.18 **Left lateral view of the pelvis of (a)** *Edmontosaurus* **and (b)** *Hypacrosaurus.* Scale = 50 cm.

ORNITHOPOD PALEOBIOLOGY AND PALEOECOLOGY

Geography

Ornithopods ranged from near the paleoequator to such high-paleolatitude occurrences as the north slope of Alaska, the Yukon, and Spitsbergen in the Northern Hemisphere; and, in the Southern Hemisphere, to Seymour Island, Antarctica, and the southern coast of Victoria, Australia (Figures 10.19 and 10.20).

Local conditions in these regions varied widely, so we can safely assume that the many ornithopods lived in quite diverse habitats with variable climates. For example, the Lower Jurassic sediments of southern Africa from which *Heterodontosaurus* and its relatives have been recovered are indicative of a semi-arid, probably seasonal (wet-dry) climate, altogether quite inhospitable.

Ornithopods elsewhere in the world and from other times are known from a vast array of terrestrial depositional environments, ranging from upper coastal plain deposits, to lower coastal plain channels and fluvial deposits. Several ornithopods, hadrosaurids mostly, are known from islands and even from rare marine occurrences, where they are thought to represent the remains of bloated carcasses swept out to sea.

Some degree of **habitat partitioning** among ornithopods has been reported. By this, it is meant that several species divide the available ecospace into domains that do not overlap with each other. For example, Horner has noted that in western North America, there are many hadrosaurine taxa that tend to be found in near-marine, deltaic sediments. In contrast, the vast majority of lambeosaurine taxa were restricted to lower coastal plain sediments that were deposited inland from these near-marine environments. Finally, *Maiasaura* has been found nowhere else but in upper coastal plain sediments and is therefore thought to have been endemic to these environments.

Habits

The wide variety of morphologies and geographic ranges of ornithopods have provoked a great deal of speculation about their habits (see Box 10.1); however,

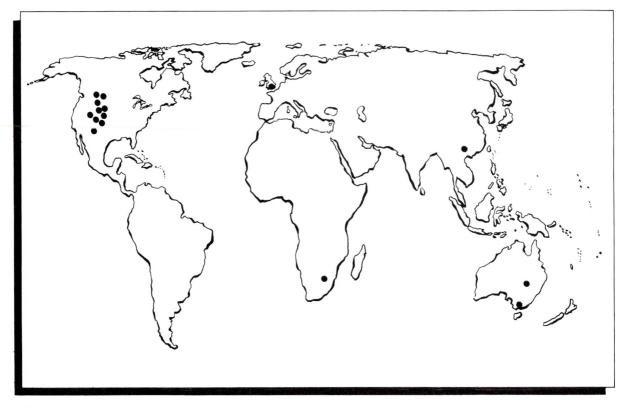

Figure 10.19
Global distribution of
Heterodontosauridae
and Hypsilophodontidae.

we now view all ornithopods as being predominantly bipedal terrestrial animals. As might be expected, many of the smaller, agile forms – heterodontosaurids, hypsilophodontids, and a few iguanodontians – must have been primarily bipedal, although they also may have adopted a quadrupedal stance when foraging or standing still. Some of the larger ornithopods, such as *Iguanodon*, may have engaged in extensive quadrupedal locomotion. These forms have a solidly built hand that clearly was capable of considerable weight support. Juveniles of these quadrupedal ornithopods may have been more bipedal than their adult counterparts. Changes in proportions of the limb elements suggest shifts in locomotion type with age. By and large, the larger iguanodontians were also not as fast-running as their smaller counterparts.

In all cases, the tail was long, muscular, strengthened by ossified tendons, and held at or near horizontal, making an excellent counterbalance for the front of the animal. In general, the powerful hindlimbs tend to be at least, and sometimes more than, twice the length of the forelimbs.

How fast could these dinosaurs have traveled? Larger iguanodontians, such as hadrosaurids, may have been able to reach 15 to 20 km/hr during a sustained run.[3] Quadrupedal galloping appears unlikely, given the rigidity of the vertebral column and the lack of movement of the shoulder against the rib cage and ster-

---

[3]Some scientists have estimated that the largest could touch 50 km/hr. Obviously this could only happen in a burst of movement and could not be sustained.

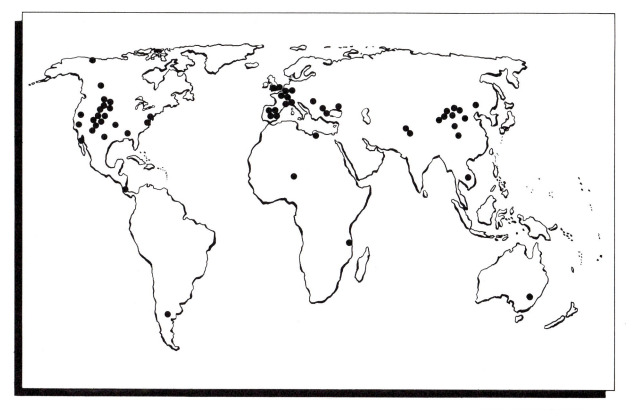

Figure 10.20 **Global distribution of Iguanodontia.**

num. For smaller ornithopods, running speeds could have been higher. Maximum speeds were possibly as fast as 60 km/hr (Figure 10.21).

Fast-running, maybe. But were they smart? According to Hopson, yes, they were. In fact, Hopson believes that they were as smart as, or smarter than, might be expected of a large archosaur! For example, *Leaellynasaura*, the hypsilophodontid from Victoria, Australia, was apparently quite brainy and had acute vision, as suggested by prominent optic lobes.[4] In general, ornithopod "brainy-ness" may relate to greater reliance on acute senses for protection, which, in the absence of the extensive anatomical defenses, may have been their only recourse. Moreover, brain size in these dinosaurs may also relate to a complex behavioral repertoire, which we will discuss.

What did the dominantly bipedal ornithopods do with their hands? *Heterodontosaurus* may have used its powerful forelimbs and its clawed hands to grab at vegetation or to dig up roots and tubers. The hypsilophodontid forelimbs and hands appear less powerful than those of either heterodontosaurids or iguanodontians. Nevertheless, because they were not usually used in weight support, the hands were free to grasp at leaves and branches, bringing foliage closer to the mouth so that it could be nipped off by the toothed beak.

Iguanodon, *Ouranosaurus*, and, to a degree, *Camptosaurus* had extraordinary hands (Figure 10.22). The first digit (thumb) was conical and sharply pointed,

[4] The animal had an estimated encephalization quotient (EQ; see Box 14.3) of 1.8; Hopson estimated that the average EQ of other ornithopods is about 1.5.

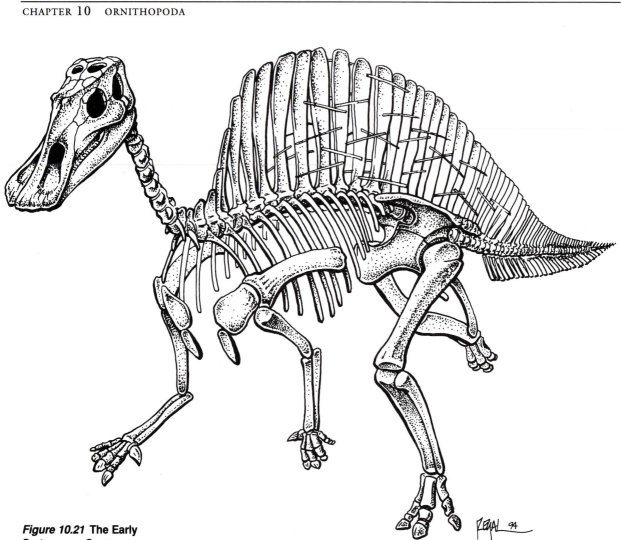

Figure 10.21 The Early
Cretaceous *Ouranosaurus*
in a hurry.

and may have been used as a stiletto-like, close-range weapon or used for breaking into seeds and fruits. In contrast, the fifth digit (pinkie) was fully opposable
(very much as the thumb is in humans)[5] and could fold against the palm of the
hand! The remaining middle three digits were all hoofed.

The same cannot be said for hadrosaurids. Their reduced hands, with three
hoofed fingers joined together in a thickened pad, hardly had any way to function other than as a support while the animal was standing. Manual dexterity was
not a hadrosaurid specialty.

Feeding and Food

No other group of dinosaurs has been the subject of as much research on feeding as ornithopods. Even ornithopod stomach contents have been fossilized

---

[5]Human success is sometimes ascribed to an opposable thumb, but as you can see in this book,
dinosaurs invented opposable digits at least twice: once in ornithopods and once in theropods.

Figure 10.22 **The hand of**
Iguanodon. Note the
spiked thumb.
Scale = 5 cm.

within so-called hadrosaurid "mummies." These spectacular specimens apparently dried before burial and replacement (see Chapter 1). Preserved are beautiful skin impressions; dried, stretched tendons and muscles; and fossilized remnants of the last supper in the gut. So what was for dinner? The hadrosaurid "mummies," at least in life, ate twigs, berries, and coarse plant matter. This correlates nicely with their size: Ornithopods are thought to have been active foragers on ground cover and on low-level foliage from conifers and, in some cases, from deciduous shrubs and trees of the newly evolved angiosperms (Box 15.3). Browsing on such vegetation appears to have been concentrated within the first meter or two above the ground, but the larger animals must have been capable of reaching vegetation as high as 4 m above the ground.

Eating coarse, fibrous food requires some no-nonsense equipment in the jaw to extract enough nutrition for survival, and ornithopods had what it took. In many cases, the group came equipped with a beak in the front for cropping vegetation; a well-developed block of cheek teeth (the dental battery) for grinding coarse plant matter (Figure 10.23); a large, robust coronoid process for serious masticatory muscles; and a tooth row that was deeply set in, indicating that, as in

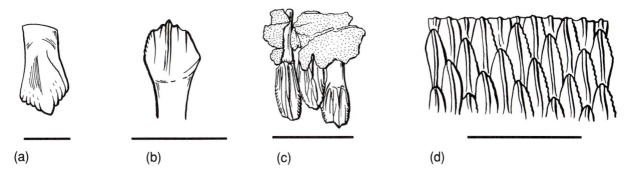

(a)   (b)   (c)   (d)

Figure 10.23 **Upper tooth of (a)** *Lycorhinus,* **(b) lower tooth of** *Hypsilophodon,*
(c) three upper teeth of *Iguanodon,* **and (d) lower dental battery of** *Lambeosaurus.*
Scale **(a)** and **(b)** = 5 mm; **(c)** and **(d)** = 5 cm.

BOX 10.1

SWIMMING IN THE TREETOPS AND OTHER DORMANT HYPOTHESES

IN THE HISTORY OF the study of ornithopods, habitats and anatomy conspired to put some of these animals in exotic places and give them unusual locomotor skills. For example, hadrosaurids were once regarded as amphibious, in part because the tail was long and deep – great for sculling in the water – the hand appeared webbed, and jaws were deemed too weak to handle anything but soft aquatic vegetation. Not true. In a similar fashion, for over one hundred years, a species of *Hypsilophodon* was regarded as a tree dweller. Upon close scrutiny by P. M. Galton, however, this animal was found to have no specializations for this particularly demanding mode of life.

The combination of a strongly seasonal African habitat and some basic heterodontosaurid anatomy created a dilemma – and ultimately a solution – for R. A. Thulborn in 1978. Heterodontosaurids, he believed, chewed by moving the lower jaw forward and backward relative to the upper jaw. Yet, evidence of tooth replacement, which he expected (given that heterodontosaurids

fed on very abrasive food), simply did not exist. To replace the teeth gradually would have impaired the ability to feed, he reasoned, so the teeth could only have been replaced en masse. How could this be accomplished? Thulborn argued that heterodontosaurids must have aestivated (lay dormant), most likely during the dry season. While the animal was dormant, the formerly functional teeth fell out and were replaced, to be worn down while the animal was active and feeding during each wet season.

Woe to fine arguments such as these, when their basic presumptions have to be corrected! Several years after Thulborn's aestivation hypothesis had appeared, J. A. Hopson reexamined heterodontosaurid jaw mechanics and tooth replacement patterns. As it turns out, heterodontosaurids chewed transversely, not forward and backward, so that tooth replacement was reduced, but not lost, in these animals. There is no compelling reason to believe that heterodontosaurids engaged in aestivation during the harshness of the southern African Early Jurassic.

---

all genasaurian ornithischians, large fleshy cheeks were present. But beyond these basics, different ornithopods had different modifications of the jaw, and different kinds of jaw motions are believed to have been used for the processing of food.

The first modern treatment of jaw mechanics in ornithopods was Ostrom's extensive 1961 study of the cranial anatomy (including the skull, musculature, vascular and nervous systems) of North American hadrosaurids. These four perspectives, which provided the basis for reconstructing the pattern of chewing in the Late Cretaceous ornithopods, suggest that hadrosaurids chewed back to front – in what is called **propalinal jaw movement** – on both sides of the mouth at the same time.

Other dinosaur paleontologists have suggested otherwise, at least for different ornithopods. Galton noted that *Hypsilophodon* may have chewed in much the same way as many mammals do today – side-to-side, on one side of the mouth at a time. And R. A. Thulborn regarded chewing in heterodontosaurids as similar to what Ostrom suggested for hadrosaurids: bilateral propalinal jaw movement.

More recently, how these herbivores chewed their food and how these jaw mechanisms evolved have been the focus of considerable research by Weishampel and Norman, as well as by Crompton and J. Attridge (Birkbeck College). These studies have been based not only on comparisons of ornithopod skulls and teeth, but also on computer analyses of cranial mobility that might translate into special kinds of movement between chewing teeth.

What emerges from these studies is yet again more ornithopod diversity, this time at the level of feeding and foodstuffs. Primitively in the clade, only the very front of the relatively narrow beak lacked teeth but was otherwise covered with a sharp, rhamphotheca. This condition – found in both heterodontosaurids and hypsilophodontids – suggests a somewhat selective cropping ability. Iguanodontians, by contrast, lose all their front teeth, broaden their snouts, and even develop a strongly serrate margin to their rhamphothecae. These animals were not selective feeders; instead, they hacked at, and severed, leaves and branches without much regard for what they were taking in.

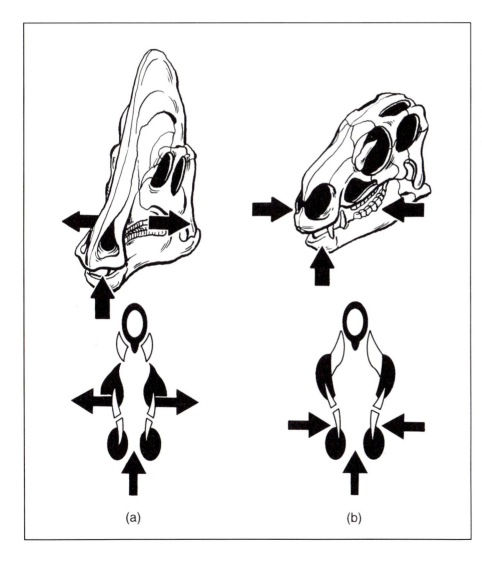

(a)   (b)

Figure 10.24 Jaw mechanics (a) in Euornithopoda, showing lateral mobility of the upper jaws (pleurokinesis), and (b) in Heterodontosauridae, showing medial mobility of the lower jaws.

Once these gulpfuls of leaves had passed the rhamphotheca into the mouth, all ornithopods chewed their food. Yet how they solved the problem of combining bilateral occlusion (where the teeth meet on both sides of the jaws at the same time) with chewing is one of the most intriguing aspects of dinosaur feeding, for heterodontosaurids and euornithopodans evolved different solutions to this problem, solutions that parallel those "invented" by ungulate mammals (such as sheep or horses) but which are uniquely distinct from those and from each other.

On the basis of skull architecture, patterns of tooth wear, and computer modeling, we know that heterodontosaurids chewed by combining vertical movement of the lower jaws with a slight degree of rotation of the mandible about their long axes (Figure 10.24b). In this way, they were able to move their upper and lower teeth in a transverse direction and thus break up the bits of plant food that the tongue had placed between them. Naturally, the fleshy cheeks prevented most of the food from falling out of the corners of the mouth.

Euornithopodans, on the other hand, evolved a distinctly different pattern of skull movement in order to solve the problem of having bilateral occlusion and still chewing side-to-side. Instead of loosening up the lower jaws to rotate about their long axes, euornithopodans mobilized their upper jaws. This kind of mechanism, which Norman called **pleurokinesis**, involved a slight rotation of portions of the upper jaw – especially the maxilla (the bone that contains the upper cheek teeth) – relative to the snout and skull roof (Figure 10.24a). When the upper and lower teeth were brought into contact on both right and left sides, the upper jaws rotated outward, the lower jaws moved inward, and the opposing surfaces of the teeth sheared past one another to break up plant food in the mouth. Impossible in humans, where the bones of the skull are solidly fused and locked together, a system such as this requires flexibility at the joints between bones of the skull. In hadrosaurids, the tight occlusion of the dental battery would have made short work of virtually all foliage. Like the situation in heterodontosaurids, pleurokinesis represented an important advance for euornithopodans, providing them the ability to chew a variety of plant foods, including those with a great deal of fiber.

As in all of the other ornithischians that have been discussed, once the food was properly chewed, it was swallowed and quickly passed to a gut whose volume suggests that fermentation took place there. Between the extensive chewing of food in the mouth and possible fermentation in the large gut, it is very likely that all ornithopods were well suited for a diet of low-quality, high-fiber vegetation.

Head Structures and Behavior

From the time of their discovery, ornithopods of all kinds have attracted a good deal of attention, particularly for their odd-appearing ornamentation. The outrageous crests on the heads – many of them highly chambered – of hadrosaurids, the tusks of heterodontosaurids, and the lumps on the forehead of *Ouranosaurus* have called out for an explanation. It is safe to say that virtually all of these features – like those odd bumps of pachycephalosaurs and the horns of ceratopsians and, for that matter, the antlers of deer and the horns of cows and antelopes – hint at sophisticated social behavior.

Hadrosaurids have attracted the most attention, in large part because they clearly stand out from the crowd with their wild headgear. Once thought to relate to the aquatic habits of the group (see Box 10.1) or to the olfactory (sense of smell) function of the nasal cavity, much of the discussion about the functional significance of hadrosaurid ornamentation now centers on combat and display, and on the reproductive consequences of such behaviors. In 1975, Hopson suggested that the unusual cranial features – principally involving the nasal cavity – that we see in hadrosaurids likely evolved in the context of social behavior among members of the same species. In particular, Hopson regarded the special cranial features in hadrosaurids as indicative of either intra- or interspecific aggression (see Chapter 9) and, more especially in the case of both solid and hollow crests, of visual and vocal display. In order to have functioned as good signals to convey information about what species, what sex, and even what rank ("Perfect '10'," "Workable," "Not!") an individual might be, the crests must have been both visually distinctive and able to make different sounds. Only then can they be regarded as having promoted successful matings by informing the consenting adults.

But as was the case for ceratopsian horns, how are we ever to make sense of these suggestions about unfossilizable behavior? Hopson made five predictions that link the fossil record of hadrosaurids to the social behaviors he anticipated were driven by sexual selection. First, to interpret incoming display information, hadrosaurids must have had both good hearing and vision. These are qualities that cannot be directly measured in extinct vertebrates, but all hadrosaurids have large eye sockets, often with sclerotic rings that encircled the outer region of the eye and give a clue to its size. In all cases, eye size was quite large, and so sight must have been reasonably acute. Similarly, we have evidence of hearing across a wide range of frequencies via preserved middle and inner ear structures.

Second, Hopson predicted that the external shape of the crest may have been as important as its internal structure if it was to act in visual display as well as a vocal resonator. Again, this prediction is upheld by hadrosaurid fossils: In virtually all cases, the profile of the crest is much more elaborate or extensive than the walls of the internal plumbing.

Prediction three was that if crests acted as visual signals, then they should be species-specific in size and shape, and they should also be sexually dimorphic. This is amply upheld in large part thanks to studies by University of Pennsylvania's P. Dodson on growth and development in lambeosaurine hadrosaurids (Figure 10.25). Using a variety of statistical techniques, Dodson was able to show that crests become most prominent when an animal approaches sexual maturity. In addition, he demonstrated that each lambeosaurine species was dimorphic, particularly in terms of crest size and shape. Could these "morphs" have been male and female? It certainly fits well with Hopson's prediction!

The two remaining predictions have to do with hadrosaurids in time and space. When several species occur together in the same area, they should exhibit great differences in the shape of their crests. Similarity would create a great deal of confusion among closely related hadrosaurids living in the same place, but distinctiveness in display structures would prevent such confusion – an obvious advantage during breeding season. Are crests more distinctive as the number of hadrosaurids living together goes up? The answer is yes. At Alberta's Dinosaur Provincial Park, where the number of hadrosaurids that have been found in the Judith River Formation (and thus are thought to have lived together) is high, there are three distinctively crested lambeosaurines: one solid-crested form (*Prosaurolophus*), and two species of hadrosaurine, each distinctive in its own right. In contrast, elsewhere – where hadrosaurid diversity is lower – the variety of flamboyant headdresses is decreased.

Hopson's final prediction, that crests should become more distinctive through time as a consequence of sexual selection (see Chapter 8), is only partially supported. In lambeosaurines, crests arguably get less distinctive over time, although in other groups (the so-called saurolophs, which consist of species of *Prosaurolophus* and *Saurolophus*), crests may show an increase in distinctiveness. In all of these forms, however, we have no inkling about the soft tissue that must have surrounded the crest.

In the end, Hopson's hypothesis of crests used for species recognition, intraspecific combat, ritualized display, courtship, parent–offspring communication, and social ranking fares very well. With such support, it is possible to put together a scenario as to how these cranial ornamentations evolved in the different hadrosaurid

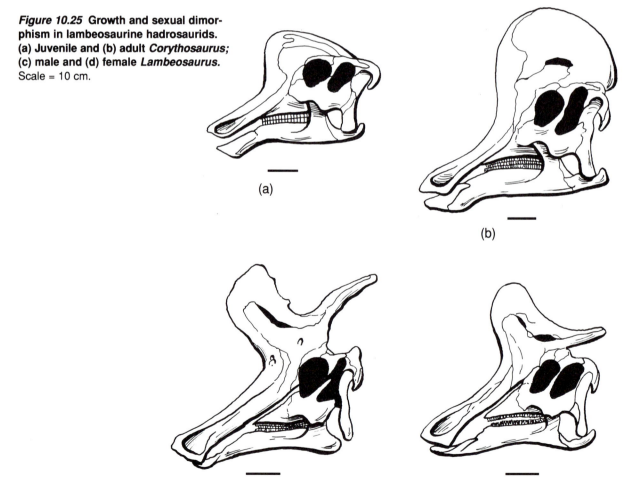

Figure 10.25 **Growth and sexual dimorphism in lambeosaurine hadrosaurids. (a) Juvenile and (b) adult *Corythosaurus*; (c) male and (d) female *Lambeosaurus*.** Scale = 10 cm.

(a)

(b)

(c)

(d)

groups. In *Gryposaurus*, *Maiasaura*, and *Brachylophosaurus*, three of the more primitive of hadrosaurine hadrosaurids, the accentuated nasal arch and stout cranial crest were likely used for broadside- or head-pushing during male–male combat. It is possible that these animals had inflatable flaps of skin that covered their nostrils and surrounding regions (Figure 10.26); these would have been blown up and used for visual display, as well as used to make some noise – a kind of Mesozoic bagpipe. In *Prosaurolophus* and *Saurolophus*, this sac would have extended onto the solid crest that extended above the eyes; in *Anatotitan* and *Edmontosaurus*, where the nasal arch is not accentuated nor is there a crest, the complexly excavated nostril region may have housed an inflatable sac. With such an exceptional development of sacs around the nostrils and up and down the crest, ritualized combat with accompanying vocal and visual display would have been the norm.

When it came to display, none did it better than the lambeosaurines. In these animals, the hollow crests perched atop the head must have provided for instant recognition. This could have been visual and/or aural (by low honking tones produced in the large resonating chamber within the crest). Either way, by sight and/or through vocal cacophony, the crests of lambeosaurines would have functioned well as species-specific display organs.

What of other ornithopods? Although not nearly so well analyzed, it appears that the evolution of canine-like teeth of heterodontosaurids may have something to do with intraspecific display and combat. Both Thulborn and Molnar have suggested that since these teeth are present only in mature "males," they would have been used not only in gender recognition, but also for intraspecific combat, ritualized display, social ranking, and possibly even courtship. A modern analogue is the tusked tragulids, living herbivorous deerlike mammals with large, canine-like teeth. Similarly, the development of a jugal boss in heterodontosaurids might also be interpreted as a form of visual display.

Following this line of reasoning, the low, broad bumps on top of the head of *Ouranosaurus* may well have similar behavioral significance (Figure 10.21). Perhaps

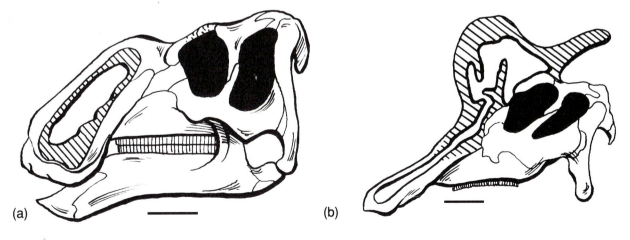

Figure 10.26 The circumnarial depression (indicated by cross-hatched region) that may have supported an inflatable flap of skin in hadrosaurines like (a) *Gryposaurus*. (b) Highly modified nasal cavity housed within the hollow crest on top of the head of *Lambeosaurus*. Scale = 10 cm.

they aided individuals in the recognition of members of the same species or members of the opposite sex. Or perhaps they were used in ritualized head-butting contests. These animals were also equipped with extremely high spines on their vertebrae, which formed a sail-like ridge down their backs. As with *Stegosaurus* (Chapter 6), it is possible that these long spines were covered with skin and used as a radiator or solar panel to warm up or cool down. Alternatively (and not mutually exclusively), they may have had a display function, providing the animal with a greater side profile than it would otherwise have had.

Display behavior in many ornithopods begins to make even more sense when considered with other aspects of their life styles. Consider communication between adults and between grownups and juveniles. There are several examples of single-species bonebeds – for example, *Dryosaurus, Iguanodon, Maiasaura, Hypacrosaurus,* and others – that support the notion that these animals were not only common, but that they formed herds of both youngsters and adults. It has even been suggested that such large aggregations required migratory movement, most likely seasonal, in order to meet the energy demands of the members of the herd.

Family Values in the Mesozoic: Bringing Up Baby

It is now known from a series of important studies by Horner (Museum of the Rockies) that at least some ornithopods nested in large colonies. For hypsilophodontids like *Orodromeus*, nesting took the form of well-organized spirals laid in relatively soft sediment. On average, 12 eggs were laid in an upright position and we think that Mom may have stuck around till they hatched. When they finally emerged from the egg, hatchlings had well-developed limb bones with fully formed joint surfaces, indicating that these young could walk, run, jump, and forage for themselves as well as any adult. Such levels of youthful activity gain support from the preservation of complete, yet hatched, eggs in the *Orodromeus* nest. Nests in which the young stuck around would have eggs that were trampled. The good condition of the hatched *Orodromeus* eggs suggests that nobody stayed at the nest very long. With the young so well formed and capable, parental care must have been minimal, if there was any at all.

Once out and about, young *Orodromeus* appear to have stayed in groups, where mutual protection and a degree of communication would have been an advantage. It is not known whether groups split up or remained together later in life, simply because these age groups have yet to be found in the fossil record. With further work, perhaps we will soon know.

Not all ornithopods took such a laissez faire attitude toward the children. *Maiasaura, Hypacrosaurus,* and probably other hadrosaurids likewise nested in colonies, digging a shallow hole in soft sediments and laying up to 17 eggs in each nest. These nests were a mother's body size apart from the next, strongly suggestive that nests were regularly tended by Mom and/or Dad. Vegetation probably covered the eggs to keep them warm. In contrast to *Orodromeus*, however, hadrosaurid hatchlings (Figure 10.27) have been found within their nests, having wreaked havoc on the eggs that once housed them. Consequently, we have an abundance of eggshell scraps but very poor information on complete hadrosaurid eggs. It is clear that hatchlings remained in the nest for extended periods of time, perhaps upwards of eight or nine months. During this nest-bound time, the off-

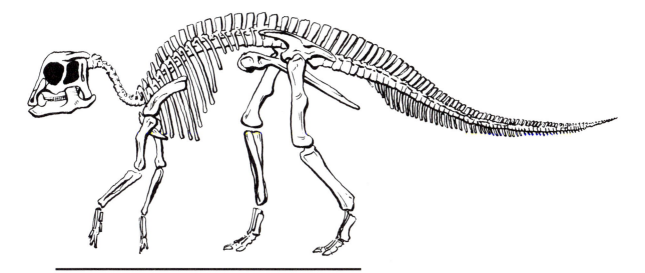

Figure 10.27 Left lateral view of the skull and skeleton of a hatchling _Maiasaura_. Like other juveniles, such as puppies and kittens, the head, eyes, and limbs are disproportionately large. Scale = 50 cm.

spring were literally helpless and must have depended on their parents to provision them with food and protection. If our estimates of the length of the nest-bound period are correct, then it appears that hatchling growth rates were exceedingly high, well within the range of fast-growing mammals and birds at approximately 12 cm in length per month (see Chapter 14)! This means that hatchlings must have channeled into growth virtually all of the foodstuff that their parents brought them.

Once these hatchlings left the nest, they appear to have stayed together at least in small groups. But during the breeding season, if not for longer periods of time, these animals gathered into exceedingly large herds; in the case of _Maiasaura_, Horner estimates that the size of the herds exceeded 10,000 individuals. This correlates nicely with the visual and vocal communication we postulated earlier and suggests complex social behavior.

With ornithopods of all kinds in and out of the nest, demanding or rejecting parental care, we enter perhaps one of the most elusive aspects of the fossil record: life history strategies. These strategies detail the ways in which particular organisms grow, reproduce, and die. Consider the mosquito, the blood-sucking blight of a warm summer day. These animals produce enormous numbers of eggs, which result in thousands of offspring, the vast majority of which do not survive to reproduce themselves even during their incredibly short lifespans. No parental care here – too many children, for one thing, and the bugs are not programmed that way anyway. Now consider us, with much longer lifespans, fewer offspring, and lots of parental care (too much, some say, when stuck with thirty-somethings returning to the parental abode!). In the former case, species survival is based on saturation – with so many mosquitoes, some are bound to survive. This kind of life strategy is referred to as an **r-strategy**, the symbol for the unrestricted, intrinsic rate of increase of individuals in a population. Because these organisms must fend for themselves, we regard them as **precocial**, which means that the young are rather adultlike in their behavior.[6] In contrast, human survival is thought to depend in large part on parental care of only a few, often slow-growing, offspring. Instead of being r-strategists, we employ a **K-strategy**, named after the symbol for the carrying-capacity of an environment.[7] Because the young are delaying their maturation – thereby requiring the extra input of care by parents – we refer to this condition as being **altricial**. So no matter how totally and maximally mature some of our young teenagers assure us they are, we all are biologically altricial, as are virtually all mammals and most birds.

How do those ornithopods for which we have information conform to either of these two contrasting strategies? With their ability to fend for themselves like adults, we consider *Orodromeus* to have been closer to an r-strategist, a claim that is based on the inferred precocial nature of these dinosaurs. In contrast, *Maiasaura, Hypacrosaurus*, and perhaps other hadrosaurids that had nest-bound hatchlings requiring parental care all appear to have been altricial and, thereby, K-strategists. Gazing elsewhere among ornithopods, it appears that precociality may be primitive for at least Euornithopoda and that altriciality likely evolved for the first time within the clade sometime prior to the origin of Hadrosauridae.

Whatever the broader meaning of these changes might be, we – and ornithopods – cannot escape from the effects of family. Among hypsilophodontids, life as a parent must have been easy – no provisioning or protection of the kids. But the toll to be paid was reduced survival of these offspring – wherever it was that they wandered off to. From a hadrosaurid perspective, however, it was a good thing to take care of the kids. For no matter how loud, squawky, and hard to handle these hatchlings might have been, Mom and Dad directly increased their chances of survival.

Important Readings

Dodson, P. 1975. Taxonomic implications of relative growth in lambeosaurine hadrosaurids. Systematic Zoology 24:37–54.
Galton, P. M. 1974. The ornithischian dinosaur *Hypsilophodon* from the Wealden of the Isle of Wight. Bulletin of the British Museum (Natural History) Geology 25:1–152.
Hopson, J. A. 1975. The evolution of cranial display structures in hadrosaurian dinosaurs. Paleobiology 1:21–43.
Horner, J. R. 1984. The nesting behavior of dinosaurs. Scientific American 250:130–137.
Horner, J. R. 1990. Evidence of diphyletic origination of the hadrosaurian (Reptilia: Ornithischia) dinosaurs; pp. 179–187 *in* K. Carpenter and P. J. Currie (eds.), Dinosaur Systematics. Cambridge University Press, New York.
Lull, R. S., and N. E. Wright. 1942. Hadrosaurian dinosaurs of North America. Geological Society of America Special Paper 40:1–242.

[6]The word "precocious," when applied to a young genius like Mozart, implied that he could, as a child, write music like an adult.
[7]The assumption embodied in this term is that resources limit the number of adult individuals that can subsist within a particular environment. The K-strategist rears its young so that all are intended to achieve adulthood. The emphasis is on quality, and not quantity, because no environment could support a situation in which all the young of an r-strategist reached adulthood.

Norman, D. B. 1980. On the ornithischian dinosaur *Iguanodon bernissartensis* from the Lower Cretaceous of Bernissart (Belgium). Institut Royal de Science Naturelle de Belgique, Memoire 178:1–103.

Norman, D. B. 1986. On the anatomy of *Iguanodon atherfieldensis* (Ornithischia: Ornithopoda). Bulletin, Institut Royal de Science Naturelle de Belgique, Science de la Terre 56:281–372.

Norman, D. B. 1990. A review of *Vectisaurus valdensis*, with comments on the family Iguanodontidae; pp. 147–161 *in* K. Carpenter and P. J. Currie (eds.), Dinosaur Systematics. Cambridge University Press, New York.

Norman, D. B., and D. B. Weishampel. 1990. Iguanodontidae and related Ornithopoda; pp. 510–533 *in* D. B. Weishampel, P. Dodson, and H. Osmólska (eds.), The Dinosauria. University of California Press, Berkeley.

Ostrom, J. H. 1961. Cranial morphology of the hadrosaurian dinosaurs of North America. Bulletin of the American Museum of Natural History 122:33–186.

Sereno, P. C. 1986. Phylogeny of the bird-hipped dinosaurs (Order Ornithischia). National Geographic Research 2:234–256.

Sues, H.-D., and D. B. Norman. 1990. Hypsilophodontidae, *Tenontosaurus*, Dryosauridae; pp. 498–509 *in* D. B. Weishampel, P. Dodson, and H. Osmólska (eds.), The Dinosauria. University of California Press, Berkeley.

Taquet, P. 1976. Géologie et paléontologie du gisement de Gadoufaoua (Aptien du Niger). Cahiers de Paléontologie C.N.R.S. Paris: 1–191.

Weishampel, D. B. 1984. The evolution of jaw mechanisms in ornithopod dinosaurs. Advances in Anatomy, Embryology and Cell Biology 87:1–110.

Weishampel, D. B., and J. R. Horner. 1990. Hadrosauridae; pp. 534–561 *in* D. B. Weishampel, P. Dodson, and H. Osmólska (eds.), The Dinosauria. University of California Press, Berkeley.

Weishampel, D. B., D. B. Norman, and D. Grigorescu. 1993. *Telmatosaurus transsylvanicus* from the Late Cretaceous of Romania: the most basal hadrosaurid. Palaeontology 36:361–385.

Weishampel, D. B., and L. M. Witmer. 1990. Heterodontosauridae; pp. 486–497 *in* D. B. Weishampel, P. Dodson, and H. Osmólska (eds.), The Dinosauria. University of California Press, Berkeley.

PART III

SAURISCHIA

SAURISCHIA:
Predators and Giants
of the Mesozoic

ORIGINALLY COINED by Cambridge University's H. G. Seeley in 1887, Saurischia consists of Sauropodomorpha (Chapter 11) and its sister-taxon Theropoda (Chapters 12 and 13). It contains both the smallest of dinosaurs and the super-giants as well. It also contains the most agile of predatory dinosaurs and the most ponderous plant-eaters.

Despite such disparate membership, Saurischia has been diagnosed by J. A. Gauthier by more than 15 derived features. These include a new opening beneath the nostril area (the **subnarial foramen**), elongation of the rearward neck vertebrae (giving a relatively long neck), the development of accessory articulations between the vertebrae (the so-called **hyposphene-hypantrum articulations**), a large thumb supported by a very broad first metacarpal bone, and a wedge-shaped splint of bone on the **astragalus** (an ankle bone) that lies flat against the shin and points upward (the **ascending process of the astragalus**). All in all, this is a well-supported dinosaurian clade (Figure III.1).

Left unmentioned in saurischian membership is the enigmatic Segnosauria, a newcomer to the lizard-hipped dinosaurs. The first inkling that these unusual beasts even existed was the surprise description of the mid-Cretaceous *Segnosaurus* (*segnis* – slow) from Mongolia, by A. Perle in 1979. This new and obviously aberrant plant-eating dinosaur has a weird pelvis and jaws unlike any sort of theropod, which is how *Segnosaurus* was originally described. What was wrong here? Was there a mix-up of two or more kinds of dinosaurs? The following year, with Perle's description of even better segnosaur material (which he named *Erlikosaurus* – Erlik was the Lamaist king of the dead), it became clear that we were dealing with an entirely new group of dinosaurs, a clade unforeseen in all prior work on dinosaur evolution. Yet there was more to come, for in 1983 Mongolian paleontologists R. Barsbold and Perle provided us with another of these unusual segnosaurs, which they called – perhaps for obvious reasons – *Enigmosaurus* (*aenigma* – riddle). These three forms, all collected from mid-Cretaceous sediments of southern Mongolia, are nearly all that we know about segnosaurs. Another form, *Nanshiungosaurus* (named for Nanziong Province) by Chinese paleontologist Dong Z.-M. in 1979, was originally thought to be an aberrant sauropod but is now known to be a very large, Late Cretaceous segnosaur from China.

It is a relatively simple matter to diagnose segnosaurs, which have been likened to an offspring of a *ménage à trois* involving a theropod, a prosauropod, and an ornithischian. All segnosaurs have six sacral vertebrae; broad hips; an ilium in which the **preacetabular process** (that piece of the ilium anterior to the hip socket) is deep, down-turned, and outwardly flared; and a retroverted pubis, in which

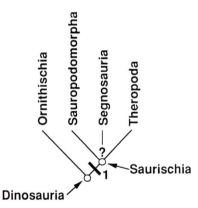

Figure III.1 Cladogram of Dinosauria emphasizing the monophyly of Saurischia. Derived characters at 1: subnarial foramen, elongation of the rearward neck vertebrae, hyposphene-hypantrum vertebral articulations, large thumb supported by a very broad first metacarpal bone, and wedge-shaped ascending process on the astragalus.

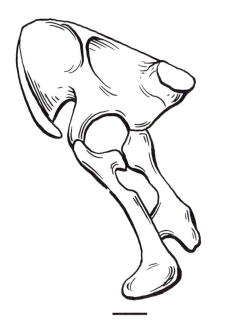

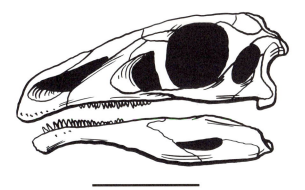

Figure III.3 Left lateral view of the skull of *Erlikosaurus.* Scale = 10 cm

Figure III.2 Left lateral view of the pelvis of *Segnosaurus*. Note that the pubis is rotated to a more rearward position to underlie the ischium. This opisthopubic condition is convergent on that seen in ornithischians. Scale = 10 cm.

the pubis actually runs downward and backward, along the lower part of the ischium (Figure III.2). This pubis looks superficially similar to that found in ornithischians, but in detail it is quite different. In addition, segnosaurs may share certain similarities of the skull and lower jaw. Unfortunately, skulls and lower jaws are known only in *Erlikosaurus* and *Segnosaurus*. But in those two genera at least, the skulls are low, long, and lack teeth in the front; moreover, they have a highly vaulted palate and braincase with an enlarged base and ear region (Figure III.3).

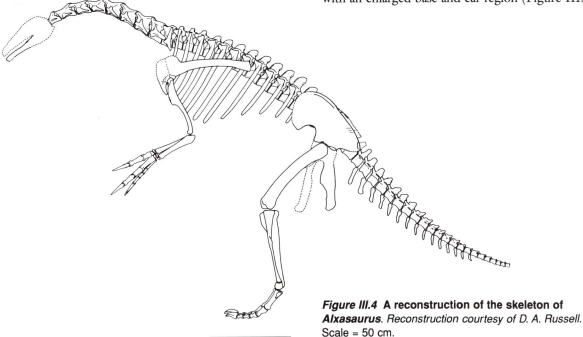

Figure III.4 A reconstruction of the skeleton of *Alxasaurus*. *Reconstruction courtesy of D. A. Russell.* Scale = 50 cm.

At present, the position of Segnosauria within Saurischia is problematic. When the clade was first recognized by Barsbold and Perle in 1980, it was placed among theropods, but more recently it has been allied with sauropodomorphs by G. S. Paul in 1986. Barsbold and Polish paleontologist T. Maryańska in 1990 placed Segnosauria in an unresolved trident with theropods and sauropodomorphs.

New work from Canadian paleontologist D. A. Russell and Dong suggests yet another – and unexpected – alternative. They described a segnosaur, *Alxasaurus* from the Alxa Desert of Inner Mongolia, China (Figure III.4), the analysis of which persuaded them that segnosaurs are actually closely related to the large, strange, and poorly understood theropod *Therizinosaurus* (see Chapter 12). Researchers from the American Museum of Natural History, restudying the beautifully preserved skull of *Erlikosaurus,* have come to the same conclusion (also see Chapter 12). Only time and more fossils will unravel the true affinities of these strange animals.

Important Readings

Barsbold, R., and A. Perle. 1980. Segnosauria, a new infraorder of carnivorous dinosaurs. Acta Palaeontologica Polonica 25:185–195.

Barsbold, R., and T. Maryańska. 1990. Segnosauria; pp. 408–417 *in* D. B. Weishampel, P. Dodson, and H. Osmólska (eds.), The Dinosauria. University of California Press, Berkeley.

Clark, J. M., A. Perle, and M. A. Norell. 1994. The skull of *Erlikosaurus andrewsi,* a Late Cretaceous "segnosaur" (Theropoda: Therizinosauridae) from Mongolia. American Museum Novitates, v. 3115, 39 pp.

Gauthier, J. A. 1986. Saurischian monophyly and the origin of birds. Memoirs of the California Academy of Sciences 8:1–55.

Paul, G. S. 1984. The segnosaurian dinosaurs: relics of the prosauropod-ornithischian transition. Journal of Vertebrate Paleontology 4:507–515.

Russell, D. A., and Dong Z. 1993. The affinities of a new theropod from the Alxa Desert, Inner Mongolia, People's Republic of China. Canadian Journal of Earth Sciences 30: 2107–2127.

CHAPTER 11

SAUROPODOMORPHA:
The Big, the Bizarre, and the Majestic

BEHEMOTHS: THERE'S HARDLY A BETTER WAY to describe many sauropodomorphs. For it was the sauropodomorphs that pushed the envelope of body size among terrestrial vertebrates to the extreme – to the tune of 75,000 kg and possibly more (Figure 11.1)! What kind of anatomy does it take to be building-sized – to grow up to 40 m long and tower 6 m at the shoulder, *and* to flourish for 160 million years?

Take the shape of your basic sauropod – that subset of sauropodomorphs epitomized by "brontosaurus." Note the long neck constructed by a quite complex system of girders and air pockets to maximize lightness and strength. At the end observe the relatively tiny skull – laughably small, until we realize that only an idiot would design a large, heavy skull at the end of an extremely long neck. The skull itself is unusual: It has relatively simple peglike or spatulate teeth, and nostrils that, instead of residing at the tip of the snout, have a phylogenetic tendency to migrate upward, toward the top of the head (Figure 11.2). Then there are the legs: four pillars that would do a Greek temple proud. The limbs were composed of bone denser than that found in the upper parts of the skeleton; again, a sophisticated adaptation placing the weight and strength in the sauropod skeleton where it is most needed. Sauropodomorphs taxed biomechanical and physiological design – weight support, neural circuitry, respiration, digestion, and the like – to the limit. It isn't easy being BIG.

Let's see who sauropodomorphs really are. Sauropodomorpha (*sauros* – lizard; *pod* – foot; *morpho* – form) is a clade composed of animals popularly known as

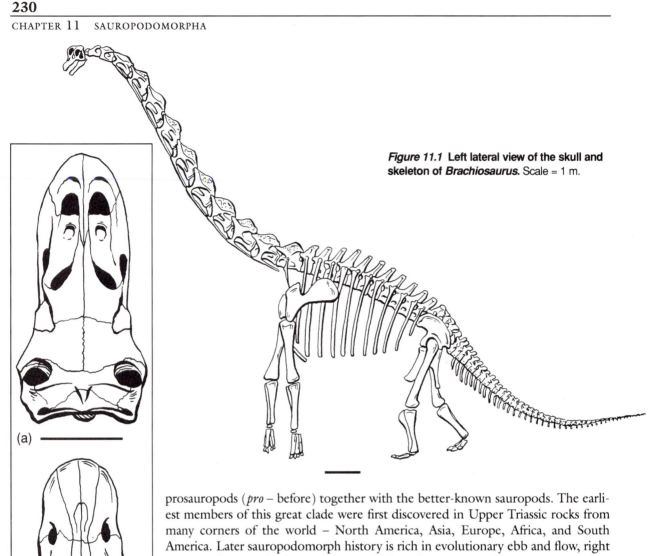

Figure 11.1 Left lateral view of the skull and skeleton of *Brachiosaurus.* Scale = 1 m.

Figure 11.2 Dorsal view of the skulls of (a) Brachiosaurus and (b) Diplodocus. Arrow points to dorsally placed external nares in *Diplodocus.* Scale = 10 cm.

prosauropods (*pro* – before) together with the better-known sauropods. The earliest members of this great clade were first discovered in Upper Triassic rocks from many corners of the world – North America, Asia, Europe, Africa, and South America. Later sauropodomorph history is rich in evolutionary ebb and flow, right up to the terminal moments of the Cretaceous. By some accounts, sauropodomorphs spawned well over one hundred different species over this interval – no small evolutionary achievement.

HISTORY OF THE STUDY OF SAUROPODOMORPHA

19th Century

The first discovered sauropodomorph was *Thecodontosaurus* (*theke* – socket; *dont* – tooth) from Great Britain, described in 1836. A year later, H. von Meyer described what has become the best known of all early sauropodomorphs, *Plateosaurus* (*plateos* – flat). Originally discovered in northern Bavaria, *Plateosaurus* (Figure 11.3) is now known throughout western and central Europe. One particular place holds special importance for this dinosaur: the southern German town of Trossingen. Otherwise famous for the manufacture of Hohner harmonicas (attention, all you blues harp players!), from the 1910s to the 1930s

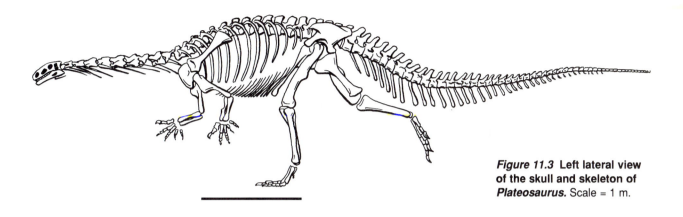

Figure 11.3 Left lateral view of the skull and skeleton of *Plateosaurus*. Scale = 1 m.

Trossingen produced one of the largest mass accumulations of *Plateosaurus* or any other sauropodomorph. In 1841, the year before he coined the name Dinosauria, Sir Richard Owen announced *Cetiosaurus* (*keteios* – whalelike), which he took to be an exceptionally large crocodile.

Yet again, it was the material from the Western Interior of the United States that cut the Gordian knot of sauropod anatomy and evolution, and yet again the two major figures in these discoveries were E. D. Cope of Philadelphia and O. C. Marsh of New Haven (Box 6.1). Among the sauropods that resulted from their rivalry were *Camarasaurus* (*kamara* – chamber; Figures 11.5c and 11.14), *Apatosaurus* (*apato* – trick or false; so named because the tail bones appeared to Marsh more like those of a lizard than a dinosaur; Figure 11.4), and *Diplodocus* (*diplo* – two; *docus* – spar or beam; Figures 11.2b, 11.5d, and 11.15).

Marsh also worked closer to his own backyard of New England. In 1885, he announced the prosauropod *Anchisaurus* (*anchi* – near; Figure 11.6a), and another – *Ammosaurus* (*ammos* – sand) – in 1891. Both are known from the northeastern United States (with an additional later record of *Ammosaurus* from Arizona).

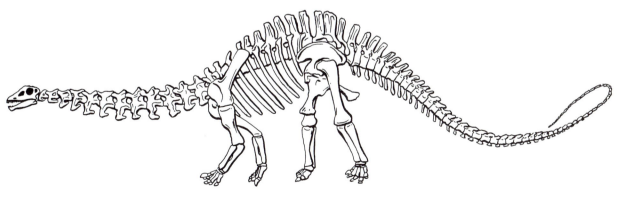

Figure 11.4 Left lateral view of the skull and skeleton of *Apatosaurus*. Scale = 1 m.

20th Century

The beginning of the 20th century was marked by continued work in the Western Interior of the United States, particularly by the American Museum of Natural History (New York) at Como Bluff and nearby Bone Cabin Quarry in Wyoming, the Field Museum of Natural History (Chicago) in western Colorado, and the Carnegie Museum of Natural History (Pittsburgh) in Wyoming, Colorado, and especially Utah. The American Museum and Carnegie Museum[1] efforts gathered one of the greatest collections of sauropods on earth. The best of the bounty includes several sauropod skeletons that have been mounted for exhibition. Go to see these staggeringly huge yet majestic skeletons if you can – *Apatosaurus* at the American Museum,[2] the Carnegie Museum, the Field Museum, the Yale Peabody Museum of Natural History (New Haven), and the University of Wyoming (Laramie); *Camarasaurus* in the Carnegie Museum and in the National Museum of Natural History (Smithsonian Institution) in Washington, D.C.; and *Diplodocus* in the Carnegie Museum, the National Museum, and Denver Museum of Natural History. Equally important to our story is *Brachiosaurus* (*brachion* – arm; Figures 11.1, 11.2a, and 11.5b) described in 1903. This sauropod, also from the Morrison Formation of Colorado, is now known from elsewhere in the western United States and, more importantly, from along the eastern coast of Africa.

Inland from the seacoast of Tanzania, in what was then German East Africa, at a place called Tendaguru Hills, huge quantities of sauropod bones, along with those of other dinosaurs, were being unearthed in what was one of the most spectacular dinosaur expeditions that ever took place (Box 11.1). When the dust from the digging had settled, the Tendaguru expeditions claimed two new sauropods (*Tornieria* [named in honor of German paleontologist G. Tornier] and *Dicraeosaurus* [*dikraios* – bifurcated]), as well as new material of *Barosaurus* and the finest specimen of *Brachiosaurus* ever found – now mounted and peering into the fourth floor balcony of the Humboldt Museum für Naturkunde.

Sauropodomorph discoveries dribbled in from the 1920s until the close of the 1930s. The most important was *Alamosaurus* – named for the Ojo Alamo Formation of New Mexico – from the Late Cretaceous of the southwestern United States (Figure 11.17).

From the late 1930s onward, a vast array of new sauropodomorphs was described from Asia. C.-C. Young (Institute of Vertebrate Paleontology and Paleoanthropology [IVPP] in Beijing) began the avalanche with his description of *Omeisaurus* (named for Mt. Emei, a sacred mountain in Sichuan Province), a basal sauropod. Shortly thereafter, there were two basal sauropodomorphs: the Early Jurassic *Lufengosaurus* (named for Lu-Feng, a locality in Yunnan Province; Figure 11.6c) and the contemporary *Yunnanosaurus* (for Yunnan Province, where this dinosaur was discovered; Figure 11.6d). And in 1954, Young described the remarkable, long-necked *Mamenchisaurus* (named for Manmenchi Ferry at Jinshajiang, Sichuan Province; Figure 11.16).

Although new sauropodomorph specimens continued to be uncovered

---

[1] One particularly rich quarry was discovered by E. Douglass in 1909 and worked successively by the Carnegie Museum, the National Museum of Natural History (Smithsonian Institution), and the University of Utah. It is now run by the United States Park Service and is famous as Dinosaur National Monument.

[2] This most famous of museums recently eclipsed its own mount of *Apatosaurus* with an exciting, free-standing mount of a *Barosaurus* mother rearing up to protect its baby against a marauding *Allosaurus*. The head of the mother reaches up five-and-one-half stories into the air (Figure 1.7)!

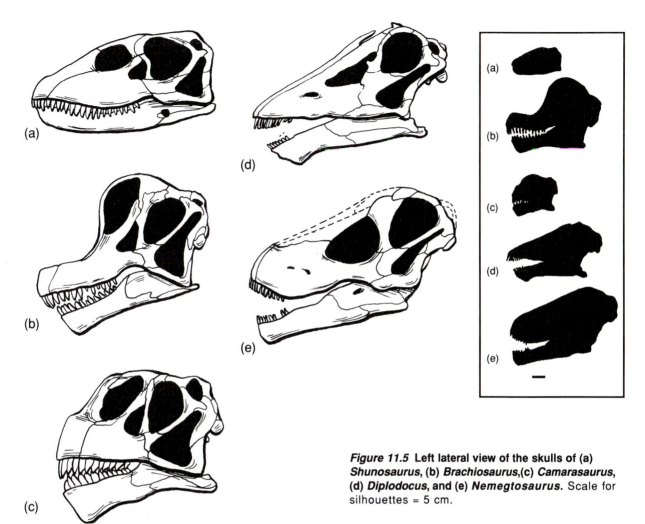

Figure 11.5 Left lateral view of the skulls of (a) *Shunosaurus*, (b) *Brachiosaurus*, (c) *Camarasaurus*, (d) *Diplodocus*, and (e) *Nemegtosaurus*. Scale for silhouettes = 5 cm.

throughout the mid-century, it was not until the late 1960s and into the 1980s that new kinds again began to be discovered in abundance, this time in South America. Initially there were *Riojasaurus* (for Rioja Province), *Coloradisaurus* (named for the Los Colorados Formation from which it was recovered; 11.6b), and *Mussaurus* (*mus* – mouse, known from one of the smallest articulated skeletons, albeit a hatchling; Figure 11.7), all three basal sauropodomorphs from the Late Triassic described by Argentinian paleontologist J. Bonaparte and colleagues. Then *Patagosaurus* (named for the Patagonia region of Argentina), *Saltasaurus* (named for Salta Province, Argentina, where it was found) and *Aeolosaurus* (named for Aeolus, Greek god of the winds, and also alluding to the windiness of Patagonia), were recovered, leading to the inference that sauropods were important members of Gondwanan faunas.

Several thousand kilometers away to the west, discoveries of two new sauropods, *Nemegtosaurus* (named for the Nemegt Formation of Mongolia; Figure 11.5e) and *Opisthocoelicaudia* (*opistho* – hind; *coel* – hollow; *caud* – tail; the name acknowledges hollow spaces in the tail), were made in Mongolia. And farther still to the west, two new basal sauropods – *Vulcanodon* (*vulcan* – volcano) from Zimbabwe and *Barapasaurus* (*bara* – big; *pa* – leg; based on several Indian languages) from India – have together provided a great deal of insight into the base of sauropod phylogeny.

The 1980s nevertheless belonged to central and eastern Asia and to the United States. Beginning with a new sauropod from Mongolia named *Quaesitosaurus* (*quaesitus* – abnormal or uncommon), the story thereafter returns to China and the work of Dong Z.-M. and colleagues from the IVPP: *Shunosaurus* (from Shuo, an old name for the Sichuan region of China; Figures 11.5A and 11.8), and *Datousaurus* (from Malay: *datou* – chieftain), which come

BOX 11.1

TENDAGURU!

THE NAME STRIKES like the boldness of a movie marquee: the hinterland of Tanzania on the eastern coast of Africa, where prides of lions and exotic herbivores slink or gallop across the wilderness. This spot, which today is monotonously formed of broad plateaus blanketed by dense thorn trees and tall grass thick with tsetse flies, was formerly the site of perhaps the greatest paleontological expedition ever assembled, and millennia before that, the place where dinosaurs came to die.

Let's go back to 1907, when Tanzania was part of German East Africa. This was the era of massive western European colonialism in Africa. With the widespread colonialism came scientists. And to then German East Africa came paleontologists in search of fossils.

The fossil wealth of Tendaguru was first discovered in 1907 by an engineer working for the Lindi Prospecting Company. Word spread quickly, ultimately to Professor E. Fraas, a vertebrate paleontologist from the Staatliches Museum für Naturkunde in Stuttgart, who happened to be visiting the region. So excited was he at the prospect of collecting dinosaurs after his visit to Tendaguru that he took specimens back to Stuttgart (including what was eventually to be called *Janenschia*) and, more especially, started drumming up interest among other German researchers to continue field work in the area.

It was W. Branca, director of the Humboldt

Figure Box 11.1 **Werner Janensch.** The driving force behind the extraordinarily successful excavations at Tendaguru, Tanzania.

from the Middle Jurassic of Sichuan Province, China. Likewise, in the United States, discoveries of new sauropods were again coming into prominence in the 1980s, thanks in large part to the work of Brigham Young University paleontologist J. A. Jensen. Working in the Upper Jurassic Morrison Formation of western Colorado, Jensen announced three new sauropods in 1985, an unprecedented number since the heyday of the late 1800s and early 1900s: *Supersaurus* (*super* – above), *Ultrasauros* (*ultra* – beyond), and *Dystylosaurus* (*di* – two; *stylos* – beam), a surprising number of new sauropods from essentially one place and one time!

The 1990s are barely half over, so who knows what sauropodomorphs await

Museum für Naturkunde in Berlin, who was the first to seize upon the opportunity presented to him by Fraas. By seeking support from a great many sources, he received more than 200,000 marks – a fortune for the time – from the Akademie der Wissenschaften in Berlin, the Gesellschaft Naturforschender Freunde, the city of Berlin, the German Imperial Government, and almost a hundred private citizens.

With money, material, and supplies in hand, the Humboldt Museum expedition set off for Tendaguru in 1909. For the next four field seasons, it was bonanza time. Under the leadership of mustachioed and jaunty W. Janensch (Figure Box 11.1) for three of these seasons (H. Reck took charge in the fourth season), the expedition in these years saw possibly the greatest dinosaur collecting effort in the history of paleontology. The first season involved nearly 200 workers, mostly natives, laboring in the hot sun as they dug huge bones out of the ground. During the second season, there were 400 workers and in the third and fourth seasons, 500 workers. By the end of the expedition's efforts, some 10 km² of area was covered with huge pits, attesting to the diligence and hard work of these laborers.

Many of these native workers brought their families with them, transforming the dinosaur quarries at Tendaguru into a populous village of upwards of 900 people. With all these people, water and food was a severe problem. Not available locally, water had to be brought in, carried on the heads and backs of porters. And with the vast quantities of food that had to be obtained for workers

and their families, it is not surprising that the funds amassed by Branca speedily disappeared.

Still, the rewards were great indeed. Over the first three seasons, some 4,300 jackets were carried back to the seaport of Lindi. A four-day walk away, this trip was made 5,400 times by native workers, each with the fossils balanced on his or her head and back, all to be shipped from there to Berlin.

Overall, work at Tendaguru involved 225,000 man-days and yielded nearly a hundred articulated skeletons and hundreds of isolated bones. When finally unpacked and studied, what a treasure-trove: In addition to ornithischians (*Kentrosaurus*, *Dryosaurus*) and theropods (*Elaphrosaurus*), and a pterosaur as well, the Tendaguru expeditions claimed not only two new kinds of sauropod (*Tornieria* and *Dicraeosaurus*), but also new material of *Barosaurus* and *Brachiosaurus*.

The Humboldt Museum never went back to Tendaguru after 1912. In 1914, World War I erupted and, with the Treaty of Versailles, German East Africa became British East Africa. This shift in the continuation of European colonialism brought British paleontologists to Tendaguru in 1924, under the direction of W. E. Cutler. This team from the British Museum (Natural History) hoped to enlarge the quarried area and retrieve some of the leftover spoils from the German effort. From 1924 to 1929, the British expedition had its ups and downs, finding more of the kinds of dinosaurs discovered earlier, but suffering severe health problems, including malaria, from which Cutler died in 1925. There has been no significant paleontological effort at Tendaguru since.

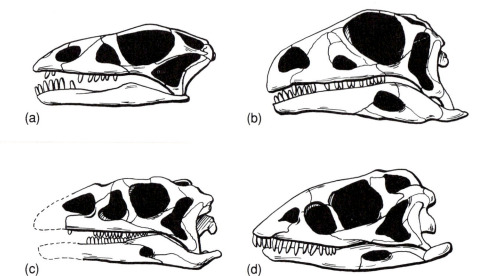

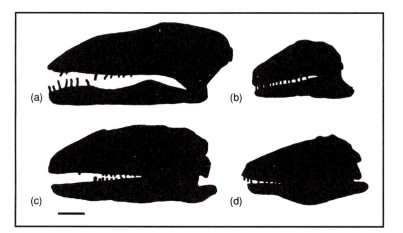

Figure 11.6 Left lateral view of the skulls of (a) *Anchisaurus*, (b) *Coloradisaurus*, (c) *Lufengosaurus*, and (d) *Yunnanosaurus*. Scale for silhouettes = 5 cm.

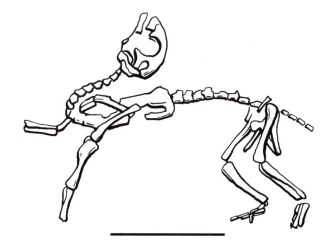

Figure 11.7 *Mussaurus*, the only sauropodomorph known from a hatchling. Scale = 5 cm.

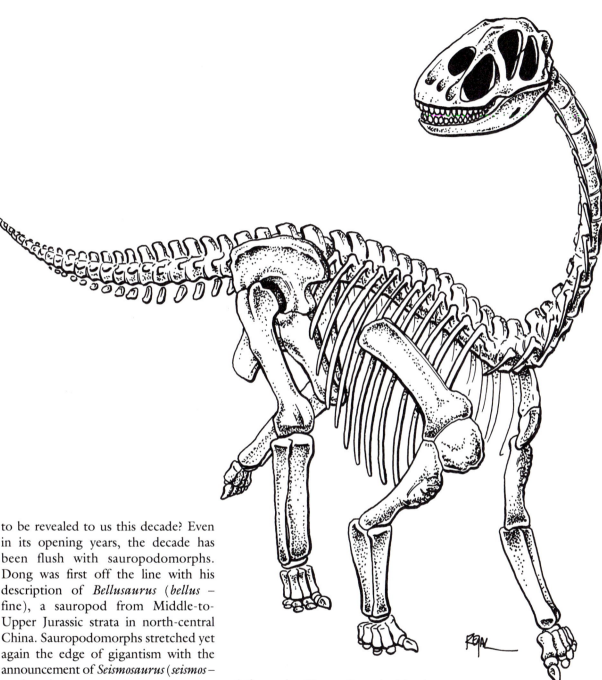

to be revealed to us this decade? Even in its opening years, the decade has been flush with sauropodomorphs. Dong was first off the line with his description of *Bellusaurus* (*bellus* – fine), a sauropod from Middle-to-Upper Jurassic strata in north-central China. Sauropodomorphs stretched yet again the edge of gigantism with the announcement of *Seismosaurus* (*seismos* – earthquake), a new, gargantuan sauropod from the Upper Jurassic Morrison Formation of New Mexico described by D. D. Gillette (Division of State History, Salt Lake City, Utah). This dinosaur, thought to be between 39 and 50 m long, is the largest animal yet known to have walked on land. The same year saw the descriptions of two new mid-to-Late Cretaceous Argentinian sauropods,

Figure 11.8 **An ambling *Shunosaurus.***

Andesaurus (named for the Andes Mountains of Argentina) and *Amargasaurus* (named for Amarga Canyon in Neuquen Province, Argentina), the latter a bizarre form with exceedingly long bifurcated neural spines.

SAUROPODOMORPHA DEFINED AND DIAGNOSED

As now should be clearly plain, Sauropodomorpha is a very diverse and long-lived clade. Sauropodomorpha consists of all the descendants of the most recent common ancestor of *Thecodontosaurus* and sauropods like *Brachiosaurus*. Sauropodomorpha is easily diagnosed by more than a dozen derived features (Figure 11.9).

Sauropodomorpha shares a close phylogenetic relationship with Theropoda (and Segnosauria?), all together forming the monophyletic clade known as Saurischia (see Part III: Saurischia).

SAUROPODOMORPH DIVERSITY AND PHYLOGENY

In recent years, dinosaur paleontologists – principally J. A. Gauthier and M. J. Benton – have argued that basal sauropodomorphs, the so-called prosauropods, are not monophyletic but instead have successively closer relationships with members of Sauropoda, universally regarded as monophyletic (Figure 11.10). This position has held sway since the mid-1980s, challenged only by two attempts – by P. C. Sereno and by P. M. Galton – at resurrecting Prosauropoda as a natural group (Figure 11.11). We present both viewpoints here. Whether a monophyletic Prosauropoda will survive further scrutiny awaits time and further study.

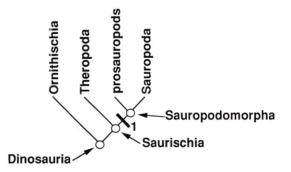

Figure 11.9 Cladogram of Dinosauria emphasizing the monophyly of Sauropodomorpha. Derived characters at 1: relatively small skull, deflected front end of the lower jaw, lanceolate teeth with coarsely serrated crowns, at least 10 neck vertebrae that form a very long neck, dorsal and caudal vertebrae added to the front and the hind end of the sacrum, enormous thumb equipped with an enlarged claw, and very large foramen (the obturator foramen) in the pubis.

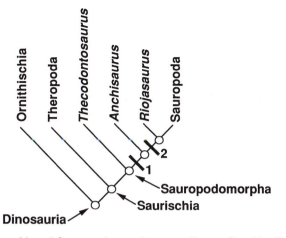

Figure 11.10 Cladogram of basal Sauropodomorpha according to Gauthier. Derived characters at 1: very robust digit I on the hand, wide-based neural spines on the front tail vertebrae, arched upper margin of the ilium, and a completely open acetabulum; at 2: compressed upper margin of the external nares, jaw joint placed well below the level of the tooth rows, strongly built forelimbs, and broad foot.

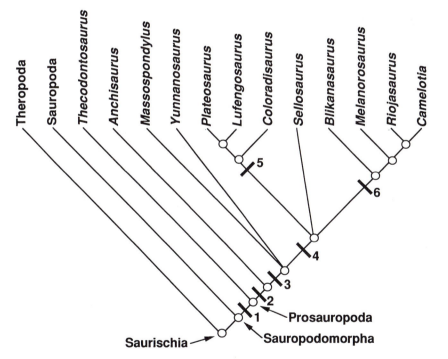

Figure 11.11 Cladogram of a monophyletic Prosauropoda according to work by Sereno and by Galton. Derived characters at 1: enlargement of one of the wrist bones, elongation of the claw on digit I of the hand, asymmetry of the joint at the end of the first bone of digit I of the hand, and shaft of the ischium subtriangular in cross section; at 2: enlargement of the external nares, decrease in size of the antorbital opening, elongation of the upper and lower jaws, and elongation of the trunk region; at 3: vertical or backwardly sloping quadrate bone, very wide and apronlike pubis, and very strongly built hands and feet; at 4: pronounced ventral offset of the jaw articulation well below the tooth rows; at 5: ventral offset in the braincase so that the occipital condyle is well below the front of the braincase; at 6: increase in the length of the trunk region.

Sauropoda

The monophyletic status of Sauropoda has never been questioned and is well supported by more than 30 derived features that unite the clade (Figure 11.12, node 1).

Despite being so well supported, the internal relationships of sauropod taxa are anything but fully resolved. The scheme that we furnish here, based upon a cladistic treatment by Gauthier, should certainly not be regarded as definitive.

In Figure 11.12, there is a sequence of isolated primitive sauropod taxa climb-

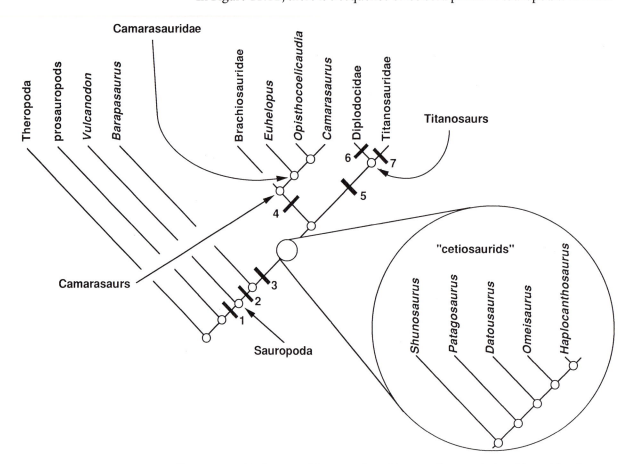

Figure 11.12 **Cladogram of Sauropoda, with more distant relationships with other sauropodomorphs and Theropoda.** Derived characters at 1: shortening of the rear portion of the skull with the lower temporal opening situated partly beneath the eye socket, reduced postorbital bone, 12 or more neck vertebrae, 4 or more sacral vertebrae, massive and vertical limbs, and solid long bones; at 2: pleurocoels, wide pelvis, and a well-defined lesser trochanter; at 3: contact between the maxilla and quadratojugal to exclude the jugal from the ventral margin of the cheek region, 5 sacral vertebrae, well-developed pleurocoels, loss of external mandibular fenestra, and vertical metacarpals; at 4: snout sharply demarcated from the rest of the skull, relatively deep contact between the pubis and ischium, and ischium twisted to become more horizontal at its end; at 5: long broad snout, elevation of the bones of the face to situate the nostrils on top of the head, elongate pencil-like teeth at the very front of the jaws, incorporation of 3 or more trunk vertebrae into the neck, and very long tail; at 6: skid-shaped hemal arches beneath the tail vertebrae, deeply cleft V-shaped neural spines in the shoulder region, and ischia that are expanded at their ends; at 7: presence of body armor, and tail vertebrae with hollow front surfaces and ball-like hind ends.

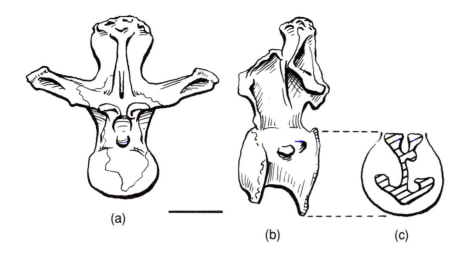

Figure 11.13 (a) Front view and (b) left lateral view of one of the back vertebrae of *Brachiosaurus*, with pleurocoels indicated in (c) cross section. Scale = 20 cm.

ing up the cladogram, with a relatively bushy crown consisting of camarasaurs and titanosaurs. The most primitive member in Gauthier's phylogeny of the group (the first to branch from the base of the cladogram) is *Barapasaurus*. On the sides of its vertebrae are well-marked depressions, not yet fully developed as large hollow spaces, or **pleurocoels**, that are so characteristic of many later sauropods (Figure 11.13). Interestingly, the isolated teeth found with *Barapasaurus* skeletal material (and therefore thought to belong to it) have coarse serrations on both edges, reminiscent of those of more basal sauropodomorphs.

Although it was not analyzed by Gauthier, there may be another, more primitive form than *Barapasaurus* within the clade. *Vulcanodon* (Early Jurassic of Zimbabwe), a medium-sized (6.5 m) form, shares all of the characters that diagnose Sauropoda. However, it lacks true pleurocoels and a wide pelvis, features that characterize remaining members of the clade, including *Barapasaurus* (Figure 11.12, node 2).

"CETIOSAURIDAE." A variety of sauropods commonly linked as "Cetiosauridae" constitutes the next batch of these massive herbivores up the tree (Figure 11.12, node 3). "Cetiosaurids" contain a wide array of forms (16 species by some counts), but it is commonly agreed that they are not monophyletic (hence the quotation marks). Among these sauropods, we're guessing that *Shunosaurus*, *Patagosaurus*, *Datousaurus*, *Omeisaurus*, and *Haplocanthosaurus* are successively more closely related to "higher" sauropods.

Shunosaurus, a 9-m-long sauropod from China, is known from several nearly complete skeletons. Its skull is relatively long and low, vaguely reminiscent of the more primitive condition (nostrils in front of the snout, many small and spatulate teeth). *Shunosaurus* likewise retains a number of primitive features on behind the head, including the absence of a lesser trochanter on the femur.

Patagosaurus had broad teeth in its jaws and relatively low-placed nostrils. As in other primitive sauropods, *Patagosaurus* lacks true pleurocoels, but has better-developed depressions on the sides of the vertebrae.

Datousaurus possesses a high, yet short, skull, with a snout reminiscent of the condition in *Camarasaurus*. The massive jaws contain fewer teeth; these are of the large and spoon-shaped variety. Similarly, *Omeisaurus* has a short, high skull with a somewhat more reduced dentition, and the number of neck vertebrae is increased over more primitive sauropods. Interestingly, here for the first time the neck vertebrae contain the well-developed pleurocoels so characteristic of many sauropods.

Haplocanthosaurus, from the western United States, may be close to the likes of *Camarasaurus* and *Diplodocus* in having a longer neck (that is, more vertebrae) and very well developed pleurocoels.

Camarasaurs and Titanosaurs

The majority of sauropods clump at the top of the family tree. This great clade, which is not yet named, is formed – according to Gauthier – of the camarasaurs and titanosaurs.

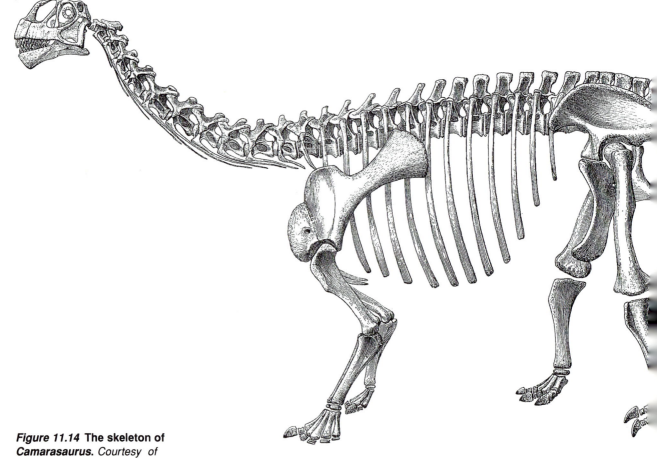

Figure 11.14 **The skeleton of *Camarasaurus*.** *Courtesy of Carnegie Museum of Natural History.* Scale = 3 m.

CAMARASAURS. Camarasaurs, one of the two descendent clades within this sauropod bush, appear to be united by five derived features (Figure 11.12, node 4). This group has a vast array of members that appear to be grouped into two smaller clades: Camarasauridae and Brachiosauridae. Best known among camarasaurids (a clade united by having important modifications of the cheek region) is *Camarasaurus* itself, the most common of all North American sauropods (Figure 11.14). Perched on the end of its 18-m-long body, its head was short, high, and powerfully built, and equipped with fewer teeth than we've yet seen in sauropods. Its vertebrae had very deep pleurocoels and, dorsally, the neural spines at the back of the neck were divided by a U-shaped cleft.

Other camarasaurids include *Euhelopus* and *Opisthocoelicaudia*. The former has only a slight bifurcation of the neural spines, whereas – like *Camarasaurus* – *Opisthocoelicaudia* has a deep cleft, although the latter is otherwise a very strange sauropod (six sacral vertebrae, very short tail, massive and relatively short forelimbs, greatly reduced foot).

Brachiosaurids, those sauropods with elevated nasal bones (further positioning the nostrils on a protrusion at the top of the head) and elongate forelimbs, are by comparison to camarasaurids must less well known. Consequently, it is fortunate that we have *Brachiosaurus* (Figure 11.1). Among the largest of all sauropods (measuring 22.5 m long and weighing in excess of 50,000 to 60,000 kg), *Brachiosaurus* captured several decades' worth of people's imaginations as the largest land-living animal of all time. Now supplanted by the likes of *Supersaurus*,

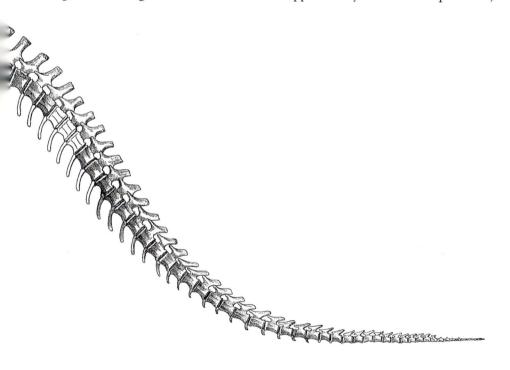

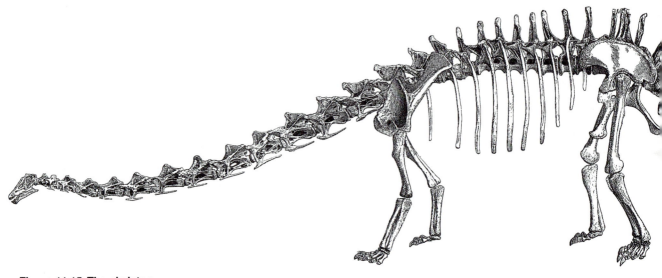

Figure 11.15 The skeleton of Diplodocus. *Courtesy of Carnegie Museum of Natural History.* Scale = 1 m.

Ultrasauros (another brachiosaurid) and *Seismosaurus*, *Brachiosaurus* is still by far the best known of all of these earthly giants.

TITANOSAURS. According to Gauthier, the other great clade of derived sauropods is the titanosaurs, united by more than a dozen derived features (Figure 11.12, node 5). The better known of the two titanosaur groups by far is the diplodocids (Figure 11.12, node 6). Completely known down to its toe bones (including new evidence of spiky skin, thanks to the work of S. A. Czerkas), the 27-m-long *Diplodocus* (Figure 11.15) has a very giraffelike skull, with an elongate snout that houses pencil-like teeth along its very front margin. The nostrils are located on the very top of the skull (see Figure 11.2b). And the tail is drawn out into a series of long cylindrical vertebrae that may have served as a kind of whip in defense against predators.

Other diplodocids are slightly less impressive in size, but no less intriguing. The 21-m-long *Apatosaurus* is known best for what it is not: "*Brontosaurus*" (Box 11.2). At about 25 m in length, *Barosaurus* hails from the western United States and from Tanzania – as does *Brachiosaurus* – providing important information on intercontinental distribution patterns among dinosaurs. Also at Tendaguru, living alongside *Barosaurus*, was *Dicraeosaurus*. The pleurocoels of this much smaller (12 m long), and relatively short-necked, diplodocid were not particularly well developed; both conditions may be primitive within the group.

Prized among sauropods, principally because of their rarity, are skulls.[3] Thus, it is especially important that two Late Cretaceous diplodocids (both from Mongolia) are known from skulls: *Nemegtosaurus* and *Quaesitosaurus*. Both are similarly long snouted, with peglike teeth restricted to the front of the jaws.

Finally in our coverage of diplodocids, we come to *Mamenchisaurus* (Figure 11.16). What stands out in this aberrant member of an already singu-

[3]After death, sauropod skulls were apparently some of the first elements to drift away with the currents, which carried sediments that buried the corpses.

lar group of dinosaurs is its very long neck. Making up nearly half of its 22-m body length, this neck gives the impression of an animal imminently in danger of tipping forward. It is remarkable and puzzling that the pleurocoels in the neck vertebrae – which would help to lighten this immense boomlike structure – are poorly developed.

At last we turn to Titanosauridae (called "antarctosaurs" by Gauthier). By far the poorest known of all sauropods, titanosaurids nevertheless may be united by several derived features (Figure 11.12, node 7). There are many named titanosaurids, but we'll concentrate on just three of the best known.

Figure 11.16 A reconstruction of *Mamenchisaurus*, the Late Jurassic sauropod from Sichuan, China. *Photo by Jim Tinios courtesy of the Ex Terra Foundation.*

Figure 11.17 Alamosaurus. What is known of this titanosaurid is shaded in black. *Reconstruction courtesy of Spencer A. Lucas, New Mexico Museum of Natural History.* Scale = 1 m.

Titanosaurus, the namesake of the group, is best known from the Late Cretaceous of India, but it is also reported from France, Spain, Madagascar, and as far away as Laos, in southeastern Asia. What we know of *Titanosaurus* are its tail vertebrae and most of the limb bones, all indicating a 12-m-long sauropod. *Saltasaurus* is known from disarticulated and somewhat incomplete material – a continuing problem for titanosaurids. However, this relatively small (12 m long), stocky-limbed form was covered with a pavement of globular and buttonlike osteoderms. *Alamosaurus* (Figure 11.17) is perhaps the best known of any member within the Titanosauridae: We have recovered a quarter of an articulated skeleton! From this and other material, it is clear that *Alamosaurus* was quite a large animal, probably measuring up to 21 m in length. A distinctive aspect of the animal is that it had relatively long forelimbs, unlike other titanosaurs and more like brachiosaurids.

SAUROPODOMORPH PALEOBIOLOGY AND PALEOECOLOGY:
Lifestyles of the Huge and Ancient

Distribution and Preservation

Let's start this section by reviewing what we know about sauropodomorph distribution (Figure 11.18). For animals the size of sauropods, you'd expect that their bones would be easily found. And to an extent, that appears to be true. These animals have a fossil record that extends from the very beginning of dinosaur history (Late Triassic) until its denouement (the close of the Cretaceous). Over this long interval, sauropodomorphs managed to walk or to be carried (on continental plates) to nearly all corners of the world except Antarctica, and they may be found there yet (see Chapter 15).

Still, the record of individual sauropodomorphs can be very poor. Except for

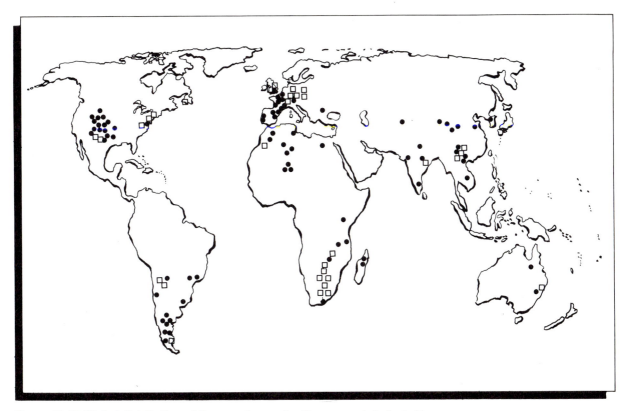

Figure 11.18 **Global distribution of Sauropodomorpha.** Prosauropods indicated by squares; Sauropoda indicated by circles.

a few basal forms (some of which are known from complete skeletons), most sauropodomorphs are known from incompletely preserved material, often missing their heads, parts of their tails, and their feet. University of Pennsylvania paleontologist P. Dodson has noted this enigma: Why are animals as large and presumably durable as sauropodomorphs not better preserved in the fossil record? In answer, he has suggested that – even in animals as large as the largest sauropods – preservation of a complete skeleton is balanced against the vagaries of both sedimentation and erosion. Perhaps many of the sauropodomorphs were just too large to be easily buried. It is one thing to bury a rat and quite another to bury a whale! This matches the observation that those body parts most likely to be lost before burial – that is, the tail, feet, and skull – are just the elements commonly missing when sauropodomorphs are collected.

Habits

Despite such problems of preservation, when we do find the remains of these magnificent animals, they come from a myriad of different depositional environments. All sauropodomorphs are found in a wide array of sediments, from fine-grained floodplain sediments to channel and even eolian sandstones. In the Upper

Jurassic Morrison Formation of the Western Interior of the United States, where sauropods are known in great abundance and where their sedimentary context has been carefully studied by Dodson, A. K. Behrensmeyer, R. T. Bakker, and J. S. McIntosh, sauropod remains are distributed across a broad spectrum of sedimentary environments, including those interpreted as lakes, rivers, and floodplains. By comparison with the distribution of the carcasses of large terrestrial mammals such as elephants and rhinos living today in places like the Serengeti of eastern Africa, we see that sauropods were similarly able to range across all available habitats found in the ancient landscapes represented by the Morrison Formation.

In addition to catholic utilization of living space, these animals may have had to cope with long dry seasons during the year. Sedimentological and paleontological evidence suggests that the ancient environment of the Morrison Formation was subject to seasonal dryness. Such annual droughts may have been severe enough to have forced sauropods into large-scale migration to more lush areas. We will have more to say about migration when considering sauropod herding behavior.

Was seasonal dryness typical of sauropod habitats? At Tendaguru in southeastern Tanzania, as well as in northern Texas at the famous sauropod trackway sites and in Maryland where the remains of the brachiosaurid *Astrodon* have been uncovered, there is strong geological evidence that these environments were once close to the sea and potentially quite humid. Perhaps these were some of the conditions that sauropods found most congenial.

IN THE SWAMPS? But wait – doesn't everybody know that sauropods frequented lakes and deep rivers, where their great bulk was buoyed up by the water? Such an idea may have originated as a consequence of Owen's mistaken notion that *Cetiosaurus* was an exceedingly large marine crocodile. It was argued that sauropods were too big and heavy to have lived on land; that their legs would not have supported them, and that, like whales, they needed the buoyancy of water to permit them to breathe. It was argued that the long tails were used for swimming. Finally, it was argued that the nostrils at the top of the skull were like the snorkel on a submarine, allowing the animal to breathe while remaining fully submerged (and presumably hidden and protected). Although E. S. Riggs, dinosaur paleontologist at the Field Museum of Natural History (Chicago), argued in 1904 that these animals were terrestrial, sauropods have persistently been viewed as having a fully aquatic existence.

Despite its stubborn longevity, the idea of aquatic sauropods is unsupported. In 1951, K. A. Kermack examined the barometric consequences of the lungs of a submerged sauropod. The air that these animals had to breathe would obviously be at atmospheric pressure (i.e., 1 atmosphere of pressure). However, because the **thorax** (in vertebrates, the part of the body between the neck and stomach) – and hence the lungs – of a submerged sauropod would be under a column of water some 6 m or more in depth, this part of the sauropod's respiratory anatomy would be under nearly 2 atmospheres (water pressure goes up 1 atmosphere of pressure every 9.8 m). This pressure would tend to collapse the thorax, pushing whatever air was in the lungs up and out of the body. How the next breath might be taken is hard to say, because the lungs would have to be expanded against high pressures, well beyond any experienced in vertebrates today. Unless sauropods had exceedingly powerful chest muscles, these animals certainly would have been unable to breathe in. For this reason, Kermack argued, it is perhaps better to envision sauropods as terrestrial animals.

Sauropods clambered out of the water for good with the studies of Bakker (then at Yale University) in the late 1960s and 1970s. This research emphasized the sturdy, pillarlike construction of the legs and feet of sauropods, surely strong enough for walking on land. Additionally, Bakker pointed to the narrow, slab-sided thorax – very unlike the amphibious hippopotamus, but like rhinos and elephants – as evidence of terrestrial habits.

So what were the long necks for, if not for keeping heads above water? From a land-based perspective, such necks made it possible for these animals to feed in tall trees. Animals like *Brachiosaurus* may have fed much like a giraffe. Not only was the neck very long, but the front limbs were longer than those in the rear (Figure 11.1). With this "extra boost," the head could apparently be raised to a height of 13 m, providing the opportunity to feed on foliage to which virtually no one else (except maybe tripodal diplodocids – as we will discuss) had access.

HIGH BLOOD PRESSURE. Animals like *Brachiosaurus* had to pay a price for looking down on the crowd. Now that the head was perched so high, the brain (the smallest – relative to body size – among dinosaurs, although certainly of the utmost importance) would have towered about 8 m above the heart. In contrast to the condition noted by Kermack for a submerged sauropod, under terrestrial conditions, the heart would have to pump blood at very high pressure to get it to the brain. Otherwise, the animal would have no recourse but to faint when it lifted its head!

What sort of blood pressure are we talking about to get blood pumped from the heart to the head of a sauropod? Recall that when you go to the doctor and your blood pressure is taken, it comes in two numbers, both of which are measures of pressure in terms of mm of mercury, abbreviated Hg. The first, the systolic pressure (usually around 120 to 130 [mm of Hg] in healthy people), indicates the pressure your blood is experiencing when the heart contracts. It is during this part of heart activity that blood is shot out to the extremities of the body; hence, it is at its highest pressure. In contrast, during heart relaxation, blood pressure is at its minimum. In healthy people, this diastolic pressure ranges between 70 and 80 (mm of Hg).

Keeping these conditions in mind, we can estimate what kinds of systolic pressures would be necessary to get blood from the heart to the head in a full-grown *Brachiosaurus*. To push blood through the arteries up its 8.5-m-long neck (or thought of another way, to counter the pressure of an 8.5-m column of blood), the heart of a *Brachiosaurus* must pump with a pressure exceeding 629 mm of Hg. Such blood pressures would have been much larger than those in any living animal. Most mammals – including ourselves – have pressures of 110 to 150 mm of Hg and even giraffes have only about 320 mm of Hg. It would indeed take a very muscular heart – some scientists estimate one weighing as much as 400 kg – to do the pumping (Figure 11.19)! And if giraffes are any measure, there may have been other features of anatomy – muscularized blood vessels with one-way valves – to force blood to the head and to keep it from flowing out again.

Other sauropods appear to have been less giraffelike than *Brachiosaurus* in their body postures. For example, animals like *Diplodocus* and *Nemegtosaurus* had shorter forelimbs than did *Brachiosaurus*, many also had much longer tails. Shorter forelimbs and longer tails have inspired two different interpretations of posture that would have had an effect on blood pressure and heart size.

The first came initially from J. B. Hatcher in 1901 and was revived by Bakker:

BOX 11.2

BRONTOSAURUS:
Beheading a Wrong-Headed Dinosaur

WITH THE DISCOVERIES OF DINOSAURS in the Western Interior of the United States during the late 19th century, boxcar-loads of brand-new, but often incomplete, sauropod skeletons were shipped back East to places like New Haven and Philadelphia. It was Yale's O. C. Marsh who described one of these new sauropods as *Apatosaurus* in 1877. With further shipments of specimens and more studies, Marsh again named a "new" sauropod in 1879 – *Brontosaurus*.

Years went by and – thanks to the burgeoning popularity of many kinds of dinosaurs – the public came to know the name *Brontosaurus* much better than it did the earlier-discovered *Apatosaurus*. Nevertheless, there was the suspicion by many sauropod researchers that *Apatosaurus* and *Brontosaurus* were the same kind of sauropod. In fact, this case was made in 1903 by E. S. Riggs. Since then, most sauropod workers have regarded *Brontosaurus* as synonymous with *Apatosaurus*. If *Apatosaurus* and *Brontosaurus* are two names for the same sauropod, the older name, *Apatosaurus*, should be applied to this Late Jurassic giant.

But the more interesting story is not in the names, but in the heads. Again we go back to Marsh. Lamenting in 1883 that his material of "*Brontosaurus*" had no head, he made his best guess as to the kind of skull this animal had: one like *Camarasaurus*, found not far away. And it was thus that *Apatosaurus* donned the short-snouted head of *Camarasaurus*.

Enter H. F. Osborn, curator of vertebrate paleontology and powerbroker of the American Museum of Natural History, and W. J. Holland,

curator of fossil vertebrates at the Carnegie Museum of Natural History and equally stalwart in his pursuit of "getting it right" about sauropods. These two men – contemporary dinosaur researchers in the early part of the 20th century – skirmished over the issue of whose head should reside on the neck of *Apatosaurus*. Osborn followed Marsh and had the American Museum of Natural History's mount of this majestic sauropod topped with a camarasaur head (which it kept for the next 90 years!), whereas Holland was strongly persuaded that *Apatosaurus* had a more *Diplodocus*-like head (based on a skull found associated with an otherwise quite complete skeleton at what is now Dinosaur National Monument). Holland's mount of *Apatosaurus* in the Carnegie Museum, therefore, remained headless in defiance of Osborn's dogma. After Holland's death, however, the skeleton was quietly fitted with a camarasaur skull, almost as if commanded by Osborn himself.

Whose head belongs to whom was finally resolved in 1978 by D. S. Berman and J. S. McIntosh. Through some fascinating detective work on the collection of sauropod specimens at the Carnegie Museum, these two researchers were able to establish that *Apatosaurus* had a rather *Diplodocus*-like skull – long and sleek, not blunt and stout as had previously been suggested. As a consequence, a number of museums that display *Apatosaurus* skeletons celebrated the work of Berman and McIntosh (and Holland) by conducting a painless head transplant – the first ever in dinosaurian history!

Figure 11.19 **Systolic blood pressures compared. (a)** A sauropod (~630 mm Hg), **(b)** a giraffe (320 mm Hg), and **(c)** a human (150 mm Hg).

Sauropod
630mm
Mercury

Giraffe
320mm
Mercury

Man
150mm
Mercury

Perhaps *Diplodocus* and the like gained access to foliage at high levels in the trees by adopting a tripodal posture, rearing up on their hindlimbs and using their tails as a "third leg" (Figure 11.20). This posture should look familiar; we discussed it in Chapter 6 when considering the possibility that stegosaurs could also adopt it. The center of gravity in sauropods was also positioned just in front of the hips, making it biomechanically rather easy to rear upward on the hindlimbs.

Recall that many sauropods have distinctive, well-developed, Y-shaped neural arches. Here is the anatomical function of this feature: The head and neck of the animal were supported by a strong and taut ligament that lay between the bifurcations of the neural spines of the dorsal and cervical vertebrae in diplodocids (apparently also in *Camarasaurus*). So raising the head and neck would not have been a constant muscular struggle; much of the weight would have been taken up by this ligament (Figure 11.21). In tripodal posture, these dinosaurs would have had to pay the same price as *Brachiosaurus* (which itself was probably not able to

Figure 11.20 A sauropod reconstructed in a tripodal posture, using the tail as a "third leg."

rear up): elevated blood pressure and a large heart to produce it.

The second interpretation requires much less blood pressure because it advocates that most sauropods generally held their necks horizontally. J. Martin (Leicester Natural History Museum) and Dodson separately have pointed out that the ligaments occupying the space between the bifurcations of the cervical neural spines of many sauropods had no possibility of contracting – that job is only for muscles. What these ligaments would have been good for was resisting the downward bending of the head and neck due to gravity, much like a very taut rubber band. In this way, the neck was more likely to have been held horizontally, because it was supported there by the ligament (Figure 11.21). And a horizontal neck would have placed the head at approximately the same level as the heart, with no outrageously high blood pressure required. In such a position, foraging for food would have included huge horizontal sweeps of the head through the branches. Even so, these animals may have occasionally reared up on their hindlimbs – during intraspecific combat, when feeding on highly placed leaves in tall trees whose lower foliage had been depleted, and certainly while mating.

Before leaving the physiological consequences of the extraordinarily long

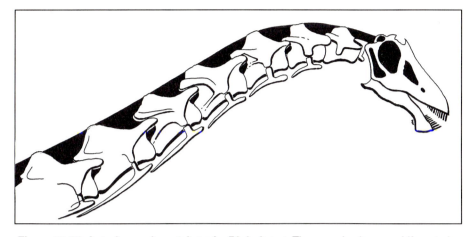

Figure 11.21 **Anterior neck vertebrae in *Diplodocus*.** The neural spines are bifurcated and are thought to have held a ligament supporting the neck – the nuchal ligament (shown in black) running from the head, down the neck, and beyond.

necks of sauropods, we must turn to the effect of sauropodomorph necks on respiration. It goes without saying that, especially for sauropods, the **trachea** (windpipe) would have been exceptionally long – approximately the same length as those arteries carrying blood from the heart to the brain but, in this case, carrying air from the back of the throat to the lungs. In all air-breathing vertebrates, this conduit is necessary to bring oxygen into contact with the physiologically active sites in the lungs (called **alveoli** in mammals and *faveoli* in birds) where oxygen is exchanged into the bloodstream for carbon dioxide, which is transmitted back to the air.

Mammals such as ourselves (as well as other mammals, lizards, crocodilians, and snakes) pass air into and out of the lungs during inhalation and exhalation, rather like a bellows. When this takes place, the trachea creates what is known as anatomical dead space: Some of the inspired air never reaches the lungs. It is simply brought into the respiratory system and returned without ever being involved in the oxygen–carbon dioxide exchange. Consequently, in the case of a single breath, as little as 20% of inhaled oxygen diffuses into the blood stream. This type of respiration is known as *bidirectional airflow*. By contrast, birds have *unidirectional airflow* through their lungs: Inhaled air travels down the trachea and then makes a single pass through a complex arrangement of air sacs and well-vascularized (i.e., efficient) lungs before returning to the trachea for exhalation. This arrangement permits about 40% of the inhaled oxygen to diffuse into the bloodstream.

But what of sauropodomorphs? We will never know for sure, but we can take clues from living long-necked mammals such as giraffes, and certain long-necked birds. Giraffes circumvent the dead space problem by possessing a narrow trachea, thus drastically limiting its volume. In fact, it is thought by some scientists that giraffes may have the longest necks that are physiologically possible for animals using bellows-style lungs. True? Perhaps – but sauropodomorphs (if their metabolic and lung ventilation requirements were less than those of giraffes) obviously exceeded this limit! And as for birds, dead space is less of a problem because of the efficiency of avian lungs. Sauropods may have had unidirectional, avian-style lungs as well, but the question of whether such lungs were necessary really boils down to the metabolic demands of being a sauropod (see Chapter 14).

Feeding

Sauropodomorphs were an important and diverse group of well-adapted herbivorous dinosaurs that almost certainly fed from the crowns of quite tall trees. Still, by comparison with modern herbivorous mammals and especially with other herbivorous dinosaurs such as hadrosaurids, the skull does not appear to be particularly well designed for powerful chewing. Yes, there is a ventrally offset jaw joint in animals like *Plateosaurus*, *Coloradisaurus*, *Apatosaurus*, and even *Camarasaurus*, but overall the skull is relatively small and lightly built, with little room for jaw muscles. In most of these animals, the teeth have simple crowns, triangular when seen in side view (as in most basal sauropodomorphs; Figure 11.22a), spatulate (primitively in Sauropoda; Figure 11.22b), or slender and pencil-like (Diplodocidae, Titanosauridae; Figure 11.22c). There is even a tendency in the clade to limit the teeth to the front of the jaws. In all cases, there is nothing like the full dental arcades or batteries seen in other dinosaurian herbivores. Interestingly, although much has been made about the lack of tooth wear in

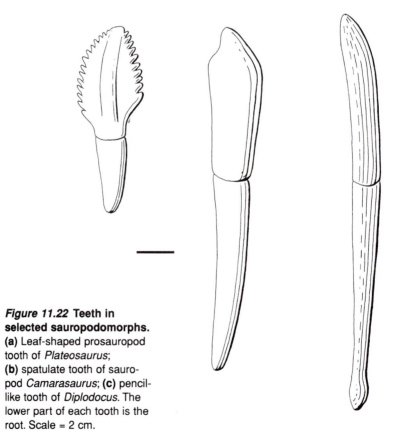

***Figure 11.22* Teeth in selected sauropodomorphs. (a)** Leaf-shaped prosauropod tooth of *Plateosaurus*; **(b)** spatulate tooth of sauropod *Camarasaurus*; **(c)** pencil-like tooth of *Diplodocus*. The lower part of each tooth is the root. Scale = 2 cm.

sauropodomorphs, this situation may be the result of lack of looking: A. R. Fiorillo (University of California at Berkeley) found that the teeth of *Camarasaurus* are heavily worn with coarse scratches and pits, whereas those of *Diplodocus* have fine scratches on their wear surfaces. Fiorillo argued that these differences in wear are a consequence of feeding on different kinds of food or possibly of feeding at different levels, with the shorter-necked *Camarasaurus* being exposed to more grit adhering to the leaves of lower levels than was the longer-necked *Diplodocus*.

From an inspection of the size and shape of the teeth and jaws, as well as from these studies of tooth wear, all of these animals appear to have nipped or stripped off foliage, unceremoniously delivering a succulent bolus to the gullet without much modification in the mouth. It is doubtful that any of these animals fed selectively, given the size of the jaws, the nature of the dentition, and most important, the gigantic size of many of these animals.

Swallowing sped the bolus down its long travel along the esophageal canal, whereupon it entered the abdomen, in particular the **gizzard**. This muscular chamber just in front of the glandular part of the stomach is thought to have been developed in sauropodomorphs; enclosed within it would have been a collection of **gastroliths**, rather large and smoothly polished stones that acted as a gastric mill. Contraction of the walls of the gizzard churned the gastroliths, thereby aiding in the mechanical breakdown of food as it passed farther along the gut. Gastroliths are known, albeit rarely, in some sauropodomorphs (including the discovery, most recently, of in-place stomach stones with the skeleton of *Seismosaurus*).[4]

In all sauropodomorphs, the gut must have been capacious, even considering the forward-projecting pubis (in contrast to all ornithischians, which rotated the pubis rearward to accommodate an enlarged gut; see Chapter 7). J. O. Farlow (Indiana University–Purdue University at Fort Wayne) envisioned these animals as having an exceptionally large fermentation chamber in the hindgut. The bacteria within the chamber would have chemically broken down the cell walls of the food plants, thereby liberating their nutritious contents.[5] Considering the size of the abdominal cavity in sauropodomorphs, these animals probably fed on low-quality food items (i.e., foliage with high fiber content); perhaps they also had low rates of passage of food through the gut in order to ensure a high level of nutrient extraction from such low-quality food. Still, these huge animals with their comparatively small mouths must have been constant feeders to acquire enough nutrition to maintain themselves. The digestive tract of a sauropod had to have been a nonstop – if slow-speed – conveyor belt.

As we earlier discussed, sauropodomorphs were the tallest browsers for their time – and for *all* time, in many cases. Forms like *Euskelosaurus* and *Plateosaurus* were able to feed at up to 3 m above the ground, particularly if they assumed a tripo-

---

[4] The term "gastrolith" was coined in 1907 by ace dinosaur collector Barnum Brown, who claimed to have evidence for gastroliths in both plesiosaurs and dinosaurs. Today, rounded, polished stones said to be gastroliths are not uncommon in curio shops. However, unquestionable gastroliths from any dinosaur are rare. To distinguish true gastroliths from stones polished by the rivers (coincidentally, the source of much dinosaur material) is difficult; therefore, most paleontologists consider stones to be gastroliths only when those stones are found preserved within the rib cages of articulated specimens.

[5] Today, endosymbiotic bacteria are ubiquitous in the animal kingdom; it is very likely that they existed in similar abundance in previous times.

dal posture. This was tall for the Late Triassic and Early Jurassic. Likewise, some later sauropods also may have been able to do so, at least until they felt faint! And the elongate necks seen in all members of Sauropodomorpha certainly extended vertical feeding ranges. Such ranges have been estimated to have been up to four or five stories! In sauropodomorphs capable of rearing up on their hindlimbs to feed, it is likely that the hands manipulated leaves and branches to the mouth or possibly assisted in balancing while the animal craned for foliage just out of reach.

From studying their anatomy, we have good reason to suspect that sauropodomorphs browsed at high levels, but were there tall trees for them to browse on? Indeed, the Late Triassic had its share of tall plants, the likes of which *Plateosaurus* and *Massospondylus* fed upon. These included ferns, conifers, seed plants, cycads, and ginkgoes. During the Jurassic, considered by some to be the heyday of sauropod evolution, a great variety of conifers, with fewer kinds of ginkgoes, cycads, ferns, and horsetails, constituted the tall plants available to a browsing *Omeisaurus*, *Diplodocus*, or *Seismosaurus* (see Box 15.3). And for the Cretaceous sauropodomorphs such as *Saltasaurus*, *Quaesitosaurus*, and *Alamosaurus*, there were the emergent angiosperms, some of which probably reached tall-tree height before the end of the period.

However, beyond recognizing this potential link between sauropodomorphs and their fodder, we have very little direct evidence of their diets. It has been suggested that a pile of carbonaceous material, old stems, bits of leaves, and other plant material collected from the abdominal region of a sauropod skeleton found in the Morrison Formation of Wyoming constituted some fossilized gut contents. Interestingly, this skeleton also had a packet of gastroliths in its belly region. Another possible example of stomach contents comes to us from the Upper Jurassic of Utah. Consisting of sections of small twigs and branches, it contains no leaves or carbonized residues. If either of these collections of plant hash are to be believed as stomach contents, they give the impression that sauropod digesta were rather coarse and fibrous.

Locomotion

The early history of locomotion in Sauropodomorpha is consistent with the primitive condition for all dinosaurs: bipedality. In basal sauropodomorphs (Figures 11.10 and 11.11), the forelimbs are relatively shorter than the hindlimbs and the trunk region is relatively short, suggesting that these animals walked principally on their hindlimbs rather than on all fours. However, *Riojasaurus* and *Blikanasaurus* appear to have become fully quadrupedal.

In all cases, basal sauropodomorphs appear to have been quite slow, perhaps the slowest of all bipedal dinosaurs. R. A. Thulborn (University of Queensland) calculated that most got around at no more than 5 km/hr, about the average walking speed of humans. Whether such an inference is true might be judged by checking it against footprints and trackways. The few trackways of these animals about which we are confident come from Lower Jurassic rocks of the eastern United States and of the American southwest, and from Upper Triassic sediments of Lesotho.[6]

[6]On the face of it, there seems to be an abundance of so-called prosauropod prints throughout the world. However, most of these – despite claims in the literature – were not made by prosauropods at all. Instead, their makers were theropods, ornithopods, primitive ornithodirans, or even crocodylomorphs!

Figure 11.23 **Five parallel trackways of Late Jurassic age.** Morrison Formation, Colorado, U.S.A. *Photograph courtesy of Martin Lockley, University of Colorado, Denver.*

What do these tracks tell us about locomotion in basal sauropodomorphs? Unfortunately, not much about walking or running rates. Nevertheless, there is information to be gleaned about limb posture from these trackways, which all come from animals walking quadrupedally. When walking on all fours, the print-maker had a rather broad trackway, with the oval prints of the hindfoot turned outward from the midline. In keeping with the rearward-positioned center of gravity, the imprints of the hands are smaller, and somewhat shallower, than the imprints of the feet. Interestingly, the large thumb claw appears to have made a mark in the ground only when the hand sank deeply into the substrate. Otherwise, it was clearly held high enough to clear the surface.

For sauropods, both skeletons and trackways reveal a great deal more about locomotion in these animals than they do for basal sauropodomorphs. Again, we look to Thulborn's estimates of speed based on skeletal information. In 1990, he calculated maximal speeds for such sauropods as *Brachiosaurus*, *Diplodocus*, and *Apatosaurus*; all ranged between 20 and 30 km/hr, a reasonable clip for animals the size of a house and weighing in excess of 3 to 10 elephants!

More to the point, though, sauropods probably walked a good deal slower most of the time, perhaps – as has been estimated – at rates of 20 to 40 km per day. These much slower rates are based on sauropod trackways, known in great quantities from the Jurassic and Cretaceous of North and South America, Europe, Asia, and Africa (Figure 11.23).

In keeping with the gigantic size of sauropods, their trackways tend to be quite narrow and their prints immense. The hand is horseshoe shaped, with no impression of the massive claw on the thumb. In contrast, the foot is ovoid, and there is an indication of toe claws and a heel pad. Most significantly, relatively few trackways include a tail-drag mark, providing strong evidence that many sauropods carried their immense tails clear of the ground.[7]

Social Behavior

The generally slow rate of their progression is certainly consistent with the long-distance movements we earlier suggested for regionally migratory sauropods. This particular aspect of sauropod – indeed, more generally of sauropodomorph – social behavior has long been of interest to dinosaur pale-ontologists. It was F. von Huene who first suggested mass movement of large groups (should we call them herds – or flocks?) of *Plateosaurus* from the more easterly highlands of Europe during the Late Triassic to account for the sub-stantial bonebed at Trossingen, Germany, where virtually only one species is preserved. Since that time, many now-famous mass accumulations in the United States, Tanzania, India, and most recently in China, together with the vast sauropod footprint assemblages described earlier, have all spoken loudly to the existence of gregariousness in at least some sauropods. Sauropods living in large groups must have been capable of wreaking severe damage on local veg-etation, either by stripping away all the foliage they could reach or by tram-pling into the ground all of the shrubs, brush, and trees that might have gotten in the way. If many kinds of sauropods actively depleted their food source, as Dodson and colleagues suggested in 1980, then it is equally likely that these herds had to migrate elsewhere for a bite to eat. In contrast, there were many other sauropods, including *Brachiosaurus* and *Haplocanthosaurus* from the Morrison Formation, and *Opisthocoelicaudia* from Mongolia, that were not so numerous. Is it possible that they lived a more solitary existence, perhaps remaining behind as the likes of *Shunosaurus* or *Camarasaurus* left them in the dust?

Beyond gregariousness, our discussion of sauropodomorph behavior culmi-nates with the time-honored subject of defense. Defense in sauropods is obvious: Large size confers the supreme deterrent against an attack. But for many basal sauropodomorphs and, of course, for the young of any of these animals, protec-tion via size was far off in both ontogenetic or phylogenetic time. Thus, we look to the large and trenchant thumb claw that is especially well developed in indi-viduals of any age and even in young and vulnerable adult sauropods (despite reduction of the rest of the digits of the hand) as the chief weaponry in these ani-mals.

Sauropodomorph Growth and Development

What do we known about young sauropods or about the general aspects of sauro-pod reproduction, growth, and life histories? Reproduction in sauropodomorphs is poorly understood. Sex in these animals assuredly involved coupling between a

---

[7] J. S. McIntosh believes that an exception to this may have been diplodocids, in whose tails half of the vertebrae have few well-developed sites for muscle or ligament attachments. He concludes that, with no obvious way for diplodocids to support their tails, the tail must have dragged.

tripodal male and a quadrupedal female; beyond this most basic of poses all else remains speculative. A suggestion by Bakker that (some? all?) sauropods gave birth to live young – an idea based on pelvic structure and the rarity of eggs attributable to sauropods – has not been given great credence by the scientific community. Therefore, it is likely that all sauropodomorphs, including sauropods, entered the world from inside an egg. There are as yet no eggs of basal sauropodomorphs (although it is possible that a nest of six eggs – with embryonic material! – from the Lower Jurassic of South Africa may have been laid by *Massospondylus*). For sauropods, the story is less desperate: Although they are very rare, in at least one case eggs have been tentatively associated with a particular sauropod (*Hypselosaurus*) from the Upper Cretaceous of southern France; eggs from Mongolia and India are thought to have been laid by unspecified sauropods. Those eggs referred to *Hypselosaurus* are nearly spherical, with a volume of 1,500–2,000 cm³ (about the size of an American football). Laid in linear pairs, these eggs may also have been covered by mounds of vegetation to keep them at optimal temperature and humidity.

It is not precisely clear how fast sauropodomorphs grew once hatched. Recent studies by A. Chinsamy concluded that growth was not rapid and continuous but, rather, episodic (Chapter 14). A number of lines of evidence suggest that it could take about 20 years or less for a sauropodomorph to become sexually mature (Chapter 14). Life spans for these animals may have been on the order of 100 years.

Despite their size and the ease that this should confer on finding their bones, sauropodomorphs raise many more questions than they provide answers. We know that sauropodomorphs were the slow-paced giants of the Mesozoic – indeed, of all time. Their high-browsing skills were unequalled throughout their 160-million-year tenure on earth. Today, they continually surprise, inspire, and baffle us with their towering qualities and the biomechanical and evolutionary consequences of immense size.

Important Readings

Alexander, A. McN. 1989. Dynamics of Dinosaurs. Columbia University Press, New York, 167 pp.

Berman, D. S., and J. S. McIntosh. 1978. Skull and relationships of the Upper Jurassic sauropod *Apatosaurus* (Reptilia, Saurischia). Bulletin of the Carnegie Museum of Natural History 8:1–35.

Coombs, W. P. 1975. Sauropod habits and habitats. Palaeogeography, Palaeoclimatology, Palaeoecology 17:1–33.

Dodson, P. 1990. Sauropod paleoecology; pp. 402–407 *in* D. B. Weishampel, P. Dodson, and H. Osmólska (eds.), The Dinosauria. University of California Press, Berkeley.

Dodson, P., A. K. Behrensmeyer, R. T. Bakker, and J. S. McIntosh. 1980. Taphonomy and paleoecology of the Upper Jurassic Morrison Formation. Paleobiology 6:208–232.

Galton, P. M. 1990. Basal Sauropodomorpha – prosauropods; pp. 320–344 *in* D. B. Weishampel, P. Dodson, and H. Osmólska (eds.), The Dinosauria. University of California Press, Berkeley.

Gauthier, J. A. 1986. Saurischian monophyly and the origin of birds. Memoirs of the California Academy of Sciences 8:1–55.

McIntosh, J. S. 1990. Sauropoda; pp. 345–401 *in* D. B. Weishampel, P. Dodson, and H. Osmólska (eds.), The Dinosauria. University of California Press, Berkeley.

Ostrom, J. H., and J. S. McIntosh. 1966. Marsh's Dinosaurs: The Collection from Como Bluff. Yale University Press, New Haven, 388 pp.

Weaver, J. C. 1993. The improbable endotherm: the energetics of the sauropod dinosaur *Brachiosaurus*. Paleobiology 9:173–182.

Chapter 12

THEROPODA I:
Nature Red in Tooth and Claw

TYRANNOSAURUS REX RULES. Everybody knows this, from the late Marc Bolan, lead singer/guitarist of the band T. rex, to Steven Spielberg, who starred the beast in the cataclysmic end of *Jurassic Park*, to the average five-year-old, who just *knows* it. An awesome amalgam of 20-cm teeth, sinewy haunches, and scimitar claws (Figure 12.1), *T. rex* assuredly could have gobbled down university professors in a single bite. *Tyrannosaurus* didn't spring out of nowhere; it turns out to be only one of many meat-eating dinosaurs. In the world of theropods, there *are* many ways to skin a cat (or other vertebrate).

Theropoda (*thero* – beast; *pod* – foot), the group created in 1881 by Yale University paleontologist O. C. Marsh to contain these dinosaurs, has had a long evolutionary history extending back to the Late Triassic and, as we shall see in the next chapter, going all the way to the present. These dinosaurs have been found on every continent, having been most recently discovered in the heart of blustery and frigid Antarctica.

Theropods are distinctive and instantly recognizable. All theropods had a bipedal stance. Many theropods are characterized as having sharp and, commonly, serrated teeth.[1] Beyond this, however, they came in various sizes and personalities. Theropod evolution is thought to have been associated in one way or another with what the animals did best: track, attack, and feed – because all theropods (alone among Dinosauria) appear to have been irredeemable carnivores.

---

[1] The teeth and bipedal stance of theropods appear to have been primitive carry-overs from earlier ornithodiran history (see Chapter 5).

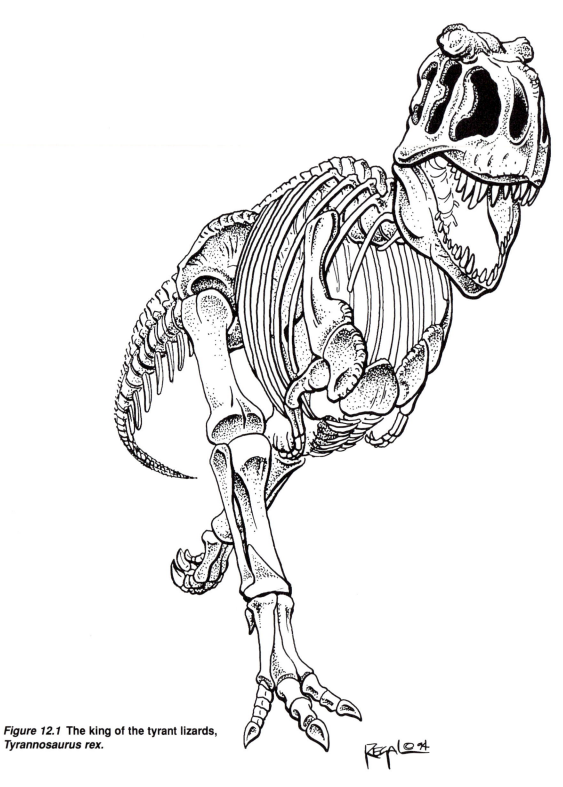

Figure 12.1 **The king of the tyrant lizards,
*Tyrannosaurus rex.***

HISTORY OF THEROPOD DISCOVERIES

19th Century

Theropod discoveries begin auspiciously enough with Oxford University cleric and professor W. Buckland's[2] work on *Megalosaurus* (*mega* – great). Based on a tooth-bearing jaw discovered in 1819 in Oxfordshire, England, and later given its name in 1824, *Megalosaurus* was one of the founding members of Sir Richard Owen's Dinosauria. Additional material of this Middle Jurassic theropod has been found in England as well as in France, but paradoxically it remains very poorly known.

Elsewhere in Europe, a very delicate skeleton of a small (less than a meter long) theropod was unearthed in Bavaria. Called *Compsognathus* (*kompsos* – delicate; *gnathos* – jaw), this animal was found in the same lithographic limestone rocks as the fossil bird *Archaeopteryx* (see Chapter 13). Interestingly, it was *Compsognathus*, rather than its feathered companion, that transfixed T. H. Huxley in his quest to understand bird origins.

Discovery of new theropods in the 19th century, however, was dominantly a North American experience. The controversial *Troodon* (*troo* – wound; *odon* – tooth) led the way. As we learned in Chapter 7, the affinity – and even the reality – of *Troodon* was a source of confusion until P. J. Currie (Royal Tyrrell Museum of Palaeontology, Drumheller, Alberta) referred a very significant skull and postcranial skeletal material to this taxon. Now *Troodon* is known in much greater detail indeed!

Elsewhere in North America, Philadelphia paleontologist E. D. Cope described *Laelaps* (named for the mythical hunting dog Lailaps ["storm wind"]) in 1866. This evocative name was dropped (it was unfortunately pre-occupied by an insect!) and replaced with *Dryptosaurus* (*drypto* – tearing) by Cope's rival, Marsh (Yale University) in 1877.

Yet the great theropod discoveries came with the western research of Marsh and Cope (Box 6.1). Marsh got off the first shot with *Allosaurus* (*allo* – other) from the Morrison Formation of the Western Interior of the United States in 1877 (Figures 12.2f and 12.3b). Then there were the second and third salvos: *Coelurus* (*coel* – hollow; *uro* – tail), again from the Morrison Formation, in 1879; and *Ceratosaurus* (*cera* – horn) from the Morrison Formation of Colorado and Utah in 1884 (Figures 12.2b and 12.3a). A final salvo came in the form of *Ornithomimus* (*ornith* – bird; *mimus* – mimic) in 1890. This theropod – from the Western Interior of Canada and the United States – heralded a large clade of ostrichlike theropods, the diversity of which we are still in the process of appreciating.

The sole theropod from Cope that comes down to us today is *Coelophysis* (*coel* – hollow; *physis* – form, a reference to air spaces within the vertebrae), recovered in 1889 from New Mexico. This small (less than 3 m long), agile theropod has become one of the best known of all predatory dinosaurs (Figure 12.2d).

20th Century

H. F. Osborn, turn-of-the-century vertebrate paleontologist and self-styled magnate of evolutionary biology, christened the first theropod in the new century –

---

[2] Like many of his Victorian peers and predecessors, William Buckland was a naturalist as well as cleric. A most unusual man, he wanted – as one of his life goals – to eat his way through the animal kingdom.

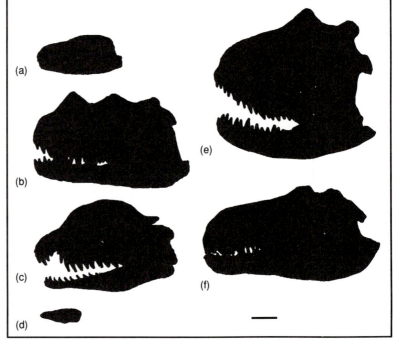

Figure 12.2 Left lateral view of the skulls of (a) *Herrerasaurus,* (b) *Ceratosaurus,* (c) *Dilophosaurus,* (d) *Coelophysis,* (e) *Carnotaurus,* and (f) *Allosaurus.* Scale for silhouettes = 10 cm.

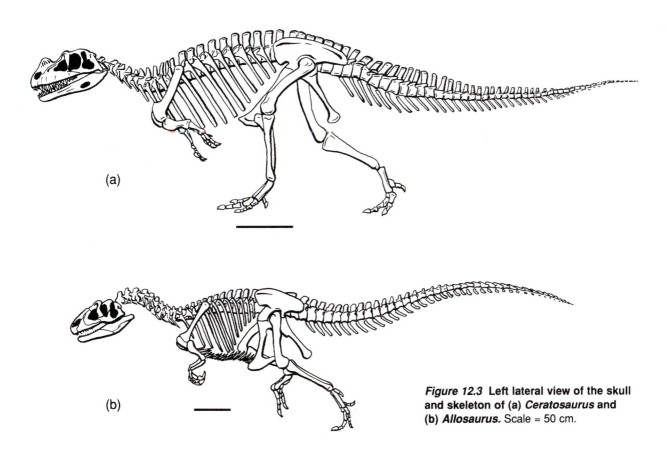

(a)

(b)

Figure 12.3 Left lateral view of the skull and skeleton of (a) *Ceratosaurus* and (b) *Allosaurus.* Scale = 50 cm.

Ornitholestes (*lestes* – robber) from the Morrison Formation of Wyoming – in 1903 (Figures 12.4a and 12.5b). Two years later, Osborn hit the record books with two additional theropods: *Albertosaurus* (named for Alberta, Canada, where the dinosaur was discovered; it is now known from elsewhere in the Western Interior of North America; Figure 12.4c) and *Tyrannosaurus* (*tyrannos* – tyrant; Figure 12.1), now known from throughout the North American Western Interior.

Then Tendaguru happened (see Box 11.1). Although the name is commonly associated with sauropods, between 1909 and 1912, W. Janensch and his co-workers were also collecting a small (3.5 m long), swift-looking animal, *Elaphrosaurus* (*elaphros* – fleet).

While the Humboldt Museum expedition was busy at Tendaguru, E. Stromer from Munich was hard at work in the Western Desert of Egypt, describing three new theropod dinosaurs from mid-Cretaceous outcrops at Baharije Oasis: *Spinosaurus* (*spina* – spine), *Carcharodontosaurus* (named for *Carcharodon*, the great white shark), and *Bahariasaurus* (named for Baharije Oasis, where it was found). It is one of the paleontological tragedies of World War II that the original specimens of all three were destroyed in the allied bombing of Munich at the close of the war.

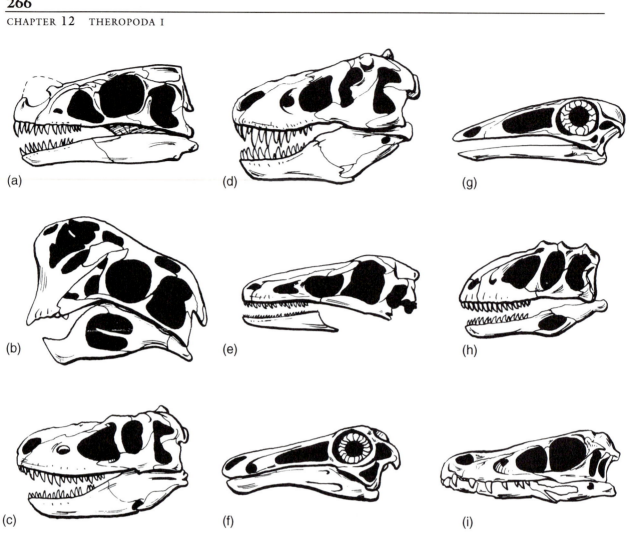

Discoveries in Alberta, Canada, eclipsed anything yet recovered. In 1917, Osborn described *Struthiomimus* (*struthio* – ostrich) and Geological Survey of Canada paleontologist L. M. Lambe described *Gorgosaurus* (*gorgos* – fearsome).[3] By the middle of the 1920s, the Upper Cretaceous strata of Alberta had yielded two small (2 m long) theropods, *Dromaeosaurus* (*dromaios* – swift) and *Chirostenotes* (*chiro* – hand; *steno* – narrow). At the same time, three new and rather exotic theropods, all about 2 m in length, were described from the Gobi: *Oviraptor* (*ovi* – egg; *raptor* – stealer; Figure 12.4b), *Saurornithoides* (*ornithoides* – birdlike; Figure 12.4e), and *Velociraptor* (*velo* – swift; Figure 12.4i). In 1933, *Alectrosaurus* (*alectros* – unmarried; in reference to the enigmatic form of this theropod) – was added to the list of Gobi theropods.

World War II exacted an extraordinary toll on humanity and human affairs, and as we have seen, paleontology was not excluded. In the 1940s and 1950s, only a few new theropods were described. Among them were *Acrocanthosaurus* (*akros* – high; *akantha* – spine), described from the Lower Cretaceous of

[3]*Gorgosaurus* has since been synonymized with *Albertosaurus*.

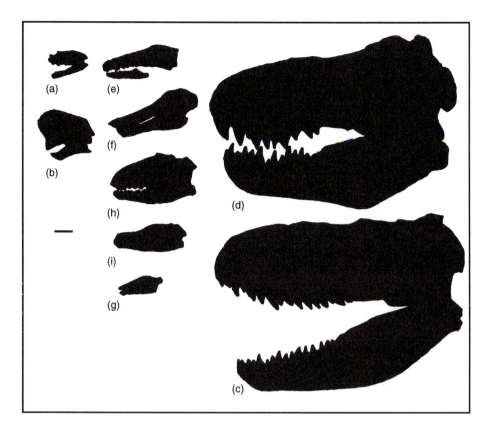

Figure 12.4 Left lateral view of the skulls of (a) *Ornitholestes,* (b) *Oviraptor,*
(c) *Albertosaurus,* (d) *Tyrannosaurus,* (e) *Saurornithoides,* (f) *Gallimimus,*
(g) *Dromiceiomimus,* (h) *Deinonychus,* and (i) *Velociraptor.* Scale for silhouettes = 5 cm.

Oklahoma (now also known from Texas), and a curious creature called
Therizinosaurus (*therizo* – reap; so-named for its large, sicklelike claws). Originally
thought to be a turtle by Soviet paleontologist E. A. Maleev, *Therizinosaurus* has
since been identified as a new, strange theropod. Maleev was on a roll; in 1955,
he announced the fearsome *Tarbosaurus* (*tarbos* – terror; Figure 12.7), the har-
binger of many new Mongolian theropods.

The 1960s began with the description of a 4-m-long carnivorous dinosaur
named *Herrerasaurus* (in honor of D. V. Herrera, landowner of the locality where
this dinosaur was discovered), a form from Argentina that is now thought by some
to be central to the origin of Theropoda (Figure 12.2a). It was 1969, however, that
was the Year of the Theropod. That year, M. A. Raath (then at the Queen Victoria
Museum, Salisbury, in what is now Zimbabwe) described *Syntarsus* (*syn* – fused;
tarsos – tarsus), a 2-m-long, gracile predator from southern Africa and, thanks to
the subsequent work of T. Rowe, from Arizona as well.

But the single most important announcement in 1969 – perhaps in the recent
history of all dinosaurs – was the watershed discovery of *Deinonychus* (*deinos* – terrible; *onycho* – claw; Figures 12.4h and 12.5a) from Montana and Wyoming.

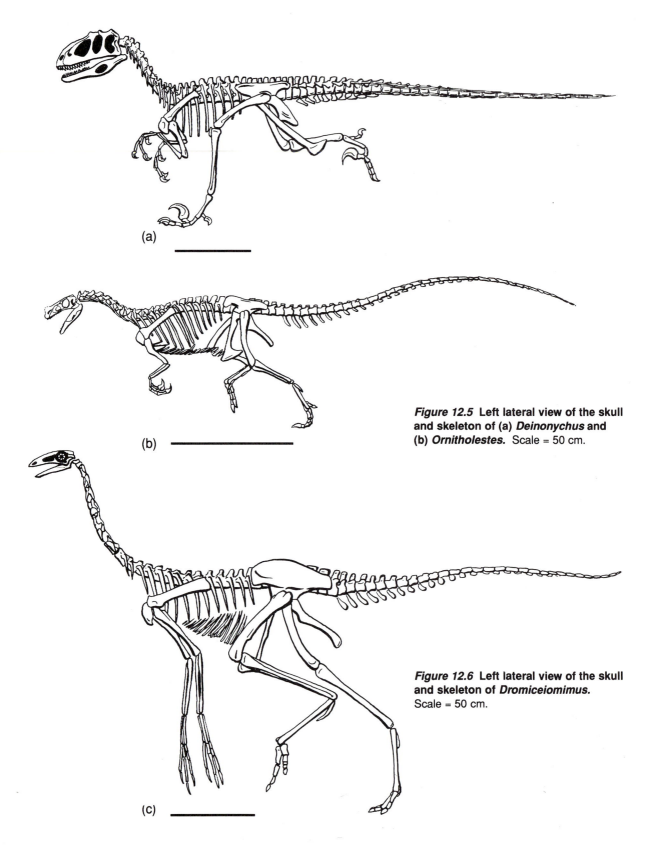

(a)

(b)

Figure 12.5 Left lateral view of the skull and skeleton of (a) *Deinonychus* and (b) *Ornitholestes.* Scale = 50 cm.

Figure 12.6 Left lateral view of the skull and skeleton of *Dromiceiomimus.* Scale = 50 cm.

(c)

Deinonychus – obviously sophisticated and lethal in its behavior – led its discoverer, J. H. Ostrom (Yale Peabody Museum of Natural History, New Haven), to reconsider claims about theropod agility, hunting behavior, evolutionary relationships, and, in a broader context, the affinities of dinosaurs to birds.

The 1970s were dominated by discoveries that came from the Polish-Mongolian Palaeontological Expeditions and the Joint Soviet-Mongolian Paleontological Expeditions in the Upper Cretaceous of the Gobi Desert. From these efforts, we have *Deinocheirus* (*chiro* – hand), known only from a set of huge forelimbs; *Gallimimus* (*gallus* – chicken), the largest-yet known ornithomimid (at 6 m long; Figure 12.4f); *Alioramus* (*ali* – other; *ramus* – branch, as in another branch of tyrannosaurids); and a new and enigmatic theropod from the Upper Cretaceous of Uzbekistan in central Asia called *Itemirus* (after Itemir, the closest village to where this dinosaur was found).

The 1970s brought *Dilophosaurus* (*di* – two; *lophos* – crest), a 6-m-long, lightly built theropod from Arizona (Figure 12.2c); a new tyrannosaurid from Alberta named *Daspletosaurus* (*dasples* – frightful); and two new ornithomimosaurs – *Dromiceiomimus* (*Dromiceius* is the generic name for the emu; Figure 12.6) from western Canada and *Archaeornithomimus* (*archaios* – ancient; *ornith* – bird) from northern China. In addition, J. H. Madsen, Jr., (then State Paleontologist of Utah) began publishing on the very abundant theropod remains from the Cleveland-Lloyd Quarry (in Morrison Formation) of central Utah. Along with nearly ubiquitous *Allosaurus* remains, he described *Stokesosaurus* (named in honor of American paleontologist and geologist W. L. Stokes) and *Marshosaurus* (named in honor of Marsh).

Finally, the 1970s brought *Saurornitholestes* (*lestes* – robber; a lizard-like bird thief), a 2-m-long predator; and *Torvosaurus* (*torvus* – savage), a 9-m-long contemporary of the Cleveland-Lloyd theropods, but from the Colorado side of Morrison Formation outcrops. *Piatnitzkysaurus* (named for Argentinian geologist A. Piatnitzky), a medium-sized (4.5 m long) dinosaur, comes from southern Argentina. In 1977, Dong Z.-M. described a nearly-2-m-long and as-yet enigmatic theropod from Xinjiang Province, which he named *Shanshanosaurus* (for Shanshan zhan, the name of the locality from which this dinosaur was collected). *Yangchuanosaurus* (named for Yangchuan County) was his next conquest; this 8-m-long theropod comes from Sichuan Province.

Mongolia was productive in the 1980s. The finds included the basal ornithomimosaurs *Garudimimus* (Garuda is a monstrous bird in Asian mythology) and *Harpymimus* (*harpyiai* – harpy); a poorly known dromaeosaurid dubbed *Adasaurus* (named for Ada, an evil spirit in Mongolian mythology); and two new oviraptorosaurs: *Ingenia* (named for Ingeni, Mongolia, near the locality that yielded this dinosaur) and *Conchoraptor* (*conch* – shell). The list goes on: a small, but as yet poorly known, theropod named *Elmisaurus* (in Mongolian, *elmyi* means hindfoot); a close relative of *Deinonychus*, called *Hulsanpes* (Khulsan is the locality in Mongolia where this dinosaur was discovered; *pes* – foot); and a new troodontid named *Borogovia* (named for the borogoves in Lewis Carroll's "Jabberwocky"[4]). But the most peculiar discovery coming out of this Gobi research is *Avimimus* (*avis* – bird), a gracile, long-legged, and perhaps toothless 1.5-m-long theropod.

---

[4] "Twas brillig, and the slithy toves/Did gyre and gimble in the wabe;/All mimsy were the borogoves,/And the mome raths outgrabe."

Figure 12.7 The "terror lizard" *Tarbosaurus,* from the Late Cretaceous of central Asia.

By comparison with the finds in Mongolia, new theropods from the rest of the world just dribbled in over the course of the 1980s. The most important include three unusual forms from Argentina. *Noasaurus* (in Spanish, NOA stands for norte-oeste Argentina) was a small theropod; the 6-m-long *Abelisaurus* (named for R. Abel, Director of the Museo de Cipolletti in Argentina) gave little

Figure 12.8 Left forelimb of (a) *Deinonychus,* (b) *Tyrannosaurus,* and (c) *Carnotaurus.* Scale = 10 cm.

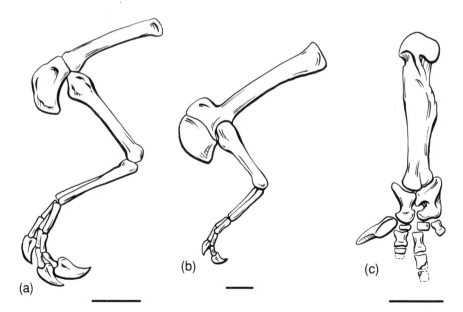

(a)

(b)

(c)

clue to its closest relative, the subsequently discovered *Carnotaurus* (*taurus* – bull; Figures 12.2e and 12.8c), a short-faced, horned, 8-m-long brute with grotesquely diminutive arms. These new theropods provided the first clear suggestion that at least some theropods – a group now called Abelisauridae – evolved in isolation within the southern land masses.

Finally, England got its first theropod to be named in nearly a quarter-century: *Baryonyx* (*bary* – heavy). This possibly piscivorous (fish-eating) theropod comes from the famous Lower Cretaceous Wealden beds, the same strata that have provided England with its greatest collection of dinosaurs.

Between 1990 and late in 1993, seven(!) new theropods appeared. The most important include the Argentinian *Eoraptor* (*eos* – dawn), considered by some the most basal known theropod; *Maleevosaurus* (honoring Maleev), a tyrannosaurid from Mongolia; and *Shuvosaurus* (named by S. Chatterjee [Texas Tech University, Lubbock] for his son, Shuvo, who discovered this dinosaur), claimed to be the earliest ornithomimosaur, from west Texas. And this was just the beginning. In November of 1993, the results of four field seasons of the Sino-Canadian dinosaur project were published. Four new theropods were recovered from Inner Mongolia, the northern part of China that borders on Mongolia. Currie and Zhao X.-J. reported *Monolophosaurus* (*mono* – one; *lophos* – ridge or crest), a large, single-crested theropod reminiscent of *Dilophosaurus*, and *Sinraptor* (*sino* – referring to China; *raptor* – thief), a big theropod a bit like the North American *Allosaurus*. D. A. Russell and Dong weighed in with a new troodontid, *Sinornithoides* from the Early Cretaceous. And as we have seen, the same region produced *Alxasaurus* (see Figure III.4). In May of 1994, W. R. Hammer and W. J. Hickerson of Augustana College, Illinois, reported the discovery of a strangely crested tetanuran theropod skull and associated postcranial material from central Antarctica, which they named *Cryolophosaurus* (*cryo* – cold). Finally, P. C. Sereno recovered *Afrovenator* (*venator* – hunter) from the Sahara. What a strange and wonderful trip it's been, and who knows what further discoveries will be made before the Millennium?

THEROPODA DEFINED AND DIAGNOSED

Theropoda is a group that contains not only *Tyrannosaurus* and *Coelophysis* (and maybe *Eoraptor* as well), but also their common ancestor as well as *all* the descendants of this common ancestor (including birds; Figure 12.9).

Theropoda is easily diagnosed (Figure 12.9). Many distinctive theropod features occur in legs and feet that are well adapted for bipedal locomotion and in an enlarged hand clearly capable of flexibility and grasping. The lowest bone in the thumb of theropods (the first phalange of carpal I) actually lies beneath (palm side of) the equivalent bones in digits II and III; its point of articulation on metacarpal I is structured so that this must occur. The result of this skeletal organization is a distinctive and unique grasping adaptation in the hand: a kind of semi-opposable thumb.

Theropods share closest relationships with Sauropodomorpha; together, the two groups form a monophyletic Saurischia. Depending on who segnosaurs turn out to be, they may have a basal relationship among saurischians as well (Part III: Saurischia).

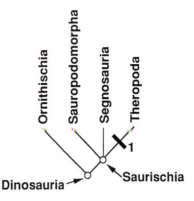

Figure 12.9 Cladogram of Dinosauria emphasizing the monophyly of Theropoda. Derived characters at 1: extreme hollowing of vertebrae and long bones; enlarged hand with vestigial fourth and fifth digits, remaining digits capable of extreme extension due to large pits on the upper surfaces of the ends of the metacarpals; lacrimal bones well exposed dorsally; another hole in front of the antorbital opening (a kind of second antorbital opening); sacrum with at least five vertebrae; ilium somewhat elongate, particularly in its forward part; and foot that is compact, narrow, and long, with the fifth metatarsal reduced to a splint.

THEROPOD DIVERSITY AND PHYLOGENY

Basal Theropods?

Our understanding of theropod relationships is currently in a state of flux – a nasty spot for the student, and disquieting for the specialist. J. A. Gauthier's 1986 view was that *Herrerasaurus* – and *Staurikosaurus* – although dinosaurs, are neither saurischians nor ornithischians (Figure 12.10a). In other words, they appear to lack those features that diagnose the most recent common ancestor of Saurischia and Ornithischia (and, of course, all the descendants of this ancestor).

Since 1986, better *Herrerasaurus* material and a new form, called *Eoraptor*, have challenged Gauthier's hypothesis. According to studies by Sereno and co-workers, both *Herrerasaurus* and *Eoraptor* appear to be theropods. This view-point is shown in Figure 12.10b. Above *Eoraptor* and *Herrerasaurus* on the cladogram are the standard gaggle of predatory theropods, the ones that we all know – and probably fear – in our dreams and now on the Big Screen.

Ceratosauria

There are two major descendant groups within Theropoda: Ceratosauria (named after one of its members, *Ceratosaurus*) and Tetanurae (*tetanus* – stiff; *uro* – tail). Ceratosauria is now well supported by a number of important derived features (Figure 12.11). One of the most distinctive is an ilium that fuses with both the pubis and the ischium in adulthood to form a strong support between the hindlimbs and the vertebral column (Figure 12.12).

Within Ceratosauria, there appear to be two major groups. The first, which

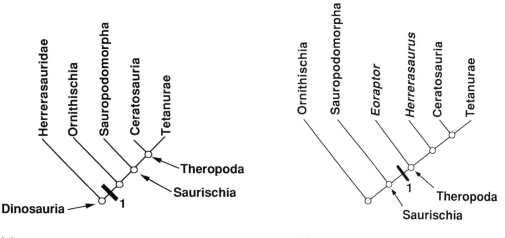

(a)   (b)

Figure 12.10 Two rival cladograms of Theropoda. In **(a)**, Herrerasauridae (*Herrerasaurus* and *Staurikosaurus*) evolved before the split between Ornithischia and Saurischia. Characters at 1: transversely expanded lower end of tibia; and lesser trochanter (a process at the top of the femur) that is shaped like a spike. In **(b)**, Herrerasauridae and *Eoraptor* are full-fledged members of Theropoda. Characters at 1: articulation in the middle of the lower jaw; vertebrae at the end of tails modified to reduce flexibility (expanded zygopophyses); and long finger bones on digits I, II, and III of hand, each with an aggressive claw.

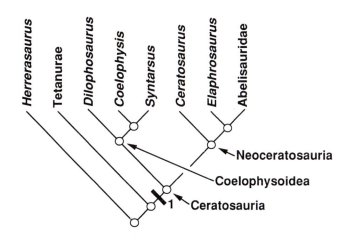

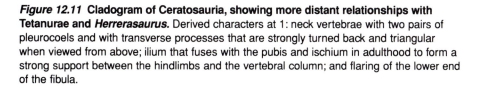

Figure 12.11 Cladogram of Ceratosauria, showing more distant relationships with Tetanurae and *Herrerasaurus*. Derived characters at 1: neck vertebrae with two pairs of pleurocoels and with transverse processes that are strongly turned back and triangular when viewed from above; ilium that fuses with the pubis and ischium in adulthood to form a strong support between the hindlimbs and the vertebral column; and flaring of the lower end of the fibula.

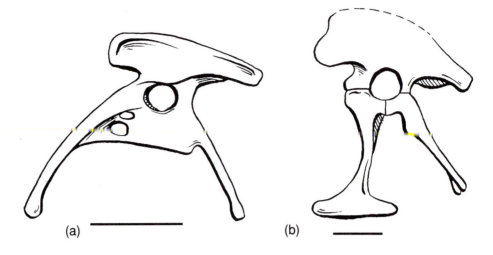

Figure 12.12 Left lateral view of the pelvis of (a) *Syntarsus* and (b) *Tarbosaurus.* (a): Scale = 10 cm; (b): scale = 50 cm.

(a)   (b)

T. Holtz (then at the U.S. Geological Survey) has called Coelophysoidea, consists of *Dilophosaurus* and its close relatives, among them *Coelophysis*, *Syntarsus*, and *Liliensternus*; the second – Neoceratosauria, first recognized and analyzed by J. F. Bonaparte and colleagues in 1990 and 1991 – is formed of *Ceratosaurus* itself, plus *Elaphrosaurus* and members of Abelisauridae, an unusual group of South American theropods (Figure 12.11).

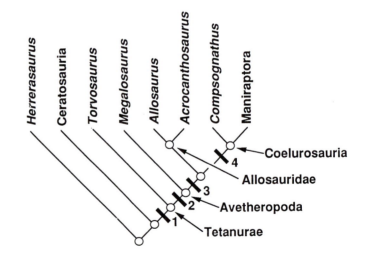

Figure 12.13 Cladogram of Tetanurae, with more distant relationships with Ceratosauria and *Herrerasaurus.* Derived characters at 1: tooth row that does not extend behind the antorbital opening, pronounced groove across the front aspect of the astragalus, procoelous dorsal vertebrae, and shortening of the vertebral column in front of the pelvis; at 2: heightening of the ascending process of the astragalus, and well-developed round accessory antorbital opening; at 3: U-shaped snout, asymmetrical front teeth, basal half of metacarpal I closely appressed to metacarpal II, and well-developed pubic "boot"; at 4: expanded circular orbit, semilunate carpal in the wrist, short ischium, further heightening of the ascending process of the astragalus, and reduction of the transverse processes of the tail vertebrae.

Tetanurae

The remaining great clade of theropods is Tetanurae (Figure 12.13), first recognized by Gauthier in 1986. The name refers to the fact that in this group of theropods, the back half of the tail is stiffened by interlocking **zygopophyses**, fore-and-aft projections from the neural arches. Members of this group, whose record extends from the Middle Jurassic to the present, share a large number of derived features (Figure 12.13, node 1).

The familiar, massive, carnivorous theropods within Tetanurae (for example, *Allosaurus* and *Tyrannosaurus*) have traditionally been united within a group evocatively named **Carnosauria**. Here, we follow the 1993 study of Holtz, who disbanded "carnosaurs" and produced the cladogram in Figure 12.13. The implication of Holtz's study is that the big bipedal carnivores epitomized by the likes of *Allosaurus* and *Tyrannosaurus* actually evolved independently several times, in several lineages of theropods (i.e., convergently).

The ex-carnosaur *Torvosaurus* is the basal member within the tetanuran clade. Thanks in large part to the recent work by B. B. Britt (Museum of Western Colorado in Grand Junction) on new material of *Torvosaurus*, we know that this dinosaur possesses all of the features of Tetanurae. The next-most-derived clade – as yet unnamed – has as its basal member that venerable theropod *Megalosaurus*. This Middle Jurassic theropod from Europe has had a checkered history,[5] but appears to have incorporated a number of significant evolutionary innovations, most notably the heightening of the ascending process of the astragalus (a splint of bone that rises up from the main articulating part of the astragalus and covers as much as the lower 25% of the shin [tibia]; Figure 12.14a).

AVETHEROPODA. First recognized by G. S. Paul in 1988, Avetheropoda (*avis* – bird; a reference to birdlike features of many members of this group) comprises the most recent common ancestor of birds and *Megalosaurus*, and their descendants. The avetheropods share a number of derived features (Figure 12.13, node 2), including a modification of the wrist and hand to allow for better grasping ability.

On our further cladal climb up Avetheropoda, we encounter, successively, Allosauridae, and then Coelurosauria, a group yet more birdlike than prior theropod taxa. With the small theropod *Compsognathus* as its most basal member, these coelurosaurs share an abundance of features uniquely evolved in their common ancestor (Figure 12.13, node 4), including a **semilunate carpal** (a distinctive, half-moon-shaped bone in the wrist; Figure 12.14b).

MANIRAPTORA. *Compsognathus* may have all it takes to be a coelurosaur, but it appears not to have the wherewithal to join the next-higher clade, Maniraptora (Figure 12.15). Maniraptora (*manus* – hand) consists of *Ornitholestes* and an unnamed taxon, which share the derived characters listed in Figure 12.15, node 1.

---

[5] The original specimen of *Megalosaurus* was one side of a theropod lower jaw. Because the jaw did not contain enough diagnostic characters, more complete specimens of large theropods found afterward were given new names, such as *Metriacanthosaurus* and *Eustreptospondylus*. *Megalosaurus*, with its historically seminal position, remains a kind of siren for paleontologists, as it has proven very tempting to refer a variety of dinosaurs to this theropod. For example, *Dilophosaurus* was first announced – headless – to the world as *Megalosaurus*.

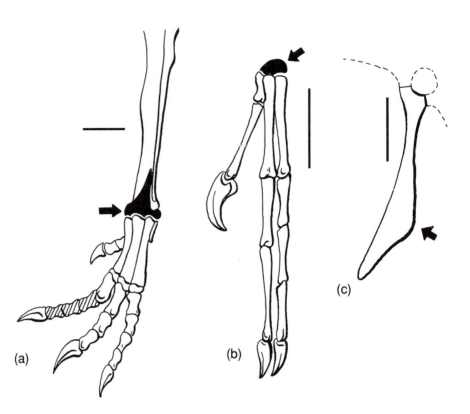

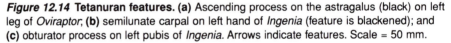

Figure 12.14 **Tetanuran features. (a)** Ascending process on the astragalus (black) on left leg of *Oviraptor*; **(b)** semilunate carpal on left hand of *Ingenia* (feature is blackened); and **(c)** obturator process on left pubis of *Ingenia*. Arrows indicate features. Scale = 50 mm.

Above Maniraptora

Ornitholestes, that lightly built, small theropod from the Late Jurassic of Wyoming and Utah, was assuredly a maniraptoran, but it diverged from the ancestor of this group prior to the origin of a number of important features that are found in our next group. So far without a name, this taxon includes, on the one hand, oviraptorids, elmisaurids, tyrannosaurids, troodontids, and ornithomimosaurs and, on the other, dromaeosaurids and birds. The common ancestor of this bundle of theropods sported a number of derived characters (Figure 12.15, node 2), including an ischium with a flange (the **obturator process**) placed down its shaft (Figure 12.14c).

After the nearly basal split of Tetanurae and Ceratosauria, the next-most-significant evolutionary event in theropod history is between the clade of oviraptorids and their relatives, and troodontids and their relatives. Oviraptorids and their nearest relatives uniquely evolved a number of additional features not found in other theropods (Figure 12.15, node 3). Within this group, the bizarre members of the Late Cretaceous Oviraptoridae (*Oviraptor, Conchoraptor, Ingenia* – all from Mongolia) are the first to diverge. This small clade shares a number of unique features that blend together to make these forms some of the most unusual among dinosaurs. These features include a deep skull with a shortened snout; a short and deep lower jaw with a short, concave, and thickened front rim;

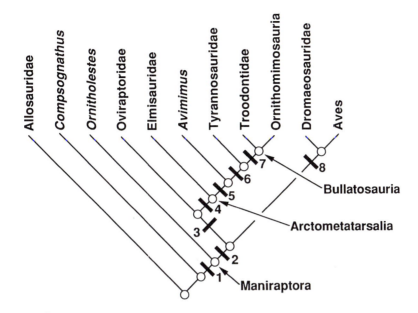

Figure 12.15 Cladogram of Maniraptora, with more distant relationships with **Compsognathus** and **Allosauridae.** Derived characters at 1: greater neck flexure, furcula, lengthening of the forelimb, bowed ulna, and elongation of the middle digit of the hand; at 2: third antorbital opening, and obturator process placed down the ischial shaft; at 3: heightening of the back of the skull, modifications to the back and sides of the braincase, more than five sacral vertebrae, and differentiation of the head of the femur from the greater trochanter; at 4: pinched upper portion of the middle metatarsal bone, and elongate tibia and foot; at 5: upward shift of the obturator process on the ischium, and elongation of the ischium; at 6: surangular opening, contact between most of the paired dorsal margins of the iliac blades, and semicircular muscle scar on the forward part of the ischium; at 7: bulbous area at forward region of the braincase; at 8: downward or down-and-backward rotation of the pubis (the opisthopubic condition), loss of the fourth trochanter on the femur, and stiffening along all but the first section of the tail.

and a palate with a pair of median, toothlike projections. At the same time, these oviraptorids lack a number of features that characterize their sister taxon, Arctometatarsalia.

ARCTOMETATARSALIA.   Arctometatarsalians, consisting of elmisaurids, tyran-nosaurids, troodontids, and ornithomimosaurs, were recognized by Holtz princi-pally because the upper portion of the middle metatarsal bone is pinched by its neighbors, giving what Holtz has called the arctometatarsal condition (*arctus* – compressed). The arctometatarsalian clade appears to be well supported by derived characters (Figure 12.15, node 4).

First off the starting line among arctometatarsalians are the as yet poorly known elmisaurids. According to Currie, this group of small theropods – consisting of *Chirostenotes* and *Elmisaurus* – appears to have uniquely evolved a number of modifications to the foot, including close contact between, or fusion of, the sec-ond to fourth distal tarsals and metatarsals to form a combined element called a **tarsometatarsus** (seen also convergently as a derived condition in birds; see Chapter 13).

TYRANNOSAURIDAE AND BULLATOSAURIA. Climbing higher still, we encounter the fearsome tyrannosaurids, plus a group called Bullatosauria (*bullatus* – inflated, referring to the braincase; Figure 12.15, node 6). Tyrannosaurids, most terrifying of carnivorous dinosaurs, have only recently been recognized as coelurosaurs, thanks to Holtz's demolition of a monophyletic "Carnosauria." Thus, the clade of *Aublysodon, Albertosaurus, Tyrannosaurus, Alectrosaurus, Daspletosaurus, Alioramus, Nanotyrannus,* and *Tarbosaurus* is comfortably nested not only within Coelurosauria, but high up in its branches![6] Tyrannosauridae itself is a natural group, based on a number of shared, derived features: front teeth that are D-shaped in cross section, very reduced forelimbs with loss of digit III, very large "boot" on the end of the pubis, and a variety of other characters.

And who are Bullatosauria (Figure 12.15, node 7)? On the one hand, we have Troodontidae; on the other, there is Ornithomimosauria. The sickle-clawed troodontids, which include *Troodon* itself, as well as *Saurornithoides, Borogovia,* and *Tochisaurus,* have evolved unique sinus systems in the base and sides of the braincase, an enlargement of the middle ear cavity, and modifications of the ankle and foot, among other characters. The ornithomimosaurs are a diverse, but distinctive, group of highly modified theropods – *Harpymimus, Garudimimus, Ornithomimus, Struthiomimus, Dromiceiomimus, Archaeornithomimus, Gallimimus,* and *Anserimimus* – that all bear a very lightly built skull with a long and low snout and with a very large orbit and antorbital opening; no upper teeth and great reduction in, or loss of, the lower teeth in a very long and low mandible; long forelimb; flat claws on the foot; and other modifications of the fore- and hindlimbs. Ornithomimosaurs bear a striking resemblance to modern ostriches (*Struthio*) and emus (*Dromiceius*). The similarities between these modern birds and ornithomimosaurs are sufficiently striking to suggest that these creatures may have behaved in similar ways.

Dromaeosaurids and Their Kin

In recent years, much attention has been devoted to the relationships between dinosaurs and birds. This subject ultimately reduces to the phylogenetic connections between dromaeosaurids and *Archaeopteryx*. Thus, the common ancestor of Dromaeosauridae and *Archaeopteryx* is thought to have newly evolved a number of important features (Figure 12.15 node 8), including a pelvis in which the pubis is rotated downward or down and back (the opisthopubic condition), loss of the fourth trochanter on the femur (a prominent ridge for the attachment of thigh muscles), and modifications to stiffen the tail vertebrae by lengthening the zygopophyses (see Figure 12.5a).

Dromaeosaurids – *Adasaurus, Deinonychus, Dromaeosaurus, Hulsanpes, Saurornitholestes,* and *Velociraptor* – are small, agile, and highly derived theropods that vie for fame even with *Tyrannosaurus*. These denizens of the Cretaceous are best characterized by their small size and their development of a large, sharp, and

[6] Holtz's radical view of tyrannosaurids as immense as coelurosaurs has not gone unchallenged. As this book went to print, J. M. Clark, A. Perle, and M. A. Norell, in a very recent issue of the American Museum of Natural History's technical publication *Novitates* (see Important Readings for this chapter), have argued strongly that Holtz misinterpreted a number of critical characters in the skull and especially in the braincase of tyrannosaurids. Clark et al. believe that tyrannosaurids sit snugly within Maniraptora – although they may not have a close relationship with bullatosaurs.

sicklelike claw on their feet (Figure 12.16), leading one to think that they might be closely related to troodontids. However, the characters on the cladogram imply that the sickle-clawed condition must have occurred two times in the history of Theropoda: once in troodontids and once in dromaeosaurids. This in turn suggests that hunting and killing styles must have been quite similar in dromaeosaurids and troodontids.

One lesson from the foregoing should be that unfolding the history of non-avian predatory dinosaurs is exceedingly complicated. We expect to see studies that modify the conclusions described here. For example, Russell and Dong's classification of *Alxasaurus* and therizinosaurs suggests some very different relationships within Theropoda. As we noted earlier, they view segnosaurs as a type of therizinosaur. But more significantly, they consider Gauthier's Maniraptora an unnatural group. Instead, they recognize two main groups of theropods: one that includes *Oviraptor* (and its relatives), troodontids, ornithomimids, and therizinosaurs, and another that rejoins large theropods as "carnosaurs." In this classification, dromaeosaurids would be recognized as near-relatives of carnosaurs. If the Russell and Dong interpretation is accepted, troodontids and dromaeosaurids, once united within Maniraptora, evolved convergently in quite disparate groups within Theropoda.

All of this reinforces the point that phylogenies are by nature *provisional*. As we saw in Chapter 3, they are explicit, testable hypotheses. Russell and Dong's – or even Gauthier's or Holtz's – phylogenies are not important because somebody finally "got it right," but because they present an entirely new way of looking at theropod relationships. Each succeeding phylogeny represents a test of the preceding ones, and as characters are debated and new organisms are integrated into what is known, we may get successive approximations to the truth.

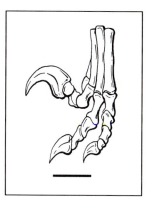

Figure 12.16 Left foot of *Deinonychus* with its disemboweling second-toe claw. Scale = 10 cm.

THEROPOD PALEOECOLOGY AND PALEOBIOLOGY

Imagine living in a world with the largest terrestrial carnivore of all time, and you have the Late Cretaceous of both western North America and central and eastern Asia. Yet beware. Living in fear of *Tyrannosaurus* or *Tarbosaurus* would not have been enough. As the enormous tyrannosaurids devoured their way through herds of ceratopsians or hadrosaurids, the small, agile troodontids and dromaeosaurids stalked the lush flowering landscape in search of their next meals as well. Hunting in packs and armed with recurved razor blades on their feet, these prehistoric killing machines must have efficiently unzipped the bowels of their unfortunate victims.

Theropods have been found on all continents (Figures 12.17 and 12.18), yet there is geographical differentiation through the history of the group. For example, tetanurans all appear to have had a Laurasian distribution, whereas most of the ceratosaurs are restricted to Gondwana. Interestingly, as pointed out by J. Le Loeuff (Musée des Dinosaures, Espéraza, France) and E. Buffetaut (University of Paris VI), the Late Cretaceous theropod faunas of Europe may have experienced dual influences – on the one hand, dromaeosaurids and bullatosaurs evolving elsewhere in Laurasia and, on the other, abelisaurids from Gondwana coming to reside as top carnivores along the European archipelago (see Chapter 15).

In whichever region they once lived, theropods have been collected from a broad range of depositional settings, from fluvial channels and overbank deposits to lacustrine environments and even eolian dunes. Habitat preference is not to be found here. In most of these instances, skeletal remains are found in isolation and often in a disarticulated state. However, several mass accumulations of single theropod species have provided important insights not only into the conditions of death and burial, but perhaps into the animal's biology as well. For the most part, these mass graveyards – which include both juveniles and adults – pertain to ceratosaurian theropods. There is *Syntarsus* in Zimbabwe, South Africa, and in Arizona. And in neighboring New Mexico, there is Ghost Ranch, one of the most profoundly rich sites of any theropod, this time yielding an extraordinary abundance – several hundred individuals! – of *Coelophysis*. Ghost Ranch is rivaled only by the Cleveland-Lloyd Quarry of Utah, from which literally tons of *Allosaurus* bones have been collected since 1927. What was the significance of these mass burials? Did ceratosaurs live in large family groups, which then perished in some sort of catastrophe? Or perhaps each accumulation represents a communal feeding site? These are important questions, and still unanswered.

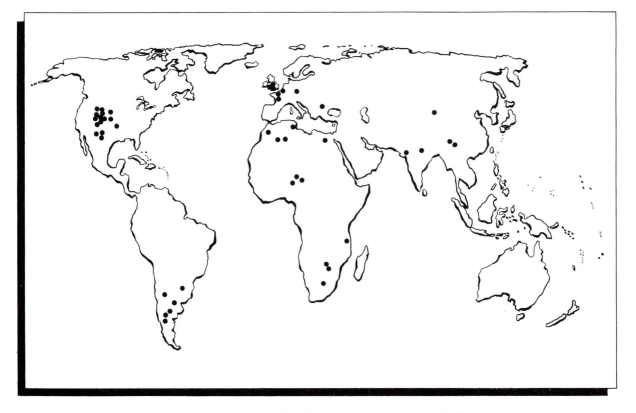

Figure 12.17 Global distribution of noncoelurosaurian Theropoda.

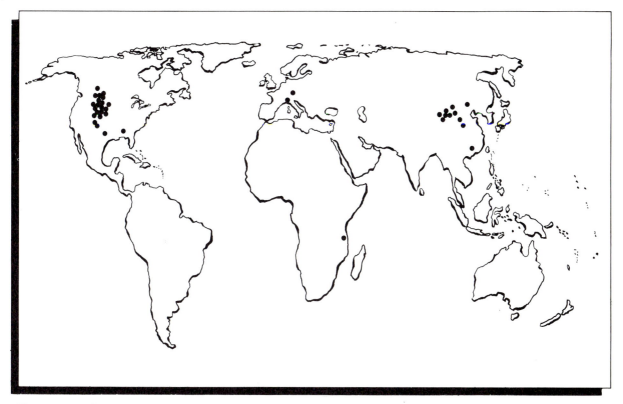

Figure 12.18 **Global distribution of Coelurosauria.**

Theropods as Living Organisms

Theropods ranged in size from about 1 m in length and about 5 kg body weight to animals nearly 40 m in length weighing 5,000 to 7,000 kg. In some cases, large theropods dwarfed their prey by nearly 35% of body length. Farlow has argued that this disparity in body size is due to both physiological and ecological causes. The immense size of tyrannosaurids and certain other theropods may have been a product of the abundance and rapid turnover of theropod food sources: the large herds of rapidly growing, abundantly reproducing hadrosaurids and ceratopsids thought to have lived concurrently with these predators. That is, the ready and continuously replenishing food source may have allowed tyrannosaurids to reach their large size. The lack of this kind of abundant resource today, Farlow argued, may restrict the size of modern mammalian carnivores.

All theropods were and are obligatory bipeds, unable to walk or run on anything but their hind legs (Figures 12.3, 12.5, and 12.6). The body was balanced directly over the pelvis, with the vertebral column held nearly horizontally. Evidence from the skeleton and trackways indicates that the hind legs were held close to the body. Theropod trackways are always narrow gauge (as are all dinosaur trackways), in some cases so narrow that one foot appears to have been placed *ahead* of the other (rather than alongside it). The trackways, as well as skeletal material, also indicate that the foot was held in a **digitigrade** stance – that is, on the toes with the ball of the foot held high off of the ground – and that the digits themselves terminated in sharp claws.

Smaller theropods (especially troodontids and *Avimimus*) have short femora compared to the great length of the rest of the hindlimb. This condition tends to emphasize powerful and long strides as the animal runs. Calculations of running speeds on the basis of hindlimb proportions (see Box 14.2) indicate that many of these animals were indeed quite rapid movers, probably clocking at 40 to 60 km/hr, metabolism permitting (see Chapter 14). Some footprint evidence bears these numbers out; for example, a trackway in Texas suggests that a theropod thundered away at upwards of 45 km/hr.

WEAPONRY.   The hindlimb, and especially the sharply clawed feet, had important functions that extended far beyond running abilities. Here we speak of the delicious topic of prey dismemberment. Nearly all theropods carried a significant portion of their weaponry on their feet (Figure 12.16). It is not coincidence that fighting cocks – living, avian theropods – use their feet to telling effect. In dromaeosaurids and troodontids, the claw on the second digit of the foot was especially huge and sharp. Because of its joint with the rest of the toe, this scimitar-like claw is capable of a very large arc of motion. During normal walking and running, this weapon was retracted to protect it from abrasion and other damage from contact with the ground. But when needed, it could be flexed back into lethal position and, with the powerful kicking motion of the rest of the leg, this razor-bladed foot could slash its way into the belly of some hapless herbivore, disemboweling the animal in one rapid stroke.

Not all of a theropod's weapons were on its feet – not by a long shot. Commonly equally well developed were the powerfully built forelimbs, equipped with strong, grasping hands. Death surely came as often and as forcefully at the hands as at the feet of many a theropod. All theropods evolved from a common ancestor that uniquely evolved an enlarged hand from the shorter-handed ornithodiran condition. In addition, the largest of digits (II, III, and IV) were capable of extreme extension, a great advantage in grabbing onto large prey. And the next evolutionary step in theropod evolution – that of lengthening the fingers and capping each with a trenchant claw – makes a great deal of sense simply from a slashing point of view. Other reorganizations to this already modified hand included changes in the wrist and hand to allow for yet better grasping ability in avetheropods and, later on, the development of the semilunate carpal bone. This modification provided a much greater-than-normal grasping ability, of obvious importance to great manual dexterity and the ability to sever flesh from bone.

The formidable armament of theropods was lethally coupled with exceptional balance. The nearly horizontal position of the vertebral column took advantage of the center of gravity being positioned near the hips. Although this was obviously important to predatory dinosaurs both large and small, it probably was not enough for the kind of agility we suspect was part of the daily routine of some of the small and highly aggressive theropods. For such animals, balance was everything. On the basis of their lightweight, yet powerfully built, skeletons, these animals must have had an extraordinary degree of agility, balance, and leaping ability.

The ability of tetanurans like *Deinonychus* and *Velociraptor* to have functioned effectively may well have come down to the design of their tail. Recall that a derived feature of dromaeosaurids is an ability to pivot a stiffened tail just behind the pelvis. This stiffening would allow the rigid tail to move as a unit in any direction and thus to function as a dynamic counterbalancing device against the motions of the long arms and grasping hands. This must have been of great importance to a dromaeosaurid as it gripped with its large, powerful hands and struck out with one of its lethal hind feet to dispatch a fleeing victim.

If grasping hands and long arms generally typify theropod skeletons, how does one explain the outrageously tiny forelimbs of *Tyrannosaurus* and *Carnotaurus* (Figures 12.8b and c, respectively)? Not easily. For example, *T. rex* could not even reach its mouth with one of its hands. It has been suggested that perhaps the arms were short in order to balance an overly large head. In bipedal animals that have exceptionally large heads (like *Tyrannosaurus*, but not so much the case in *Carnotaurus*), increasing head size may require downsizing other aspects of the front half of the body to remain balanced with the back half at the hips. Still, there's the issue of what functions (if any) these dinky limbs may have had. The bones of the arms and hands are stout, and there is good reason to believe that the shoulder and upper part of the limb were quite powerfully muscled. In fact, Denver paleontologist K. Carpenter and M. Smith of the Museum of the Rockies, Bozeman, Montana, have calculated that the arm of *Tyrannosaurus* may have been able to lift 200 kg! In the 1960s, English paleontologist B. Newman suggested that extending a small forelimb might be enough to catapult one of these large animals up from its presumed belly-down sleeping position. Not much accepted today, Newman's idea requires that tyrannosaurids slept belly down, a matter that remains pure speculation.

Theropod skulls are formidable affairs. In the case of the big tyrannosaurids, they can be upwards of 1.75 m in length. In the case of smaller theropods, skulls commonly reveal a brace of serrated bladelike teeth, and they hint at acute vision capabilities coupled with serious brain power.

In general, theropod skulls are rather primitive. By comparison with ornithopod skulls, for example, they have few of the specializations that suggest that much processing of food took place up in the head. Still, because of the well-rounded occipital condyle and its articulation with the first part of the cervical vertebrae, it is thought that the skull had considerable mobility on the neck. So too did many of the joints of the skull with each other. Their appearance of mobility, however, may be an illusion; instead, they may have been constructed in such a fashion to lessen or dissipate the stresses passing through the skull as the animal bit, subdued, and dismembered its struggling prey.

TEETH. The sharply pointed, often recurved and serrated teeth in the upper and lower jaws ultimately served the animal best in handling its prey. It has never been an issue that an animal equipped with sharply keeled and pointed teeth was anything but meat eating (Figure 12.19). Yet it took till 1992 for someone to look at the details of theropod tooth construction and its significance to the cutting of flesh. Chicagoan W. L. Abler compared different theropod teeth with an assortment of serrated steel saw blades. Abler concluded that the teeth of smaller

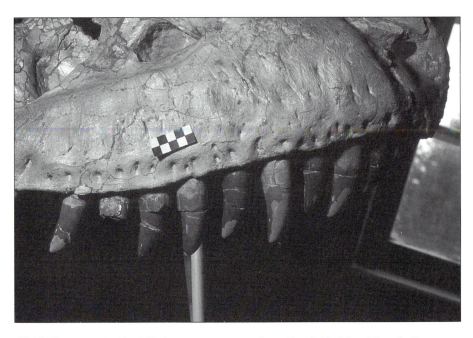

12.19 The upper teeth of *Tarbosaurus* as seen from the right side of the skull. Scale in cm.

theropods such as *Troodon*, with their prominently pointed serrations, functioned effectively in a "grip-and-rip" fashion, a bit like a serrated hacksaw blade. Tyrannosaurids, on the other hand, with their banana-shaped teeth with thickly rounded serrations, had weaker cutting ability. Based upon creases and striations in the teeth, however, Abler speculated that the teeth were subjected to complex, strong, and violent forces, such as might occur with a powerful, actively struggling prey. Beyond this, the precise function of tyrannosaurid teeth remains somewhat enigmatic (Box 12.1).

In addition, Abler observed pockets at the base of the serrations along the front and back of each tooth. These he suggested may have acted as havens for colonies of bacteria. As is the case in the living Komodo dragon (a large, predatory lizard), bacterial infections would have provided a double whammy to the victims of a theropod biting attack.

At least twice in their history, particular groups of non-avian theropods have drastically lost many or all of their teeth. Through their phylogenetic history, ornithomimosaurs lost first their upper dentition while retaining only a few of the forwardmost lower teeth. All more-derived members of the clade lost even this dental modicum. In oviraptorids, there is not even this brief glimpse of the transformation in the number of teeth in the jaws: All of these unusual-looking dinosaurs are toothless at the most basal level of the history of the clade. Instead, between their shortened upper jaws is a pair of peglike projections dead center in the middle of the palate.

What these animals ate has been widely debated. For the case of ornithomimosaurs, the muscles were oriented for rather rapid jaw closure. The margins of the jaws were almost certainly covered with cornified skin or scales that may have been relatively sharp, much like the condition in living birds. On the basis of these kinds of observations, most researchers in recent years – beginning with Russell and with H. Osmólska and her colleagues in 1972, and ending with R. Barsbold and Osmólska in 1990 – have suggested that ornithomimosaurs may have had a rather broad diet, feeding not only on such small and often soft food items as insects and their larvae, small vertebrates, eggs, and fruit, but also on larger items (medium-sized vertebrates?) that might easily slide down their throats.

Many scientists have also regarded oviraptorids as having a similar diet. However, in 1977 and again in 1983, Barsbold provided a detailed analysis of the mechanics of the skull in these animals. Unlike the situation in ornithomimosaurs,

---

BOX 12.1

TRICERATOPS AS SPOILS OR SPOILED TRICERATOPS?

IN 1917, L. M. LAMBE suggested that *Albertosaurus* (then called *Gorgosaurus*) was not so much an aggressive predator, but instead maintained its sustenance by scavenging. The basis for his remarks was the absence of heavy wear on the teeth of this theropod – these animals therefore must have fed primarily on the softened flesh of putrefying carcasses. This interpretation has appeared on and off again in discussions of theropod diet and hunting behavior.

The notion of theropod scavenging rests on the assumption that tooth wear is usually absent and that carcasses were readily available. It is further bolstered by the lack of a convincing account of why the forelimbs of these animals are so small. Contrary to Lambe's claims, tooth wear, as we have seen, is present on the teeth of nearly all large theropods. Not that that "proves" that tyrannosaurids and other large theropods had to have been active predators; both modern scavengers and active predators alike can have a high degree of wear on their teeth. The commonness of carcasses, putrefied or otherwise, was probably dependent on the season – dry, stressful seasons probably claimed their share of dead hadrosaurids, ceratopsians,

sauropods, and the like. The carcasses would have been in the form of tough, dry bodies – dinosaur jerky – not the kind of soft, predigested carrion postulated by Lambe.

Recently, the suggestion of tyrannosaurids as carrion eaters has come from the observation that these theropods had rather broad, bulbous teeth. Modern scavengers such as the hyena likewise have broad teeth, which are used to crush the bones of carcasses. Indeed, the bulbous teeth of tyrannosaurids could be a reflection of carrion eating, but they could as easily be due to the fact that as the animals increased in size, the teeth likewise increased in size, but their proportions changed as large size was attained.

In the end, we think it likely that animals like *Tyrannosaurus* would have been a more than adequate, indeed terrifying, active predator, yet one that wouldn't turn up its nose at a lunch of carrion. Indeed, it is probably better to view all theropods – from the exceptionally large tyrannosaurids down to much smaller troodontids, dromaeosaurids, and coelophysoids – as equal opportunity consumers of large herbivores, smaller flesh eaters, and even the occasional carcass.

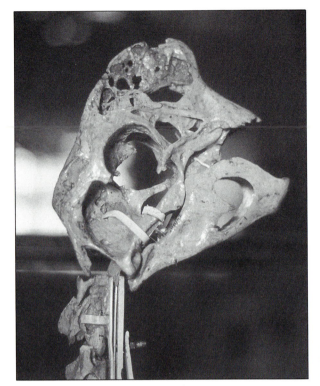

Figure 12.20 **The boxlike, toothless skull of *Oviraptor*.**

where the skull is very long and the jaw muscles relatively weak, in oviraptorids the skull is very short, boxlike and strong, and the jaw musculature is very well developed (Figure 12.20). As a consequence, Barsbold rejected a diet of eggs that had earlier been ascribed (in name and in deed) to these Late Cretaceous theropods, arguing instead that their jaws were designed to feed on hard objects that required crushing. According to Barsbold, these animals cracked the shells of bivalves open by the brute force of their jaw muscles acting on the thick horny bill covering the margins of the mouth and the palate, and especially on the stout pegs in the center of their palates.

SENSES.   To locate and track their prey, theropods of all kinds needed a keen sense of vision, so it is not surprising that eye size in these animals is large. Indeed, relatively large eyes is a shared, derived character of coelurosaurs, and within this group, ornithomimosaurs have exceptionally immense eyes. In some cases, in the more-derived tyrannosaurids and in troodontids, the eyes have migrated to a more forward-looking position, most probably conferring a degree of binocular vision to these animals. What better way to maintain good eye contact with a fleeing meal?

Likewise, hearing is exceptionally important to predatory animals, and so it is not surprising that many theropods probably had good sound perception. Indeed, the middle-ear cavity of troodontids and ornithomimosaurs was greatly enlarged, suggesting that these theropods were especially able to hear low-frequency sounds. L. M. Witmer (Ohio University) has taken analyses of hearing in troodontids even further. He noted that in these animals, the two ears communicate with each other via a pneumatic channel that passes through the base of the braincase. Because of such connections, this channel might have been used – as it is in modern birds, which also have them – in sound localization. From a purely biomechanical standpoint, airborne sound waves hitting one of the two ear drums would have been transmitted through what Witmer calls the interaural ("between the ears") pathway to the internal surface of the other ear drum. In this way, each ear conveys the same vibration to its respective middle, and then inner, ear, but one ear would have done so out of synch with the other. Out-of-synch perception of the same sound wave provides the cue for perceiving sound directionality. Put another way, the snap of a twig by, say, a baby *Maiasaura* in hiding or the grunt of an incautious solitary *Centrosaurus* would have arrived asynchronously at the outside *and* inside of a troodontid's two ears. It is a relatively simple matter (all hard-wired into the central nervous system) to evaluate the phase differences in sound, turn the head in the right direction, and bolt off after one's next meal.

PREY. Exactly who these "next meals" were is no trivial matter. In our chapters of herbivorous dinosaurs, we occasionally spoke about stomach contents and coprolites, which provide a more or less direct association of feeders and fodder. For theropods, we similarly have only a rare glimpse of this kind of information. Some of the adult skeletons of *Coelophysis* from Ghost Ranch have the remains of juveniles in their bellies. Are these cases of cannibalism? Most scientists think so. Two additional examples can be found among coelurosaurs. The first, the nearly complete skeleton of *Compsognathus* that led Huxley to his views on the origin of birds, contains much of the skeleton of a fast-running lizard. Not only did *Compsognathus* swallow whole and headfirst this delectable lacertilian meal, but it must have captured this victim through its own speed and maneuverability. Finally, the prize specimens to have come out of the years of field work in the Gobi Desert also attest to dromaeosaurid diets. For it is a particular specimen, a *Velociraptor* with its hind feet jutting half into the belly of a subadult *Protoceratops* and its hands grasping, or being held in, the jaws of the soon-to-be victim, that provides the most dynamic and irrefutable evidence about the preferred prey of *Velociraptor* (see Figure 15.1).

Beyond these few direct, and often astounding, observations of dietary preferences among non-avian theropods, we are left to speculate in a more roundabout way on the issue of who ate whom. These arenas of guesswork are often the faunal associations of the predators themselves. For example, it may be that *Marshosaurus* fed on contemporary small- to medium-sized, fleet-footed ornithopods like *Othnielia* and *Dryosaurus*; after all, they are members of the same fossil assemblages. Likewise, *Liliensternus* may have fed on *Plateosaurus*, *Alectrosaurus* may have fed on hadrosaurids, and so forth. Whether true or not, these are often the best available data that can be used to address the question of theropod diets (see also Chapter 15).

That theropods ate flesh – whether from the backs of contemporary brethren, or from mammals, lizards, snakes, or turtles, or from the inside of mollusk shells – is not a controversial issue. Whether they did it with any great style, is. The degree to which dinosaurs did *anything* in a sophisticated fashion may well be reflected in their brain size (Box 14.3). And in this measure, theropods are not at all badly off. According to J. A. Hopson (University of Chicago), all theropods for which there is brain-size information have surprising cerebral powers – their brains are every bit as large as one would expect of a crocodilian or lizard "blown up" to the proper body size. And for some, brain size is fully within the realm of a scaled-up bird. Among this latter group, troodontids had the largest brains for their body size of any of the "conventional" theropods. What this may mean in terms of lifestyle is debatable to a degree, but it is certainly suggestive that these animals probably had more complex perceptual ability and more precise motor-sensory control than some of the smaller-brained dinosaurs. It certainly implies relatively high activity levels and perhaps sophisticated inter- and intraspecific behavior.

For predatory animals, complex intraspecific behavior may have taken the form of pack hunting. In his study of *Deinonychus*, Ostrom noted that the first remains of this predator to be recovered consisted of three partial skeletons that were found in close association with a skeleton of the iguanodontian ornithopod *Tenontosaurus*. Because of the great disparity in size between these two dinosaurs (*Tenontosaurus* is nearly 7 m long, much larger than a 3.5-m-long *Deinonychus*), Ostrom argued that it was unlikely that a single *Deinonychus* could bring down a *Tenontosaurus* alone. However, add in a few more *Deinonychus*, each with their "terrible claws" unleashed, and the tables are obviously turned. A coordinated attack by two, three, or more individuals would have easily overpowered even the largest of tenontosaurs, with more than enough spoils going to pack hunters.

Although *Deinonychus* and other small predatory dinosaurs may have pursued their meals in packs, it is likely that the larger forms hunted in a solitary fashion. How this was accomplished has been a matter of some debate. Baltimore's G. S. Paul has suggested that these large predators made sharklike hit-and-run attacks, crippling or killing their victims with a deep bite to flanks, belly, or neck. They would then wait for weakened prey to succumb from loss of blood and resulting shock. Then they swooped in for the *coup de grace*.

Alternatively, large and solitary theropods may have dispatched their prey in a more active way, much like large cats do today. According to J. O. Farlow and R. E. Molnar, it may be that they suffocated their victim by seizing its snout or neck between their jaws, thus holding closed its nostrils and mouth or the trachea. All that remained was to hold on until the prey weakened or until it bled to death. This kind of attack, which is consistent with the large gape of these predators (large enough to wrap around the snout or neck of any prey animal), has more action and terror than theropods-as-sharks, but in both cases these speculations have been made in the absence of real data supplied by the fossil record. We know that theropods were active predators, but the exact means by which they brought down their prey in the majority of cases is lost in the mists of time.

As fixated as we might be on how theropods tracked, attacked, and fed, there are other aspects of their biology that are equally fascinating. And here we finally turn to intraspecific social behavior. What can the skeletons of non-avian theropods tell us about how these animals related to each other socially? Like our discussion of crests in hadrosaurids or frills and horns in neoceratopsians, we turn first to the adornments on the skulls of numerous theropods. Quite a number of these predatory dinosaurs – from *Syntarsus*, *Dilophosaurus*, *Proceratosaurus*, and possibly

Ornitholestes, to *Ceratosaurus*, *Alioramus*, *Cryolophosaurus*, *Monolophosaurus*, and *Oviraptor* – sported highly visible cranial crests. Some are made of thin lamellae of bone; others are hollow – presumably part of the cranial air–sinus system. Beyond these, theropods such as *Yangchuanosaurus*, *Allosaurus*, and *Acrocanthosaurus*, and the tyrannosaurids bore slightly elevated upper margins on the snout and the raised and roughened excrescences over the eyes. These structures are believed to have been cores for hornlets (small horns) made of keratin, which sheathed these roughened bumps (in the way that a cow's horn sheaths a bony base beneath) and which must have given the face a slightly spiky look (i.e., Figure 12.2e).

The prominent yet quite delicately built crests and the stouter hornlets assuredly functioned in display and – at least for the latter – may also have been occasionally used in head-butting squabbles over territories or mates. But how can we test such a form–function relationship? As with Hopson's analysis of the function of hadrosaurid crests (see Chapter 10), we can look to a few adjuncts of display in social animals. If crests and hornlets functioned in visual display, particularly in those theropods that lived in large groups (see the foregoing), we might expect them to be species-specific and probably sexually dimorphic so as to signal a given animal's identity and sex. And likewise we might expect crests to show their greatest development in reproductively mature individuals; youngsters should have small, poorly developed crests and hornlets.

Are these expectations met? To a degree, yes. Sexual dimorphism appears to be present in the ceratosaurs *Syntarsus* and *Coelophysis*. In these two theropods, one of the two morphs is characterized by having a relatively long skull and neck, thick limbs, and powerfully developed muscles around the elbow and hip. The other, a more gracile form, retains a number of juvenile features, including a shorter skull and neck, and slender limbs. Unfortunately, it is unclear whether the prominent cranial crest is sexually dimorphic in size and shape. In one other theropod, however, the case has been made for sexually dimorphic ornamentation: In 1990, Carpenter hypothesized sexual dimorphism in *Tyrannosaurus*, based in part on the development of the bony horn cores on the head. Interestingly, it is the larger, more robust morph that has been identified as female. Is the same true for *Syntarsus* and *Coelophysis*?

We know little about the growth and development of theropods. We again have information from the famous bonebeds (for example, Ghost Ranch and Cleveland-Lloyd), as well as from some of the Upper Cretaceous localities of the Gobi Desert. For *Coelophysis* and *Syntarsus*, apparently there was a 10- to 15-fold increase in body size from hatchling to maturity, but A. Chinsamy (see Chapter 14) suggests that this took place over as long as eight years. Accompanying this growth were proportional changes in the skull (relatively smaller eye socket, and enlargement of the jaws and the areas for muscles), relative lengthening of the neck, and relative shortening of the hindlimb. Similar changes – when they can be identified – are thought to occur in *Tarbosaurus* and *Albertosaurus* as well. Recently, eggs that for 70 years were thought to belong to *Protoceratops* have now been attributed to *Oviraptor* (Figure 12.21; see also Chapter 9, footnote 2). The embryo preserved in these is the first egg-bound embryo known for Theropoda.

There is one more question that we can ask about theropod behavior before closing this chapter: Could theropods growl, howl, grunt, and groan? The obvious answer is that we don't yet know. Could they hiss? More likely, since any ani-

Figure 12.21 **A nest of "*Protoceratops* eggs," now believed to belong to *Oviraptor*.** *Photograph courtesy of the American Museum of Natural History.*

mal that can exhale can (and usually does) hiss. In any event, it is certainly true that, were theropods capable of making any noises, the great cavernous nature of many of their skull bones, from pneumatic lacrimal bones to sinuses of the maxillae, would have been conducive to vocal resonation. Thus we might expect deep and sonorous bellows from the likes of *Albertosaurus* and *Tarbosaurus*, and perhaps higher-frequency screeches from *Troodon* and *Dromiceiomimus*. Not surprisingly, these kinds of vocal frequencies roughly match those thought to have been perceived by the theropod ear (see our earlier discussion of low-frequency acuity). Moreover, we know that sound plays an important part in the social interrelationships of living theropods (birds). Is it too radical to propose that non-avian theropods likewise relied upon sound for social interactions?

We have come full swing, from the initial discovery of the tooth-bearing theropod jaw in 1819 (given the name *Megalosaurus* by Oxford don Buckland) to our re-creation of theropods as complex, effective predators. Moreover, we have set the stage for their ascent into the heavens. But that – that story's for the birds!

Important Readings

Barsbold, R., and H. Osmólska. 1990. Oviraptorosauria; pp. 249–258 *in* D. B. Weishampel, P. Dodson, and H. Osmólska (eds.), The Dinosauria. University of California Press, Berkeley.

Barsbold, R., and H. Osmólska. 1990. Ornithomimosauria; pp. 225–244 *in* D. B. Weishampel, P. Dodson, and H. Osmólska (eds.), The Dinosauria. University of California Press, Berkeley.

Clark, J. M., A. Perle, and M. A. Norell. 1994. The skull of *Erlikosaurus andrewsi*, a Late Cretaceous "segnosaur" (Theropoda: Therizinosauridae) from Mongolia. American Museum Novitates, v. 3115, 39 pp.

Colbert, E. H. 1989. The Triassic dinosaur *Coelophysis*. Museum of Northern Arizona Bulletin 57:1–160.

Currie, P. J. 1990. Elmisauridae; pp. 245–248 *in* D. B. Weishampel, P. Dodson, and H. Osmólska (eds.), The Dinosauria. University of California Press, Berkeley.

Currie, P. J., Dong Z.-M., and D. A. Russell. 1993. Results from the Sino-Canadian Dinosaur Project. Canadian Journal of Earth Sciences 30:1997–2272.

Gauthier, J. A. 1986. Saurischian monophyly and the origin of birds. Memoirs of the California Academy of Sciences 8:1–55.

Horner, J. R., and D. Lessem. 1993. The Complete *T. rex*. Simon and Schuster, New York, 239 pp.

Holtz, T. R. 1994. The phylogenetic position of the Tyrannosauridae: implications for theropod systematics. Journal of Paleontology 68:1100–1117.

Molnar, R. E. 1990. Problematic Theropoda: "carnosaurs"; pp. 306–319 *in* D. B. Weishampel, P. Dodson, and H. Osmólska (eds.), The Dinosauria. University of California Press, Berkeley.

Molnar, R. E., and J. O. Farlow. 1990. Carnosaur paleobiology; pp. 210–224 *in* D. B. Weishampel, P. Dodson, and H. Osmólska (eds.), The Dinosauria. University of California Press, Berkeley.

Molnar, R. E., S. M.Kurzanov, and Dong Z. 1990. Carnosauria; pp. 169–209 *in* D. B. Weishampel, P. Dodson, and H. Osmólska (eds.), The Dinosauria. University of California Press, Berkeley.

Norell, M. A., J. M. Clark, D. Dashzeveg, R. Barsbold, L. M. Chiappe, A. R. Davidson, M. C. McKenna, A. Perle, and M. J. Novacek. 1994. A theropod dinosaur embryo and the affinities of the Flaming Cliffs dinosaur eggs. Science 266:779–782.

Norman, D. B. 1990. Problematic Theropoda: "coelurosaurs"; pp. 280–305 *in* D. B. Weishampel, P. Dodson, and H. Osmólska (eds.), The Dinosauria. University of California Press, Berkeley.

Novas, R. E. 1993. New information on the systematics and postcranial skeleton of *Herrerasaurus ischigualastensis* (Theropoda: Herrerasauridae) from the Ischigualasto Formation (Upper Triassic) of Argentina. Journal of Vertebrate Paleontology 13:400–423.

Osmólska, H., and R. Barsbold. 1990. Troodontidae; pp. 259–268 *in* D. B. Weishampel, P. Dodson, and H. Osmólska (eds.), The Dinosauria. University of California Press, Berkeley.

Ostrom, J. H. 1990. Dromaeosauridae; pp. 269–279 *in* D. B. Weishampel, P. Dodson, and H. Osmólska (eds.), The Dinosauria. University of California Press, Berkeley.

Paul, G. S. 1988. Predatory Dinosaurs of the World. Simon and Schuster, New York, 464 pp.

Rowe, T., and J. A. Gauthier. 1990. Ceratosauria; pp. 151–168 *in* D. B. Weishampel, P. Dodson, and H. Osmólska (eds.), The Dinosauria. University of California Press, Berkeley.

Russell, D. A., and Dong Z. 1993. The affinities of a new theropod from the Alxa Desert, Inner Mongolia, People's Republic of China. Canadian Journal of Earth Sciences 30:2107–2127.

Russell, D. A., and D. E. Russell. 1993. Mammal–dinosaur convergence. National Geographic Research and Exploration 9:70–79.

Sereno, P. C., C. A. Forster, R. R. Rogers, and A. M. Monetta. 1993. Primitive dinosaur skeleton from Argentina and the early evolution of Dinosauria. Nature 361:64–66.

Sereno, P. C., and F. E. Novas. 1993. The skull and neck of the basal theropod *Herrerasaurus ischigualastensis.* Journal of Vertebrate Paleontology 13:451–476.

Sues, H.-D. 1990. *Staurikosaurus* and Herrerasauridae; pp. 143–147 *in* D. B. Weishampel, P. Dodson, and H. Osmólska (eds.), The Dinosauria. University of California Press, Berkeley.

CHAPTER 13

THEROPODA II:
The Origin of Birds

This textbook: "A cow is a fish."
Reasonable student: "Right. And a bird is a dinosaur!"

LET'S CUT TO THE CHASE: Birds *are* dinosaurs. As we shall see, this is not quite so radical a statement as it sounds, but it does involve many of the principles that we applied in Chapter 5 when we established the relationships among vertebrates.

Suppose we said to you, "Humans are mammals." You wouldn't be likely to argue. Mammalia, after all, is a relatively large group that encompasses all kinds of creatures, including humans. But in the end, the issue boils down to a much simpler point (discussed in Chapter 3): Humans are mammals because humans have the diagnostic features that pertain to all mammals (i.e., hair, mammary glands, a unique ear apparatus and jaw joint, and double tooth replacement, to name but a few of the characters diagnostic of Mammalia) and because the most recent common ancestor of humans and other mammals was itself a mammal.

Claiming that birds are dinosaurs is no more radical than saying that humans are mammals. If birds are dinosaurs, then birds must have the characters that are specific to Dinosauria. If they do, and such characters are really unique to the dinosaurs (so we argued in Chapters 4 and 5), then birds are *by diagnosis* dinosaurs.

In applying the cladograms to understand the ancestry of birds, we will be able to develop answers to three fundamental questions about birds. These questions are

1. Where do birds come from?
2. Where do feathers come from?
3. How did avian flight evolve?

Before we can address these questions, however, we must first establish what features characterize birds. Having established these, we can go back within Archosauria in general, and Dinosauria in particular, and consider the diagnostic characters the groups share.

WHAT IS A BIRD?

Living birds are an easily identifiable group of organisms (Figure 13.1). It would be very difficult to confuse any other creature with a bird, and it would be equally improbable to confuse a bird with anything else. Birds are obviously vertebrates; they possess a nerve cord and a backbone. Moreover, they are clearly tetrapods (they bear four limbs) and amniotes (their eggs have amniotic membranes). It turns out that bird skulls belong to the great diapsid clan (they have upper and lower temporal openings) and, within diapsids, they are indubitably archosaurs by their possession of an antorbital opening. Birds don't superficially resemble crocodiles (the other living archosaurs), and it is helpful to think of birds as possessing a unique suite of highly evolved features superimposed upon the archosaur body plan.

FEATHERS. All birds – and only birds – have feathers. Feathers are complex in terms of both their structure and their development (Figure 13.1a). All feathers are composed of fingernail-type (keratinized) material. They consist of a hollow, central **shaft** that decreases in diameter toward the tip. Radiating from the shaft are **barbs**, processes of feather material that, when linked together along the length of the shaft by small hooks called **barbules**, form the sheet of feather material called the **vane**. Feathers with well-developed, asymmetrical vanes called **flight feathers;** such feathers are essential for flight. Feathers in which the barbules are not well developed tend to be puffy, with poorly developed vanes; such feathers are called **downy feathers** or **down** and, as we know from sleeping bags, comforters, and ski parkas, are superb insulation.

The importance of all this in an evolutionary sense is that *feathers evolved (or originated) only once.* After all, what are the chances of so complex a structure's having evolved more than once? Using parsimony, we must conclude that feathers evolved only one time, which means that the organism that bore the first feathers (a bird) is the ancestor of all other feathered creatures. In turn, this means that all birds share a common avian ancestor, which means that on the basis of feathers, birds are a monophyletic group.

ABSENCE OF TEETH. No living bird has teeth. The jaws of birds are covered with sheaths of fingernail-like material called a *beak* (rhamphotheca). The absence of teeth implies that no bird chews its food in the way that mammals do; processing of food is left to organs farther down the gut.

LARGE BRAINS AND ADVANCED SIGHT. Although we speak derogatorily of "bird brains," modern birds have well-developed brains, certainly by comparison to their living diapsid brethren (lizards, snakes, and crocodiles), and even by comparison to many mammals. It is believed that birds originally evolved high

295

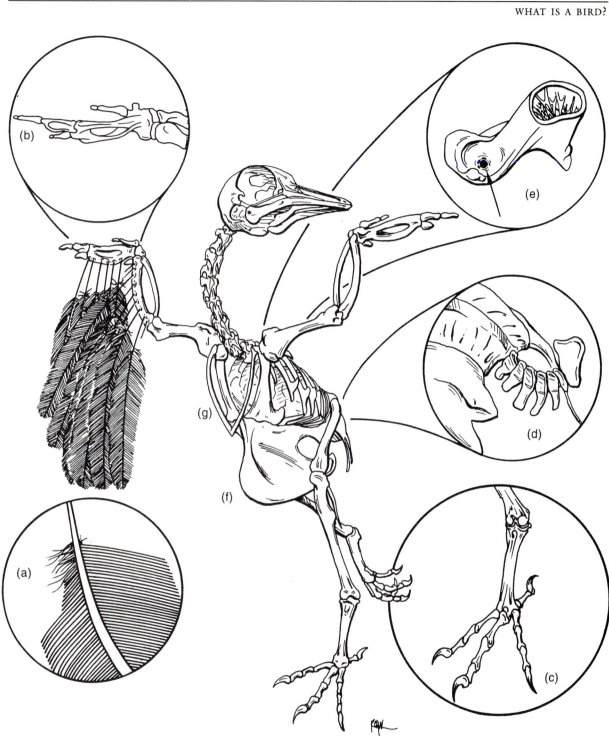

Figure 13.1 **The skeleton of a pigeon, showing major features of its skeletal anatomy.** **(a)** Detail of feather structure; **(b)** carpometacarpus; **(c)** tarsometatarsus and foot; **(d)** pygostyle; **(e)** hollow bone with pneumatic foramen (arrow); **(f)** sternum with large downward projecting keel; and **(g)** furcula.

encephalization through possession of endothermy (and the complex controls necessary to maintain it; see Chapter 14) and because of the sophisticated loco-motor control necessary for flight. Along with relatively large brains, birds tend to have well-developed vision. This also is thought to have been associated with sophisticated motor activities, such as flight.

CARPOMETACARPUS.   The wrist and hand bones in the hands of modern birds are fused into a unique structure called the **carpometacarpus** (Figure 13.1b). Most of us are more familiar with the carpometacarpus than we might realize; it is the carpometacarpus that forms the flattened tips of "buffalo wings" that we savor as appetizers. Precisely which fingers are involved in the formation of the car-pometacarpus remains up for grabs. All agree that the fifth finger is lost in birds. Embryologists, studying the development of bird embryos, have argued that the first finger is lost as well, making the digits that compose the carpometacarpus II, III, and IV. The evidence from paleontology (based upon the ancestry of birds), however, suggests that the carpometacarpus is composed of digits I, II, and III. Thus, a strong case can be made for the loss of the fourth finger as well as the fifth.

FOOT.   The feet of all birds are distinctive. They consist usually of three toes in front and one toe toward (or at) the rear. The fifth digit is lost; in front are digits II, III, and IV, and toward the rear is digit I (homologous with our big toe). Birds have well-developed, strong claws on their feet. The three metatarsals (foot bones, to which the toes attach; see Chapter 4) – digits II, III, and IV – are fused to one another and with some of the ankle bones, forming a structure called a **tar-sometatarsus** (Figure 13.1c).

PYGOSTYLE.   No living bird has a long tail. Instead, the bones are fused into a small, compact, pointed structure called a **pygostyle** (*pygo* – rump; *stylus* – stake; Figure 13.1d). The flesh surrounding this tail remnant rejoices under the nick-name of "pope's (or parson's) nose" at Thanksgiving time. At least two groups of birds do not possess a pygostyle, but neither do they possess well-developed tails.

PNEUMATIC BONES.   Living birds have an extremely complex and sophisticated system of air sacks throughout their bodies. In many cases, the bones are extremely light, with thin walls and a minimal series of splintlike buttresses brac-ing them internally. This type of bone structure is called **pneumatic**, reflecting the fact that the bulk of the volume in bird bones is taken up by air spaces. The key feature of pneumatic bones is **pneumatic foramina**: openings in the wall of the bone for the air sacks to enter the internal bone cavities (Figure 13.1e). Traditionally, pneumatic bones have been interpreted by ornithologists as an adaptation for flight: Tremendous strength is achieved with minimal weight. Indeed, soaring birds, like the albatross and vultures, have pneumatic bones throughout the body, whereas diving birds, such as the loon, have much less pneumaticity in their bones. Although there is no doubt that the lightness associ-ated with pneumatic bones is important in flight, we shall see that hollow bones, at least, have a quite ancient history unrelated to flight.

RIGID SKELETON. Bird skeletons have undergone a series of bone reductions and fusions to produce a light, rigid platform to which the wings and the muscles that power them can attach.[1] The chest region of birds is a kind of semiflexible basket, in which fused vertebrae in the back are connected with a well-developed breastbone, or sternum, by ribs with upper and lower segments. The pelvic region is fused together into a **synsacrum** – a single, locked structure consisting of pelvic vertebrae. The sternum itself is commonly quite large and, particularly in flapping flyers, has developed a broad, deep **keel**, or downward-protruding bony sheet, for the attachment of flight muscles (Figure 13.1f).

Rigidity is maximized in the shoulder region, where the powerful flight muscles attach. Pillarlike coracoid bones buttress against the front of the sternum, the scapulae (shoulder blades), and against the **furcula** (Figure 13.1g), formed from the paired, fused **clavicles** (collarbones). The furcula is a rather distinctive feature in birds, particularly in flying ones (it can be reduced or absent in nonflying birds). Arching upward and outward, the furcula connects with the coracoids and scapulae at the shoulder joint (which is the attachment site for the wing). At the dinner table, we know the furcula as the "wishbone." Virtually no other organism has a furcula; it occurs in all birds and is known in two oviraptorids (maniraptoran theropods), *Oviraptor* and *Ingenia* (Chapter 12).

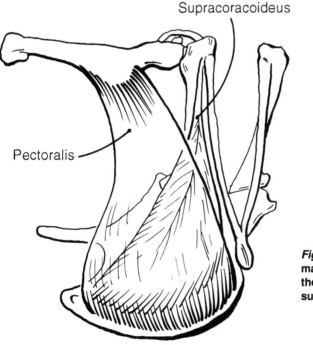

Supracoracoideus

Pectoralis

Figure 13.2 **The two major muscles for flight: the pectoralis and the supracoracoideus.**

---

[1]The situation is not unlike that found in racing bicycles. Racing bicycles are designed to maximize the power output of the rider. For this reason, they are light (so that the weight that the rider has to lug around is minimal) and the frames are stiff. Stiff frames translate an optimal amount of power from the rider's legs to the back wheel. A frame that flexes effectively absorbs some of the energy that could otherwise be delivered to the rear wheel.

Finally, modern flying birds have a unique arrangement of muscle to effect the wing recovery stroke. The *downward* stroke is obtained by the pectoralis muscle, which attaches to the front of the coracoid and sternum and to the furcula and humerus. The *recovery* stroke, however, is carried out by the supracoracoideus muscle. This muscle inserts at the keel of the sternum, runs up along the side of the coracoid bone, and inserts a tendon at the top of the humerus through a hole formed by the coracoid, furcula, and scapula (Figure 13.2). Modern birds can fly if that tendon is cut; however, as noted by ornithologist R. Raikow, take-off is greatly impaired. The hole through which the supracoracoideus tendon reaches the top of the humerus is an important, unique adaptation of modern birds to a particular type of flight stroke.

ARCHAEOPTERYX LITHOGRAPHICA AND THE ANCESTRY OF BIRDS

Archaeopteryx (*archaeo* – old; *pteryx* – wing) began first as a feather (Figure 13.3). Recall that names are applied to complete fossils if possible, but any part of a fossil organism can be named if the material that is found is sufficiently different from existing taxa to warrant a new name. After all, if more material is found later, it can be *referred* to the original name. So it was with *Archaeopteryx*. The name was first applied to a feather impression, found in 1860 in fine-grained deposits of carbonate mud located near the town of Solnhofen, Bavaria (southern Germany).[2] These deposits are the remnants of a Late Jurassic–aged, stagnant, poorly oxygenated coastal lagoon and are so fine grained that for hundreds of years they had been quarried as a source of lithographic plates: The smooth, light-tan-colored rock split beautifully into thin slabs, which could then be delicately carved to be used as printing plates (hence the species name *lithographica*). Periodically, well-preserved, delicate fossils had been found at this locality – ancient crabs, lobsters, fish, insects, and even fragile pterosaurs. The low energy of the lagoon had apparently allowed things to settle undisturbed, and the lack of oxygen hindered microorganisms that would otherwise have destroyed the creatures. Now Solnhofen produced a fossil that extended the fossil record of birds back to the Late Jurassic![3]

The excitement caused by the isolated feather was nothing compared to that generated a year later, when a specimen – feather impressions *and* bones – was quarried out of Solnhofen. The half-meter-long fossil was rather strange, because it had bird feathers contentedly residing with clearly "reptilian" features, such as

---

[2] There is some controversy about the name *Archaeopteryx*. J. H. Ostrom suggests that the original feather was unnamed and that the name was really intended for the fossil bones (with feather impressions) now housed in the British Museum (Natural History), the so-called London specimen.

[3] S. Chatterjee of Texas Tech University claims that the fossil record of birds runs considerably farther back than Late Jurassic time. In 1988, he discovered a Late Triassic–aged creature that he interpreted to be a bird. A large amount of publicity preceded the publication of his study describing the specimen, which he named *Protoavis* to commemorate its supposed position as the oldest known bird. Unfortunately, other paleontologists studying the *Protoavis* material could not reach Chatterjee's conclusions. Most thought the animal was a primitive theropod (or parts of several different creatures, including a theropod!), without the necessary diagnostic characters to securely identify it as a bird. Chatterjee, however, still adheres to his original claim that it is the oldest known bird; specialists still generally consider *Archaeopteryx* to be the oldest known bird.

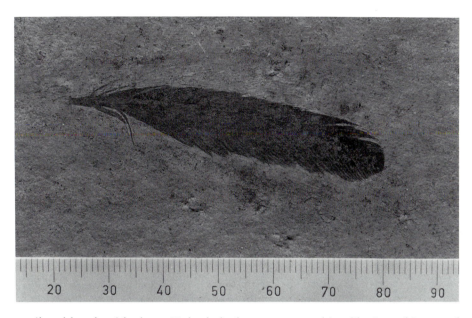

Figure 13.3 **The first evidence for Jurassic- aged birds: the feather of *Archaeopteryx litho- graphica,* described in 1861, from the Solnhofen quarry.** *Photograph cour- tesy of J. H. Ostrom.*

a tail and hands with claws. Nobody had ever seen anything like it, and it caused a furor.[4]

This second specimen of *Archaeopteryx* quickly got into the entrepreneurial hands of a local doctor, who made it his business (literally) to sell the thing to the highest bidder. So the first *Archaeopteryx* with bones was sold to an astute Sir Richard Owen (and the Trustees of the British Museum [Natural History]; hence it is now called the London specimen), who made it his mission to obtain the fos- sil; it was ironic that as late-19th-century nationalism underwent the nascent rum- blings that eventually led to World War I (and beyond), the prized fossil of Germany ended up in Britain.

The next *Archaeopteryx* specimen, retrieved from Solnhofen in 1877, ended up (after no small amount of negotiating and money) securely in German hands (this is called the Berlin specimen, which is where it presently resides)![5] This third specimen was a gorgeous thing: complete, with feathers spread out in natural position, and with a skull clearly showing teeth (Figure 13.4). Since then, five other specimens have been found, all from Solnhofen: one in 1951 (the Eichstätt specimen), one in 1956 (the Maxberg specimen), one that was collected in 1855 but not identified as *Archaeopteryx* until 1970 (the Teyler specimen), one in 1989 (the Solnhofen specimen), and an unnamed one reported in August 1993. To date, the taxon *Archaeopteryx* consists of seven specimens and one feather. The "reptilian" ancestry of *Archaeopteryx* was established to virtually everybody's satisfaction.

---

[4] The timing was spectacular; Darwin had just published *On the Origin of Species* in 1859, propos- ing that species evolved into other species. Here, a mere two years later, was discovered just that: an apparent "missing link" that had both "reptilian" and avian features. Interestingly, Darwin never referred to *Archaeopteryx* in later editions of *On the Origin of Species.*

[5] The bartering that distributed the various specimens of *Archaeopteryx* to where they currently reside is beautifully described in A. Desmond's book *The Hot-Blooded Dinosaurs,* as well as in A. Feduccia's book *The Age of Birds.*

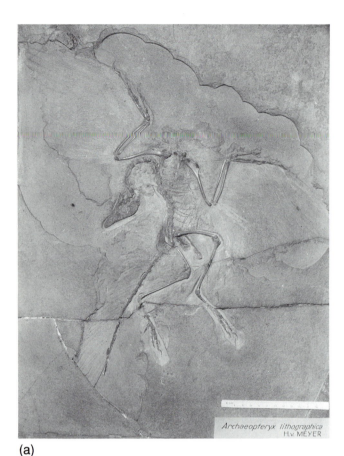

(a)

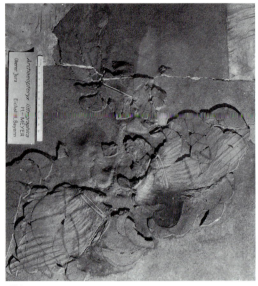

(b)

Figure 13.4 **The beautifully preserved, complete Berlin specimen of *Archaeopteryx*.**
(a) Main slab preserving most of specimen;
(b) counterslab, preserving opposite side of specimen, primarily impressions. *Photographs courtesy of J. H. Ostrom.*

Anatomy of *Archaeopteryx*

Learning who or what *Archaeopteryx* is requires a close look at some of its features (Figure 13.5).

SKULL. A beautiful skull of *Archaeopteryx* is preserved in the Eichstätt specimen (Figure 13.5a). The skull is approximately triangular in shape, with nasal, antorbital, and eye openings present. The eye opening is rather large and round and possesses a **sclerotic ring**, a series of plates that supported the eyeball. *Archaeopteryx* clearly has teeth; they are conical and recurved.

ARMS AND HANDS. The arms are quite long ($\geq$ 70% of the length of the legs) and, by comparison with modern birds, relatively unspecialized. The hands on *Archaeopteryx* are about as large as the feet, and each hand bears three fully moveable, separate fingers. Each finger is tipped with a well-developed, recurved claw. The wrist of *Archaeopteryx* bears a semilunate carpal (see Chapter 12).

LEGS AND FEET. The foot has three toes in front and a fourth toe behind (Figure 13.5b). The three in front are more or less symmetrical around digit III, and all the toes have well-developed claws. Analysis of the strong curvature of the rear claw has suggested to University of North Carolina ornithologist A. Feduccia that *Archaeopteryx* was a tree-dwelling animal; however, this view has since been criticized several times through comparisons with obviously ground-dwelling (nonavian) theropods. The jury remains out.

The ankle of *Archaeopteryx* is a modified mesotarsal joint (see Chapter 5). Although there has been some disagreement, it is generally agreed that a small splint of bone rises up from the center of the astragalus (the ankle bone) to form a tall ascending process. The three foot bones (metatarsals II, III, and IV) are elongate, narrow, and pressed closely together, but clearly unfused. The legs are slightly longer than the arms, and the thighbone has a gentle S-shape with the ridgelike head turned at 90° to the shaft. The thighs are considerably shorter than the shins (tibia and fibula), although the fibula itself is attenuate; that is, it becomes smaller and smaller as it approaches the ankle.

LONG BONES. *Archaeopteryx* has thin-walled, long bones with large hollow spaces within.

TRUNK AND TAIL. The axial skeleton of *Archaeopteryx* seems to lack many of the highly evolved features that characterize modern birds. The body is relatively long and shows none of the foreshortening or fusion that one sees in the vertebrae of birds. The sternum is relatively small. A large, strong furcula is present (Figure 13.5c). Also present are **gastralia**, or belly ribs, which primitively line the belly in vertebrates (13.5d).

Archaeopteryx lacks a synsacrum and, instead, has a more typical (i.e., generalized) "reptilian" pelvis. In the Berlin specimen, the pubis is directed backward; however, the bone has apparently been broken where it connects with the ilium. In other specimens it points downward. A footplate on the pubis is well developed, although the front part is absent.

Archaeopteryx has a long, straight, well-developed tail. Projections from the neural arches (zygopophyses) are elongate, meaning that the tail is "locked up" and has little potential for movement.

FEATHERS. *Archaeopteryx* has well-preserved feather impressions.[6] The best-preserved feathers are clearly flight feathers (Figure 13.5e). Those on the arms are in number and arrangement very much like those seen in modern birds and, as in modern birds, the vanes are asymmetrical. Moreover, as in modern birds, the

---

[6] The feathers of *Archaeopteryx* have caused no small amount of comment, much of it frustratingly incorrect. The most egregious example of this was a series of publications in the *British Journal of Photography* in the middle 1980s on the possibility that the feathers of *Archaeopteryx* were forged. The authors were two distinguished astronomers, F. Hoyle and N. C. Wickramasinghe, and some collaborators. Observation by low-angle photographic techniques led them to the conclusion that the feather impressions were actually made by carving and by the addition of a gypsum paste. There were some obvious misidentifications (such as claiming the tail to be a single feather), and the charges were thoroughly refuted by a variety of scientists. The gist of one of the most important refutations goes as follows: *Archaeopteryx* specimens come from slabs of Solnhofen limestone that have been split, revealing the specimen. The fossil is found on the so-called main slab, but impressions and bone fragments are also found on its opposite, the "counterslab." It would have to be one extraordinary forger to produce the precise, but opposite, feather pattern on the counterslab. P. Wellnhofer, in describing the 1988 Solnhofen specimen, banged the final nails in the coffin by noting that the Solnhofen specimen has feather impressions; it would have to be a spectacular forger, indeed, who could carry on a case of superhumanly skillful forging for 130 years!

Figure 13.5 **A reconstruction of *Archaeopteryx*. (a)** Skull, including teeth; **(b)** foot; note the arcuate claws, suggestive of arboreal adaptation; **(c)** robust, theropod-like furcula; **(d)** gastralia; **(e)** a close-up of the feathers; and **(f)** hand. Note semilunate carpal (compare with the theropod *Ingenia* in Figure 12.14b). *Photographs courtesy of J. H. Ostrom.*

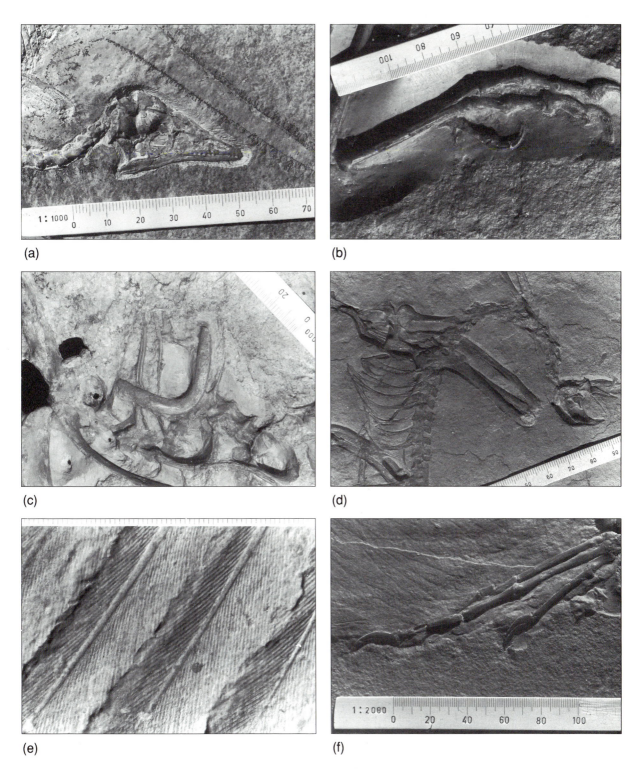

(a)

(b)

(c)

(d)

(e)

(f)

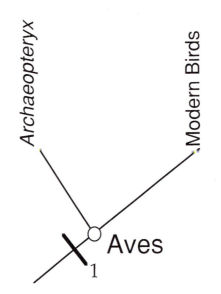

Figure 13.6 Cladogram of Aves (all birds). A diagnostic character uniting Aves at 1 is the presence of feathers.

asymmetrical vanes produce an airfoil cross section in the wing. Unlike in modern birds, however, there are feathers also lining the well-developed tail. These radiate out from the vertebrae and form an impressive tail plume.

So what, then, is *Archaeopteryx*? Obviously, if the creature has feathers, and all birds have feathers, then *Archaeopteryx* must be a bird. This is not only a matter of diagnosis; it also has important evolutionary consequences. Earlier, we said that feathers evolved only once. That being the case, it is clear that *Archaeopteryx* and modern birds must share a common feathered ancestor that is itself a bird. Based upon the presumed uniqueness of feathers, birds are a monophyletic group (Figure 13.6).

Birds as Dinosaurs

The relationship of birds to dinosaurs as outlined here is hardly new. The famous early Darwinian T. H. Huxley, as well as a variety of European natural scientists from the middle and late 1800s, recognized the connection between the two groups. As noted in 1986 by J. A. Gauthier of the California Academy of Sciences, Huxley outlined 35 characters that he considered "evidence of the affinity between dinosaurian reptiles and birds," of which 17 are still considered valid today.

So what happened? Why is it news that birds are dinosaurs? During the very early part of the 20th century, Huxley's ideas fell into some disfavor, because it was proposed that many of the features shared between birds and dinosaurs were due to convergent evolution.

What evidence was there to argue for convergence in the case of dinosaurs and birds? Really, not terribly much. But in light of limited knowledge of dinosaurs at the time, the group just seemed too specialized to have given rise to birds. Moreover, clavicles were not known from coelurosaurs (then, as now, the leading contender as the most likely dinosaurian ancestor of birds). Thus, the fused clavicles (furcula) in birds had to be explained. What was needed was a more primitive group of archosaurs that did not seem to be as specialized as the dinosaurs.

In the early part of the 20th century, such a group of archosaurs, the ill-defined "Thecodontia," was established by the Danish anatomist G. Heilmann as the group from which all other archosaurs evolved (see Chapter 4). Because this was by definition the group that gave rise to all archosaurs, and because birds are clearly archosaurs, it was concluded that birds must have come from "thecodonts." Heilmann had in mind an ancestor such as *Ornithosuchus* (note the name: *ornith* – bird; *suchus* – crocodile), a 1.5-m carnivorous bipedal archosaur that would have looked a bit like a long-legged crocodile. For over 50 years, Heilmann's detailed and well-argued analysis held sway over ideas about the origin of birds.

Several events caused the thecodont-ancestry hypothesis to fall into general disfavor. The first was that clavicles were found on coelurosaurs. Moreover, it later came to be recognized that "Thecodontia" is not monophyletic. How could one derive birds (or anything else) from a group that had no diagnostic characters?!

The renaissance of the dinosaur–bird connection must be credited to J. H. Ostrom of Yale University. In the early 1970s, through a series of painstakingly researched studies, he spectacularly documented the relationship between *Archaeopteryx* and coelurosaurian theropods. His ideas inspired R. T. Bakker and P. M. Galton, who in 1974 published a paper suggesting that birds should be included within a new Class Dinosauria. The idea didn't catch on, in part because it involved controversial assumptions about dinosaur physiology (see Chapter 14) and because the anatomical arguments on which it was constructed were not completely convincing. In 1986, however, Gauthier applied cladistic methods to the origin of birds, and with well over one hundred characters demonstrated that *Archaeopteryx* (and hence, birds) are indeed coelurosaurian dinosaurs.

Before we construct the relationships of birds to other dinosaurs, a subtle point must precede our analysis. We have just established that birds are a monophyletic group. If this is true, then the primitive characters that *Archaeopteryx* bears must have something to do with the ancestry of Aves. Because birds are a monophyletic group, we do not need to go to the highly evolved living taxa to learn about their origins; we can (and should) go through the most basal member of the group known – *Archaeopteryx* – to investigate bird ancestry. When we show that *Archaeopteryx* is an archosaur and that *Archaeopteryx* is a bird – remembering that birds are monophyletic – we have shown that birds are archosaurs.

HIGHER TAXONOMIC RELATIONSHIPS OF BIRDS. We can now identify some of the larger (more encompassing) taxa to which *Archaeopteryx* belongs. First, because *Archaeopteryx* and indeed all modern birds have an antorbital opening, the avian clade – including *Archaeopteryx* – must be archosaurs. In addition, a glance at your Thanksgiving turkey (or any bird) should convince you that all birds (again, including *Archaeopteryx*, but you won't see *that* on the table!) have a fully erect posture, in which the shaft of the femur is 90° to the head. The head of any bird femur does not have a ball (as in Mammalia); instead the head is elongate. Moreover, the ankle of *Archaeopteryx* (and all birds) is a modified mesotarsal joint.

Consider the foot of ornithodirans. Ornithodiran feet are four-toed. Three toes point forward (digits II, III, and IV), and the fourth (digit I) is reduced. This last-mentioned toe lies partway down the bone adjacent to it and sticks out either to the side or toward the back. Ornithodiran feet are symmetrical around digit III,

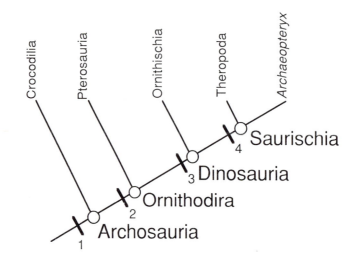

Figure 13.7 Cladogram depicting the position of *Archaeopteryx* within Archosauria. Diagnostic characters for each node include at 1: antorbital opening (Archosauria); at 2: four-toed, clawed foot, with symmetry around digit III; and digit I reduced, lying closely appressed to, and along side of, digit II (Ornithodira); at 3: semiperforate acetabulum (Dinosauria); and at 4: ascending process on the astragalus (Saurischia). The cladogram shows that *Archaeopteryx* and, therefore, birds share the diagnostic characters of Dinosauria. Birds are dinosaurs. Within Dinosauria, the character at 4, among others (see Part III: Saurischia), clearly indicates that *Archaeopteryx*, though a bird, is also a saurischian dinosaur.

and all toes bear large, arcuate claws. This condition is found in all birds (living and extinct). Birds are ornithodirans (Figure 13.7).

Within Ornithodira, Dinosauria is diagnosed by the possession of three or more vertebrae in the pelvis, a reorientation of the shoulder, reduction of the fourth finger in the hand, and a semiperforate acetabulum. *Archaeopteryx* clearly has five pelvic (or sacral) vertebrae, the reorientation of the shoulder (from the primitive archosaurian condition), and a perforate or semiperforate acetabulum; in the hand, digit IV is absent (i.e., *very* reduced; Figure 13.7).

In summary, then, the characteristics that distinguish Dinosauria within Ornithodira clearly apply to *Archaeopteryx*. The inescapable conclusion, then, is that *Archaeopteryx*, because it bears the diagnostic characters of Dinosauria, is itself a dinosaur. This is not so revolutionary; look at the illustrations within this chapter (and the preceding one) and you will see that *Archaeopteryx* indeed looks like a dinosaur. What is more revolutionary is the clear consequence of avian monophyly: Birds are dinosaurs. That being the case, Aves must be a subset of Dinosauria, and both of them should be included within an expanded Reptilia (Figure 13.7, and Figure 13.8, node 1)!

BIRDS AS THEROPODS.   Birds bear the diagnostic characters of theropods. As we have seen, since the late 1800s it has been recognized that the vertebrae and the leg and arm bones in all theropods are thin walled and hollow. This distinctive characteristic makes theropods easily recognizable in the field, even when very little fossil material is preserved. *Archaeopteryx* and all modern birds have hollow bones; this is

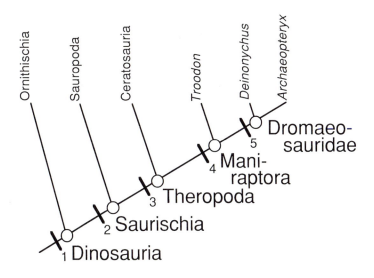

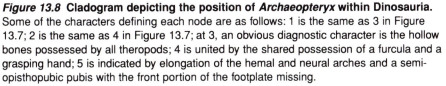

Figure 13.8 Cladogram depicting the position of *Archaeopteryx* within Dinosauria. Some of the characters defining each node are as follows: 1 is the same as 3 in Figure 13.7; 2 is the same as 4 in Figure 13.7; at 3, an obvious diagnostic character is the hollow bones possessed by all theropods; 4 is united by the shared possession of a furcula and a grasping hand; 5 is indicated by elongation of the hemal and neural arches and a semi-opisthopubic pubis with the front portion of the footplate missing.

an inheritance from the original condition found in theropod leg and arm bones. Moreover, *Archaeopteryx* bears an enlarged three-fingered hand with the deep pits at the end of the metacarpals diagnostic of Theropoda (Figure 13.8, node 3).

If we conclude that *Archaeopteryx* is a theropod, it must also be concluded that birds, too, are theropods, although highly derived ones.

BIRDS AS BASAL THEROPODS. Within Theropoda, *Archaeopteryx* is clearly situated above *Herrerasaurus*. Recall that the group above *Herrerasaurus* is united by a suite of features, including the semiopposable thumb, a semicircular antorbital opening, and expanded muscle attachment sites on the shoulder blade and humerus. *Archaeopteryx* shares these characters with other theropods above the level of *Herrerasaurus*.

As we have seen, theropods are divided into two great groups above *Herrerasaurus*. One is Ceratosauria, and the other is Tetanurae. *Archaeopteryx* has undoubted tetanuran affinities, with its stiffened tail, shortened tooth row, astragalar groove, and shortened vertebral column (ahead of the pelvis).

Archaeopteryx is clearly not at the base of Tetanurae. It is now generally agreed that *Archaeopteryx* has a high ascending process on its astragalus; however, historically, the feature has been a "bone" of contention: S. Tarsitano and M. K. Hecht in a detailed 1980 study claimed that the feature was a bit of extra mineral matter precipitated on the specimen, and not true bone. Few observers accept this interpretation.

BIRDS AS COELUROSAURS. Tetanurans, as we have seen, come in a fantastic variety of shapes and sizes. Within Tetanurae, however, *Archaeopteryx* possesses diagnostic characters of the coelurosaur clade, placing it squarely within the group. Coelurosaurs, it will be remembered, all have the distinctive semilunate carpal in their wrist. This half-moon-shaped bone is unique in Theropoda. Also seen in *Archaeopteryx* are the shortened ischium (far shorter than the pubis) and large, circular orbits. *Archaeopteryx* is a coelurosaur.

BIRDS AS MANIRAPTORANS. As befits their name, all maniraptoran coelurosaurs have a grasping, three-fingered hand that is a modification of the ancestral theropod condition. The distinctive maniraptoran addition to that hand is an elongation of the middle digit (II). *Archaeopteryx* has this feature. In those maniraptorans in which they are known, the clavicles are fused to form a furcula. That furcula is a large, robust bone and very unlike the splinterlike wishbone with which we are familiar. *Archaeopteryx* has a thick, U-shaped furcula, rather like that found in *Oviraptor* and *Ingenia*. Other maniraptoran features found in *Archaeopteryx* include a highly flexed neck, elongate forelimbs, and the distinctive bowed ulna. *Archaeopteryx*, and thus birds, are clearly members of Maniraptora (Figure 13.8, node 4).

BIRDS AND DROMAEOSAURIDS. Within Maniraptora, an unnamed[7] monophyletic group consisting of dromaeosaurids and *Archaeopteryx* is united by a variety of features. The most obvious of these is to be found in the pelvis, which is semi-opisthopubic and missing the front portion of the pubic footplate. Moreover, the tail vertebrae of this group all show elongation of the zygopophyses. Recall that in *Deinonychus*, this tendency takes a most extreme form, in which the tail becomes completely inflexible (Figure 13.8, node 5).

What can we conclude from all this? The bird, *Archaeopteryx*, is also the theropod dinosaur, *Archaeopteryx*. Within Theropoda, *Archaeopteryx* apparently is diagnosable as a maniraptoran coelurosaur. Finally, within Maniraptora, *Archaeopteryx* is probably most closely allied with dromaeosaurids. This means that the closest relatives of *Archaeopteryx* (and thus, the closest relatives of birds) were the highly predaceous and active dromaeosaurids, such as *Deinonychus* and *Velociraptor*.[8]

DISSENT. The concept of birds as saurischians has historically provoked controversy because of the position of the pubis. As we have seen, the pubis points backward both in modern birds and in ornithischian dinosaurs (a similarity reflected in the name "Ornithischia"). A number of workers, as early as the late 1800s, observed this shared character, but by the early part of the 20th century, thanks to the exhaustive studies of Heilmann, the similarities between ornithischians and birds were recognized as convergent. This leaves the pubis pointing forward in saurischians, and backward in birds. Although this might appear to be a stumbling block in the hypotheses presented here, two facts strongly support the bird–dinosaur relationship: (1) Recall from Chapter 4 that the pubis pointing forward is a primitive character within Tetrapoda; it should not be used, therefore, to falsify the monophyly of dromaeosaurids and *Archaeopteryx*; and (2) the pubis

[7]Obviously, not every branching point on a cladogram needs a name. In this case, the lack of a name simply means relief from the already too complex nomenclature of dinosaur paleontology.

[8]Here, then, is the rationale for the paleontological answer to the question (raised earlier), of which fingers form the bird carpometacarpus. If, in fact, birds are maniraptoran coelurosaurs, then the fingers in the bird hand must be I, II, and III, since those digits are unmistakably present in non-avian maniraptoran coelurosaurs.

in dromaeosaurids does not point forward, but instead downward; as we have seen, this is now believed to be the position of the pubis in *Archaeopteryx*. Our analysis suggests that in the history of dinosaurs, the pubis rotated backward three times: once in ornithischians, once in segnosaurs, and once in the dromaeosaurid-*Archaeopteryx*-Aves clade. It is significant that in modern bird embryos, the pubis initially points forward (the primitive condition) and rotates backward as the embryo develops.

A few scientists still adhere to a modified form of the thecodont-ancestry hypothesis – that some group of early archosaurs is actually responsible for the origin of birds. The approach taken to sustain this view is not a phylogenetic one; it involves nonparsimonious character distributions and convergences. Its most vocal adherent, Tarsitano of Southwest Texas State University, notes that convergence is rampant within Archosauria, and for this reason he rejects parsimony as a criterion for distinguishing phylogenetic hypotheses. Without parsimony, it is not clear how hypotheses can be falsified (see Chapter 3). For this reason, we do not find these arguments terribly convincing.

Another minority of specialists derives birds from crocodiles. This idea was first articulated by A. D. Walker of the University of Newcastle-Upon-Tyne at about the same time as Ostrom formulated his bird–dinosaur hypothesis. Subsequent study demonstrated to virtually everybody's satisfaction that Walker's hypothesis was developed on misleading or convergent characters; indeed, about 15 years after it was first proposed, even Walker himself rejected the idea. Regardless, a few specialists still adhere to modified forms of the idea, basing the hypothesis of crocodile–bird affinities on aspects of the braincase. If the crocodile–bird hypothesis were true, then the remarkable number of characters shared between birds and coelurosaurs would have to be ascribed to convergent evolution, a prospect that is generally viewed as unlikely.

We are convinced that there is much to recommend the dromaeosaurid–bird relationship and that this is the most refined statement of bird origins available. But a lesson worth remembering is that this is science, and older hypotheses (as we have seen!) are sometimes abandoned when newer and better ones come along. And sometimes even *older* hypotheses are vindicated (such as Huxley's idea that birds descended from theropods). This is not a problem, an embarrassment, or bad news; rather, this is simply the way science proceeds. The phylogenetic relationships of theropods are very much in a state of flux, and how birds fit into the overall picture of theropod systematics could change as our understanding of theropods changes.[9] In the last analysis, time, further specimens, and more study will test the dromaeosaurid–bird relationship that we have reiterated here.

Pneumatic Bones and Feathers: Adaptation vs. Inheritance

The hypothesis that birds are dinosaurs has some important implications with regard to the questions posed at the outset of this chapter. We can now answer

---

[9]For example, a recent publication by A. Elzanowski (Max-Planck-Institut für Biochemie) and P. Wellnhofer suggests that *Archaeopteryx* and birds may have a closer affinity with the theropod *Archaeornithoides*, a juvenile skull fragment from the Late Cretaceous of Mongolia, than with dromaeosaurids. *Archaeornithoides* seems to have a close relationship with troodontids, *Spinosaurus*, and *Baryonyx*, not with dromaeosaurids. Still, the relationships of *Spinosaurus* and *Baryonyx* within Theropoda are themselves poorly understood, and *Archaeornithoides* is but a juvenile skull fragment.

the first question, Where do birds come from? by stating that they appear to have evolved from the common ancestor of dromaeosaurids and *Archaeopteryx* that bore the diagnostic features of that clade. This maniraptoran was almost certainly a typical member of its clade: a highly predaceous, gracile, active biped, of small to medium size, that extensively used its grasping, dextrous hands in a variety of prey-catching and manipulation functions.

If this is true, it leads to several interesting speculations. Birds are commonly cited as supreme examples of adaptation. Indeed, there is no doubt that pneumatic bones maintain avian lightness, and that feathers are a superb adaptation for flight. The question is, however, did pneumatic bones and feathers evolve for lightness and flight, respectively? The paleontological answer is, probably not. As we have seen, hollow bones are a theropod character (recall that even the name *coelurosaur* contains a reference to the hollow bones in these dinosaurs). However, true pneumatic bones (with pneumatic foramina) are not found in any Mesozoic dinosaur; indeed, no Cretaceous bird has pneumatic bones, yet many are believed to have been excellent fliers. Truly pneumatic bones may have developed from an active, hollow-boned, cursorial ancestor, rather than having appeared in birds as an entirely new adaptation just for flight.

As regards the origin of feathers, the fossil record is unfortunately not terribly informative. In the early 1900s, Heilmann postulated, almost by default (because he had demonstrated that birds were more closely related to "reptiles" than to mammals), that feathers are an outgrowth of "reptilian" scales. His (and later) embryological work supports this hypothesis; however, all of this really begs the question of when feathers arose, and in whom. A Triassic archosauromorph, *Longisquama*, bears elongate featherlike structures that were once interpreted as potential feathers; although these enigmatic structures are not yet fully understood, they are no longer widely believed to have any relationship to the origin of feathers (Figure 13.9). In 1978, the Russian paleontologist A. S. Rautian described a Jurassic "feather" that he interpreted as an early state in the evolution of feathers. He called the fossil *Praeornis sharovi*. Other workers, examining the specimen, thought it could be a plant fossil! With the historical record so equivocal, it has to be concluded that at present, virtually nothing is known of the evolution of feathers and the natures of the earliest creatures bearing them.

We can, however, speculate on the potential functions of the earliest feathers. As noted in Chapter 14, the maniraptoran ancestor of birds and dromaeosaurids is an extremely good candidate for "warm-bloodedness." The structure of its bones shows that here was a highly active, running, predatory beast of relatively small size. Given this behavior, "warm-bloodedness" could have been an asset to this creature; however, heat loss calculations suggest that it would have needed an insulatory covering to maintain a "warm-blooded" metabolism. The insulatory properties of down are well known. It is possible that feathers originated not as a flight adaptation, but as an insulatory mechanism in a nonflying, cursorial, "warm-blooded" dinosaur. This idea becomes less farfetched when it is remembered that all modern birds are "warm-blooded." It simply becomes a matter of how far back in the bird lineage "warm-bloodedness" occurred. If it occurred with *Archaeopteryx*, then *Archaeopteryx* would have been the first dinosaur to require insulation (and thus bear feathers). But it probably occurred well before *Archaeopteryx*, because *Archaeopteryx* has feathers that are beautifully differenti-

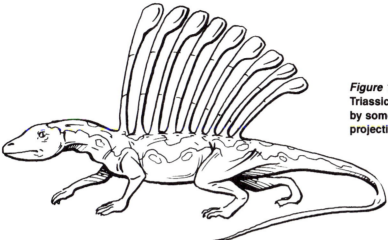

Figure 13.9 **A reconstruction of the Triassic tetrapod *Longisquama*, claimed by some to possess a row of featherlike projections along its back.**

ated into down and symmetrical and asymmetrical flight feathers; the morphology and distribution of the feathers in *Archaeopteryx* is essentially modern. Much evolution (and, presumably, time) must have elapsed between the invention of the first feather and the condition found in *Archaeopteryx*. It is possible, therefore, that the earliest feather-bearing vertebrates were nonflying, cursorial, light-bodied, small theropods, which developed feathers not for flight, but rather for insulation.

All of this of course is speculation, and a rather different view may be taken. J. A. Ruben of Oregon State University has suggested that *Archaeopteryx* could have been "cold-blooded" (see Chapter 14) and still could have generated the necessary energy available for limited flight. Ruben envisions a "cold-blooded" *Archaeopteryx* that might have been capable of limited, short-duration flapping flight, and his calculations indicate that the animal could have taken off from the ground or from the trees (see following section). The implication of Ruben's view is that feathers truly are first and foremost an adaptation for flight, and secondarily came to serve an insulating function.

Ruben's proposal, though staunchly defended by him on solid theoretical grounds, has not been widely accepted, because it is uncertain why a "cold-blooded" animal that depends upon external sources of heat energy would develop insulation that would shield it from those very sources. Indeed, for that very reason, no modern "cold-blooded" creature is insulated.

Flight

Given all of the preceding, it is unclear how intimately being a bird (i.e., being a feathered creature) is linked with flight. Feathers were clearly a prerequisite to flight in theropods, even if flight may not have been a prerequisite to feathers!

In general, two opposing endpoints on a continuum of hypotheses exist regarding the origin of bird flight (Figure 13.10). One is the so-called **arboreal** (or trees down) **hypothesis**: that bird flight originated by birds' gliding down from trees (Figure 13.10a). In this hypothesis, gliding is a precursor to flapping (powered) flight; as birds became more and more skillful gliders, they extended

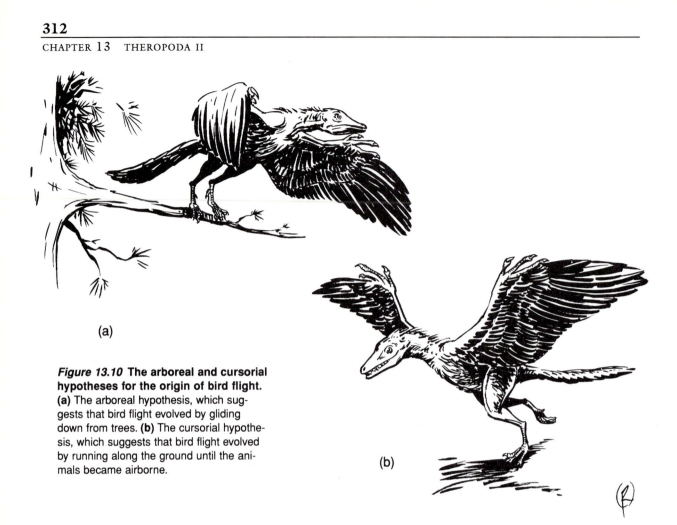

Figure 13.10 The arboreal and cursorial hypotheses for the origin of bird flight. (a) The arboreal hypothesis, which suggests that bird flight evolved by gliding down from trees. (b) The cursorial hypothesis, which suggests that bird flight evolved by running along the ground until the animals became airborne.

their range and capability by developing powered flight. Perhaps flapping developed as a modification of the motions used in controlling flight paths. In this scenario, a quadrupedal ancestor is usually postulated, and a "proavis," or early-stage gliding quadruped, is also implicated. *Archaeopteryx* is viewed as a second stage in the process; it had a perching foot, and the unfused fingers of the hand were used for grasping tree limbs. It would have flown primarily as a glider from tree to tree, using flapping flight for short bursts and only when necessary. The arboreal hypothesis has had many adherents, the most consistent and vocal of whom is probably W. J. Bock of Columbia University.

Antithetical to the arboreal hypothesis is the **cursorial** (or running) **hypothesis** for the origin of flight (Figure 13.10b). The cursorial hypothesis states that bird flight originated by an ancestral bird's running along the ground. In this scenario, perhaps as obstacles were avoided, the animal became briefly airborne. Flapping (powered) flight appeared early on, as the animal strove to overcome more fully the force of gravity. This idea obviously requires a highly cursorial ancestor, in which feathers were already present. In this hypothesis, the legs, feet, and hands of *Archaeopteryx* are viewed as an inheritance from a cursorial maniraptoran ancestor. The cursorial hypothesis has been strongly advocated by Ostrom,[10] Gauthier, and K. Padian of the University of California at Berkeley.

[10]Ostrom originally suggested that flapping flight evolved as a prey-catching mechanism. He imagined a small, insectivorous theropod "harvesting" insects from the air.

WHICH TO CHOOSE? Both the arboreal and cursorial hypotheses have much to recommend them. The arboreal hypothesis is intuitively appealing, for it requires little imagination to produce a flying creature from an arboreal-dwelling one. The advantages to be gained are obvious, and potentially difficult evolutionary challenges such as getting airborne are easily solved – as easy as jumping from a branch. The cursorial hypothesis is strongly supported because, based upon the detailed anatomy reviewed in this chapter, ultimately the ancestor of birds had to have been a cursorial creature. Moreover, indications of cursorial ancestry are present in all birds. One has but to observe the characteristic bipedality in all birds (including *Archaeopteryx*) and the gracile features in the limbs – including the ratios of thigh length to shin length, and the digitigrade stance – to recognize the hallmarks of a running ancestor. Indeed, it is not coincidence that bird limbs are little changed from the nonflying coelurosaurian condition.

The strong bipedality in modern birds affords another clue about their ancestry. As previously noted, tree-borne gliding animals of quadrupedal origin commonly develop a flap of skin linking the front with the rear legs. In other words, to a greater or lesser degree, the ancestral quadrupedality is preserved even in the flying animal. This is clear enough in gliders like the flying squirrel, but links between the hind- and forelimbs are also present even in well-developed flapping fliers like bats. Modern birds, however, are clearly bipeds, and there is no flap of skin connecting the wings with the legs. This is another clear indication of the ultimate bipedal ancestry of birds; such bipedalism is more indicative of cursorial than of arboreal behavior in the ancestor.

Fundamentally, however, it has so far proven nearly insurmountable to "design" a cursorial theropod that developed flight by running along the ground. Functional morphologists – that is, scientists who study the function of particular anatomical structures – have been unable to satisfactorily model the running-to-flight transition in early birds. Adherents maintain, however, that if the animal ran, leapt, and took a few flight strokes at a time, flight could have evolved from leaping and being briefly airborne, to the sustained, sophisticated flight we observe today.

It has been argued that perhaps the earliest birds scaled trees, and from that position learned to fly. There is, however, no evidence for an arboreal proto-bird, no evidence for climbing adaptations, and no evidence in the skeleton of any non-avian theropod for arboreal habits.

FLIGHT IN *ARCHAEOPTERYX*. It might be supposed that the earliest known bird would shed some light on the origins of flight. In this regard, *Archaeopteryx* has been a disappointment. *Archaeopteryx* clearly has wings and a feather arrangement typical of flying birds, producing an airfoil. In spite of this, many of the features that are significant in strengthening, rigidifying, and lightening the body for flight – features that are found in modern birds – are absent in *Archaeopteryx*. Instead, the creature has a primitively elongate trunk, gastralia, no synsacrum, no carpometacarpus, weakly developed coracoids, a small sternum without much of a keel, and none of the supracoracoideus adaptations of modern birds.

Still, that *Archaeopteryx* flew – in some at least limited flapping capacity – is clear enough. How well or poorly remains uncertain. J. M. V. Rayner of the University of Bristol has attempted to analyze flight in *Archaeopteryx*. His conclusion is that *Archaeopteryx* could flap its wings, attaining moderately high speeds, but could not perform the kind of slow flight that a running take-off might require. For this reason, Rayner reasoned that *Archaeopteryx* had to be primarily a tree dweller that initiated flight by gliding. The conclusions of Feduccia (see preceding) that *Archaeopteryx* was a tree-dwelling animal correlate well with those of Rayner.

If *Archaeopteryx* retains in its morphology much of its ancestry (as indeed we have claimed throughout this chapter), then the arboreal adaptations of *Archaeopteryx* suggest that the transition from ground to trees occurred very early in bird evolution – perhaps even *before* flight. On the other hand, although *Archaeopteryx* retains many ancestral skeletal features, its arboreal adaptations may not tell us much about the origin of flight in birds, because these are clear indications that it has undergone much evolution since the cursorial maniraptoran condition. For the present, we just do not know which of these interpretations is correct.

The plumage of *Archaeopteryx* suggests an animal with a good deal of evolutionary history behind its clear flight adaptations. If we wish to learn something of the *origin* of flight, it would serve us best to consider not the morphology of an animal that already can fly, but rather the features of that animal's closest ancestor. The preservation and discovery of fossils is in large part a matter of luck, and though *Archaeopteryx* takes us breathtakingly close to the ancestry of flying birds, it is clearly not that ancestor. What is known of that ancestor suggests a cursorial animal; however, the ambiguity of *Archaeopteryx*'s mosaic of primitive and advanced characteristics leaves unresolved the question of the arboreality vs. the terrestriality of the origin of flight.

THE EARLY EVOLUTION OF AVES

Getting to Be a Modern Bird

Archaeopteryx is obviously not a modern bird. It has many features – including teeth, an unfused hand, a tail, gastralia, and no synsacrum – that are far from the condition found in modern birds. How and when did the changes take place that distinguish modern birds from *Archaeopteryx*? At present, the scantiness of the Mesozoic bird record and the delicacy of specimens have led to some confusion; however, there are aspects of Mesozoic bird evolution about which the fossil record, despite its limitations, is sending a clear signal.

Early Cretaceous

Next to *Archaeopteryx*, the oldest unambiguous birds are from the Early Cretaceous. The oldest of these is the most recently discovered: In 1992, P. C. Sereno of the University of Chicago and C. Rao of the Beijing Museum of Natural History reported on a new bird from an Early Cretaceous lake bed in China. The bird, *Sinornis* (*sino* – China), is but 15 million years younger than *Archaeopteryx* but already shows a mosaic of features closer to modern birds than to *Archaeopteryx* (Figure 13.11). Most obviously, the body is shortened and a pygostyle is present. The wrist joint is rather more like that of modern birds than like that of *Archaeopteryx*. Modern birds have a modified wrist joint to allow the wing to fold tightly against the body: *Sinornis* has this; *Archaeopteryx* does not. Finally, *Sinornis* is clearly a flying animal, with pillarlike coracoids and a moderately large sternum. *Sinornis* has adaptations that suggest perching. In particular, the first toe is opposite the others, a morphology interpreted by Sereno and Rao

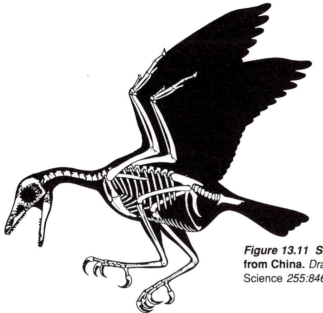

Figure 13.11 *Sinornis,* **an Early Cretaceous bird from China.** *Drawing from Sereno and Rao, 1992,* Science *255:846.*

to be clearly indicative of perching. If so, *Sinornis* assuredly lived in trees – again, evidence that flight was an integral part of its life habits.

Sinornis is hardly a modern bird, however, and in the same animal are clearly ancestral features. Like *Archaeopteryx, Sinornis* has gastralia. Moreover, it has an unfused hand (no carpometacarpus). So, although *Sinornis* was clearly a well-developed flier, it was very unlike any modern bird.

More poorly known – but equally important – is *Iberomesornis* (the name is a reference to the Iberian Peninsula), a bird from Las Hoyas, Spain. This bird was described in 1988 by J. L. Sanz, J. F. Bonaparte, and A. Lacasa. It, too, has a mixture of primitive and advanced characters. The advanced characters include a pygostyle and a strutlike coracoid. Primitive characters consist of relatively numerous back vertebrae (a number intermediate between the 13–14 found in *Archaeopteryx* and the 4–6 found in living birds), an unfused tarsometatarsus, and an unfused pelvis. Gastralia are not preserved, and it is thought that they were not present. As with *Sinornis*, flight is clearly indicated for this creature.

Enaliornis from the Early Cretaceous of England has been known since the late 1800s from well-preserved skull material and from an ankle with a shinbone attached (**tibiotarsus**). This bird appears to have been highly specialized for diving (much like the modern loon) and has been suggested as an early representative of Hesperornithiformes (see following section).

Two other Early Cretaceous birds are known: the Australian bird *Nanantius* (*nan* – dwarf) and *Ambiortus* (*ambi* – surrounding; *orta* – young bird), a flying bird from Mongolia. *Nanantius*, first described in 1986 by Australian paleontologist R. E. Molnar, is known from just a tibiotarsus. Nevertheless, this key bone identifies it as the earliest known member of the very diverse Late Cretaceous enantiornithiform clade, whose members we will discuss subsequently. The Early Cretaceous *Ambiortus* is a reasonably complete animal with feather impressions, missing only a rear end and a skull. It has derived features such as pillarlike coracoids, a keeled sternum, and a furcula. Its retention of primitive characters (e.g., teeth and an unfused hand) mark it as a primitive bird that cannot easily be assigned to any known clade.

Each of these birds shows a mosaic of primitive and advanced characters. Nevertheless, using *Archaeopteryx* as representative of the primitive condition and modern birds as representative of the derived condition, we can begin to establish the sequence through which birds acquired a modern aspect (Figure 13.12). Among the primitive characters that are retained are teeth, an unfused hand, an unfused pelvis, and gastralia. Advanced characters include a pygostyle, a decrease in the number of vertebrae, a well-developed bony sternum, and pillarlike coracoids (implying a supracoracoideus musculature like that found in modern birds). These features make up the bulk of the adaptations necessary for the kind of sophisticated flight found in modern birds. Not yet developed in this regard was the carpometacarpus, although as we shall see, this would be de rigueur in birds by Late Cretaceous time. Moreover, although the bird ancestor may have been cursorial, perching and arboreal habits were indubitably established by Early Cretaceous time. The fact that virtually all the necessary ingredients for modern bird flight were present by just 15 million years after *Archaeopteryx* lived (i.e., by Early Cretaceous time) has sparked some observers to speculate that either the rate of evolution in birds was greatly accelerated during the Late Jurassic–Early Cretaceous time interval, or that *Archaeopteryx* itself was kind of a retrograde creature, a throwback that, although it lived in the Late Jurassic, was rather primitive even for its time.

In fact, both viewpoints may be true. Because the oldest bird known is *Archaeopteryx*, only finds of equivalent age or older can resolve without question whether or not *Archaeopteryx* was an anachronism. On the other hand, the presence

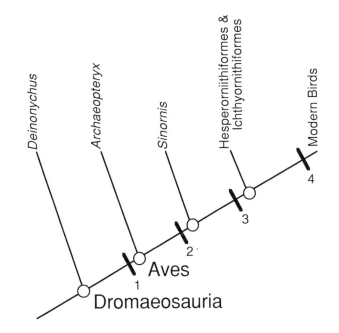

Figure 13.12 Cladogram depicting some of the steps in the early evolution of birds.
1 defines Aves and includes, among other features, feathers; 2 includes a reduction in the number of vertebrae in the trunk (from 14 to 12), development of a pygostyle, the development of a flexible furcula, the development of a strutlike coracoid, a fully-folding wing, and a reduction in the grasping ability of the hand; 3 is marked by the development of the carpometacarpus and the loss of gastralia; 4 is marked by loss of teeth.

of a specialized diving bird, *Enaliornis*, suggests that even as far back as the Early Cretaceous, birds were capable of exploiting a variety of environments (in this case, arboreal and marine environments) and had evolved into specialized forms.

Late Cretaceous Birds

The fossil record of birds kicks in again in the Late Cretaceous, with at least two diverse, intimidatingly named clades, Enantiornithiformes and Hesperornithiformes. Enantiornithiformes is a poorly known, but apparently numerous, group of flying birds (many specimens are known only from isolated limb elements, and a complete skull is as yet unknown for any member of the group!); Hesperornithiformes is a well-known group of marine birds that were highly specialized for diving. Both groups are believed to have retained the primitive feature of teeth (although this is an educated guess in the case of Enantiornithiformes) but both have a well-developed carpometacarpus, suggesting that this important character, so typical of modern birds, developed some time between the Early and Late Cretaceous.

ENANTIORNITHIFORMES. Enantiornithiform birds were all sparrow-sized, arboreal birds with well-developed flight capabilities. They were recognized as a monophyletic group only in 1992, when Argentinian paleontologist L. Chiappe diagnosed and analyzed the group. As far as is known, enantiornithiform birds all possess a perching foot; a well-developed, keeled sternum; a relatively slim furcula, similar to that found in modern birds; a strutlike coracoid; and powerful, well-developed arms (robust humerus, radius, and ulna). These advanced flight features, however, are combined with a primitive foot: The tarsometatarsus is not fully fused (it looks comparable to that found in *Archaeopteryx*, *Sinornis*, and *Iberomesornis*).

Enantiornithiform birds apparently had a worldwide distribution. *Nanantius* is from Australia. The others, all described after 1975 by a variety of authors, include *Kizylkumavis* and *Sazavis* from Uzbekistan (Asia), *Alexornis* from Mexico, *Enantiornis* and *Patagopteryx* from Argentina, and another from the United States. Most are rather fragmentary; however, distinctive aspects of the unfused tarsometatarsus allow them to be recognized as enantiornithiform birds.

HESPERORNITHIFORMES. Hesperornithiform birds are a monophyletic clade of relatively large, long-necked, flightless diving birds that used their feet to propel themselves through the water (Figure 13.13). In many respects the group is quite close to modern birds, in that all members have the shortened, fused trunk, a carpometacarpus, a pygostyle, a completely fused tarsometatarsus, and a synsacrum. Moreover, no member of this group bears gastralia. All differ from modern birds, however, in that they retain teeth in their jaws. As with modern diving birds, some of the pneumaticity in the bones has been lost. Presumably because of the loss of the flight adaptation, the furcula, coracoid, and forelimb bones are highly reduced.

Hesperornithiformes has been known for some time. Three species of *Hesperornis* (*hesper* – western; *ornis* – bird), as well as *Coniornis* (*coni* – cone-shaped; however, this bird may in fact be referable to *Hesperornis*) and its smaller relative, *Baptornis* (*bapt* – dipped or submerged), were described by Yale paleontologist O. C. Marsh in the late 1800s. Within the last decade, L. D. Martin of the University of Kansas has extensively studied hesperornithiform birds and has described *Parahesperornis* (*para* – near) and restudied *Neogaeornis* (*neo* – new; *gaea* – earth). All hesperornithiform birds are from North America.

ENIGMATA. A variety of birds from the Late Cretaceous are difficult to place in any reasonable phylogenetic scheme. This is in part because of scrappy preservation, but also because some of them were as highly specialized in their way as *Hesperornis* was in its. That Late Cretaceous birds could be quite specialized and were able to exploit a variety of niches was driven forcibly home by the 1993 report of the running bird *Mononykus* (*mono* – one; *onyx* – claw; Figure 13.14). *Mononykus* was first unearthed in the Gobi Desert of Mongolia in 1987 by Soviet-Mongolian field parties; specimens were subsequently (1992–3) recovered by field parties from the American Museum of Natural History. In life, *Mononykus* apparently lived in a desert as well; the fossils came from a Late Cretaceous sand dune sea (or erg). *Mononykus* is a strange beast. From the center of the back toward the tail, it looks like a typical theropod dinosaur: strong, elongate, well-developed hindlimbs, and a long, straight tail. The arms, however, have been described as those of a bird, albeit a highly aberrant one: the digits are fused into a short, stout carpometacarpus, and the arms are also stout and short, with a large process (the olecranon process) for developing power at the elbow joint. Along with these apparently avian characters, *Mononykus* has a keeled sternum, again suggesting a close relationship with modern birds.[11] Although no feathers were preserved with the specimen, *Mononykus* is probably a bird, if an unusual one.

Considerably more flightworthy than the confusing *Mononykus* were ichthyornithiform birds, a group of long-necked, toothed, gull-sized birds with probably equivalent flying skills. Two genera are known, *Ichthyornis* (*ichthy* – fish) and *Apatornis* (*apat* – false) from the Late Cretaceous of North America. These birds had a massive keeled sternum and an extremely large process at the head of the humerus, the **deltoid crest**, that was probably an adaptation for powerful flight musculature. In other respects, they share many of the adaptations of modern birds, including a shortened, fused trunk, a carpometacarpus, a pygostyle, a completely fused tarsometatarsus, and a synsacrum. Ichthyornithiform birds have been found exclusively in North American latest Cretaceous marine deposits; they are inferred to have behaved rather like a modern gull.

Gobipteryx, described in 1974 by A. Elzanowski, is known from a couple of incomplete, crushed skulls and some embryo fragments from the Gobi Desert in Mongolia. The embryos suggest that the (adult) animal flew. Martin has proposed that *Gobipteryx* has affinities to enantiornithiform birds and, although not a member of Enantiornithiformes, is closer to these than to any other known bird, either living or fossil.

Finally, fragments of several genera of Cretaceous birds can be assigned to the living orders Charadriiformes (shorebirds such as sandpipers, gulls, and auks) and (possibly) Procellariiformes (wing-propelled divers such as the modern petrel). Only two Mesozoic birds, however, can be referred to any modern family of birds,[12] and thus it must be assumed that most were early dead-end radiations.

---

[11] These characters are not without attendant controversy. Several workers have claimed that although *Mononykus* bears a fused hand, it is unlike the carpometacarpus found in birds. Moreover, it is claimed that the keeled sternum is rather unbirdlike.

[12] The Chilean Cretaceous bird *Neogaeornis* and a Cretaceous bird from Antarctica have been referred to the modern loon family (Gaviidae).

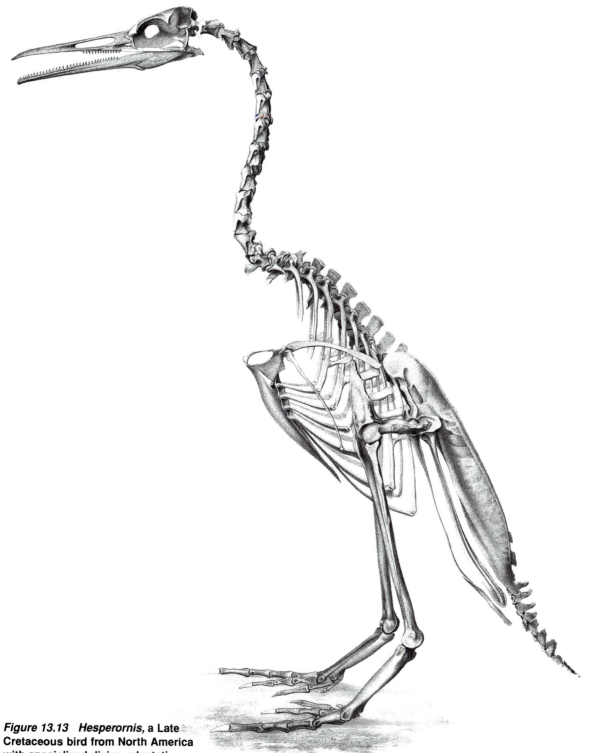

Figure 13.13 *Hesperornis,* a Late Cretaceous bird from North America with specialized diving adaptations. *Print courtesy of J. H. Ostrom.*

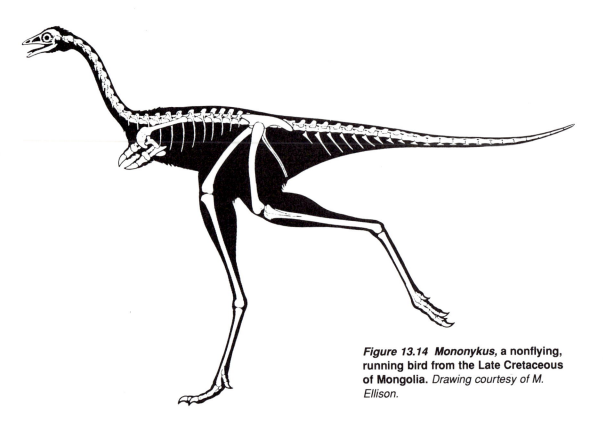

Figure 13.14 *Mononykus,* **a nonflying, running bird from the Late Cretaceous of Mongolia.** *Drawing courtesy of M. Ellison.*

For all of its limitations, the Mesozoic record of birds still provides insights into the transition from the condition in *Archaeopteryx* to that found in modern birds. The cladogram in Figure 13.12 outlines the character sequence.

We began this chapter by promising to answer three fundamental questions about birds. To the question, Where do birds come from? our answer is, of course, dinosaurs. With so much of this book behind you, however, you can now see that a better way to phrase this question would be, What *are* birds? because who they are dictates where they come from. The transition from primitive theropod dinosaurs through *Archaeopteryx* to modern birds is, despite a very imperfect fossil record, remarkably complete. Nobody expects any fossil to be *the* ancestor of a particular group, and *Archaeopteryx* does have some specializations that bar it from being the perfect ancestor to all birds. Regardless, in its mosaic of primitive theropod and advanced avian characters, it is just about as satisfactory an intermediate as one could envision.

Our second question, Where do feathers come from? is not easily answered in a mechanistic way. As we have seen, they are presumed to be an outgrowth of archosaurian scales, but to date the fossil record is silent on how or when that event took place. Nevertheless, the idea that the origin of feathers might initially have been an adaptation for a "warm-blooded" metabolic state, and not for flight, bears serious consideration in light of who birds are.

Finally, we asked, How did flight evolve? Again, the answer is not fully laid out in the fossil record; however, it is clear now that the ultimate ancestors of birds were highly adapted for cursorial behavior. We are commonly tempted to search for some kind of intermediate on the way to flight – some sort of "proavis" creature – but the remarkable thing about the dromaeosaurid-*Archaeopteryx* clade is that many of the adaptations necessary for flight were already present in that clade. For example, one might be tempted to search for a proavis with arms of increased length – on their way to becoming wings, as it were. But dromaeosaurids – which didn't fly – were animals that already had long arms and large hands, in the proportions also found in *Archaeopteryx*. Their use for those arms and hands was not for flight, but rather for the manipulation of prey. Regardless, the structure was there. Birds simply co-opted these primitive limb proportions for their own purposes.

This reminds us of a fundamental property of evolution. Evolution modifies existing structures. The primitive grasping arms were modified – remarkably little – to make wings in avian dromaeosaurids. It was a breathtaking evolutionary sleight of hand!

So, in fact, birds are not some separate biological entity, distinct and apart from "reptiles." Birds are dinosaurs. And not all the dinosaurs have gone extinct; one group, the birds, survives. What did dinosaur meat taste like? A trip to Kentucky Fried Dinosaur, or a mouthful of Dinosaur McNuggets can answer that question!

Important Readings

Bock, W. J. 1986. The Arboreal origin of avian flight; pp. 57–82 *in* K. Padian (ed.), The Origin of Birds and the Evolution of Flight. California Academy of Sciences Memoir no. 8.

Feduccia, A. 1980. The Age of Birds. Harvard University Press, Cambridge, 196 pp.

Gauthier, J. A. 1986. Saurischian monophyly and the origin of birds; pp. 1–56 *in* K. Padian (ed.), The Origin of Birds and the Evolution of Flight. California Academy of Sciences Memoir no. 8.

Hecht, M. K., J. H. Ostrom, G. Viohl, and P. Wellnhofer, eds. 1984. The Beginnings of Birds: Proceedings of the International *Archaeopteryx* Conference Eichstätt. Freunde des Jura Museums Eichstätt, Willibaldsburg, 382 pp.

Martin, L. D. 1983. The origin and early radiation of birds; pp. 291–353 *in* A. H. Brush and G. A. Clark, Jr., Perspectives in Ornithology. Cambridge University Press, New York.

Norell, M., L. Chiappe, and J. M. Clark. 1993. New limb on the avian family tree. Natural History 9/93:38–43.

Olson, S. L. 1985. The fossil record of birds. Avian Biology 8:79–239.

Ostrom, J. H. 1974. *Archaeopteryx* and the origin of flight. Quarterly Review of Biology 49:27–47.

Ostrom, J. H. 1976. *Archaeopteryx* and the origin of birds. Biological Journal of the Linnaean Society 8:91–182.

Rayner, J. M. V. 1988. The evolution of vertebrate flight. Biological Journal of the Linnaean Society 34:269–287.

Ruben, J. A. 1991. Reptilian physiology and the flight capability of *Archaeopteryx*. Evolution 45:1–17.

Schultze, H.-P., and L. Trueb, eds. 1991. Origins of the Higher Groups of Tetrapods: Controversy and Consensus. (See "Section IV: Birds," pp. 427–576, for articles by L. D. Martin, J. H. Ostrom, L. M. Witmer, and S. Tarsitano.) Cornell University Press, Ithaca, N.Y.

Sereno, P., and C. Rao. 1992. Early evolution of avian flight and perching: new evidence from the Lower Cretaceous of China. Science 255:845–848.

PART IV

ENDOTHERMY, ENVIRONMENTS, AND EXTINCTION

Chapter 14

DINOSAUR ENDOTHERMY:
Some Like It Hot

IT WAS ONCE SAID OF ELVIS PRESLEY that he didn't invent the pelvis, he only popularized it. You could make an equivalent claim for R. T. Bakker, author, paleontologist, and free spirit: He didn't invent "warm-bloodedness" in dinosaurs, but he certainly popularized it.

By now, thanks to the vigorous efforts of Bakker and many other workers, the idea of dinosaurs as sluggish, cast-iron clunkers has been abandoned by all. Our story could begin in many places; however, as befits his aggressive role in the evolution of the concept of dinosaur "warm-bloodedness," let us begin with the work of R. T. Bakker.

THE GENESIS OF AN IDEA

One of many observations that motivated Bakker's conclusion that dinosaurs were "warm-blooded" was based upon an apparently unexplainable fact of the changes in terrestrial vertebrate biotas during the Triassic.

The Therapsid–Dinosaur Transition

Recall the "wedge" of Chapter 5. Throughout the late Paleozoic Era, **therapsids**, the clade of synapsids to which mammals and their ancestors belong, evolved into a wide variety of forms, filling a broad range of niches. By Early to Middle Triassic times, therapsids gave every appearance of being poised to dominate terrestrial vertebrate faunas. Instead, dinosaurs, which appeared at approximately the same time as the earliest animals that paleontologists consider mammals (Late Triassic), dominated terrestrial vertebrate ecosystems for the next 160 million years (see

Figure 5.8). Indeed, it was only after the extinction of the dinosaurs that mammals assumed the importance among vertebrate faunas that they now occupy.

On the basis of this seemingly paradoxical pattern, Bakker asked the question, What happened? His conclusion, expressed in a 1975 article, was startling:

> One is forced to conclude that dinosaurs were competitively superior to mammals as large land vertebrates. And that would be baffling if dinosaurs were "cold-blooded." Perhaps they were not. (R. T. Bakker, 1975, "Dinosaur Renaissance," *Scientific American*, April 1975, p.61)

This statement embodies two assumptions. The first is that dinosaurs directly competed with mammals (and/or their therapsid ancestors), and the second is that – particularly in the case of large vertebrates (and by large, Bakker meant greater than 10 kg) – "warm-bloodedness" is superior to "cold-bloodedness." These assumptions require close consideration.

COMPETITION.   The first assumption – that the earliest dinosaurs *out-competed* Middle and Late Triassic therapsids (including mammals) – is problematical (see the discussion of the "competitive edge" in Chapter 5). Were mammals and dinosaurs in direct competition for the same resources?

The earliest mammals were tiny creatures, estimated at less than 30 g in weight (Figure 14.1). Features of the skulls, teeth, limbs, vertebral column, and hands suggest that the early mammals were insect-eating, at least in part tree-dwelling, and possibly nocturnal. These creatures were obviously a far cry from the earliest dinosaurs or their near-ancestors, which appear to have been large (approximately 2 m), carnivorous, ground dwelling, and diurnal (functioning during the day). The earliest mammals do *not* seem to have been competitive with the earliest dinosaurs; each group was behaving in rather different ways.

And what of the therapsid ancestors of the earliest mammals? Could these have somehow competed and lost against dinosaurs and/or their archosaurian forbears? This is a more difficult question to answer. The larger, more clearly carnivorous therapsids – ones that might have been truly competitive with the ancestors of dinosaurs – may or may not have been "warm-blooded." Regardless, few paleontologists would claim that the archosaurs against whom the therapsids could potentially have competed were themselves "warm-blooded." In any case, the competitive advantage afforded by "warm-bloodedness" is by no means obvious.

Considered another way, why must we assume that there was a one-to-one correspondence between therapsid and archosaur behavior and lifestyles? There is more than one way to be a large, terrestrial carnivore. For example, the behavior of *T. rex* and a lion are not comparable; these two large vertebrates, with very different morphologies, must have behaved in rather different ways.

IS "WARM-BLOODEDNESS" SUPERIOR TO "COLD-BLOODEDNESS?"   The second assumption inherent in Bakker's statement is that "warm-bloodedness" in large land vertebrates is clearly a superior metabolic condition to "cold-bloodedness." After all, if it were otherwise, the success of the dinosaurs would not be so "baffling."

A look at the global distribution and sizes of "cold-blooded" tetrapods apparently makes a case for the advantages of "warm blood." Virtually all "cold-blooded"

Figure 14.1
***Morganucodon*, a Mesozoic mammal.**
Painting courtesy of Margaret Colbert.

tetrapods live within 45° north or south of the equator; most of the diversity of "cold-blooded" tetrapods is concentrated within 20° of the equator. Some scientists infer from this that "warm-blooded" tetrapods are freer of their environment than "cold-blooded" tetrapods. But is this really meaningful? The idea that any organism is "free" from its environment has little validity in biology. It is true, however, that temperature can play a central role in directly controlling behavior: Until a lizard can, by basking or other warming strategies, bring its core temperature to that necessary for optimum physiological functioning, it cannot perform its activities (see "Metabolism" in "Endothermy and Ectothermy" section of this chapter).

Along with having restricted global distributions, modern "cold-blooded" tetrapods tend to be relatively small creatures. At least in part, this derives from the relationship between the *surface area* of the body and its *volume*. Elementary arithmetic tells us that surface area is a square function and volume is a cubic function. For organisms, this has important implications: As an organism gets bigger and bigger, its volume becomes proportionally greater than does its surface area (Figure 14.2). Thus, to maximize its surface-to-volume ratio, the organism must be very small. "Cold-blooded" animals, however, depend upon their *surface area* to elevate their body temperatures: They bask in the sun, or lie on warm substrates to obtain heat. Thus, as "cold-blooded" animals become larger, the more difficult it is for their surface areas to warm their proportionally much vaster interiors. The largest "cold-blooded" tetrapods, such as seagoing turtles and crocodiles, tend to be aquatic or marine; that is, they inhabit fresh or salt water. Crocodiles can get upwards of 6 m long and weigh as much as 1,000 kg; the turtles are more squat affairs, but can weigh as much as 1,000 kg. As we saw in Chapter 2, the temperatures of water bodies show less fluctuation than do temperatures on land. Because of the temperature stability of bodies of water, aquatic "cold-blooded" tetrapods can attain larger size without simultaneously roasting externally and yet freezing internally.

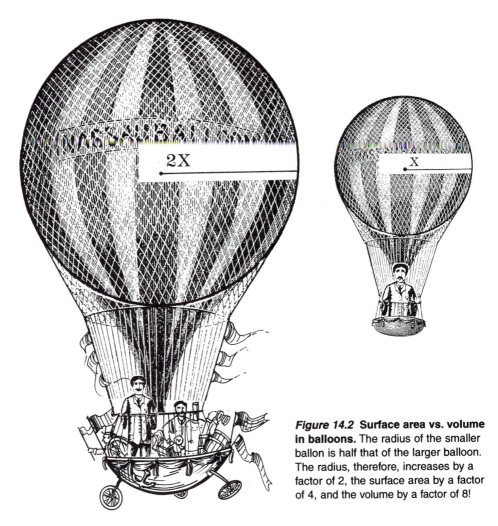

**Figure 14.2 Surface area vs. volume
in balloons.** The radius of the smaller
ballon is half that of the larger balloon.
The radius, therefore, increases by a
factor of 2, the surface area by a factor
of 4, and the volume by a factor of 8!

Well, we have seen that in terms of geographic distribution and size range,
"warm blood" confers some apparent advantages. Still, the fact is that *the vast
majority of living vertebrates are "cold-blooded."* "Warm" blood is only characteristic of birds and mammals (but see Box 14.1). Here is a conundrum: If "cold-
blooded" vertebrates are supposed to be inferior, why are they so much more
numerous (that is to say, more *successful*) than "warm-blooded" vertebrates?

An obvious solution to this problem is *not* to equate "warm-bloodedness" with
superiority. An alternative hypothesis to the superiority of "warm-bloodedness" is
that "warm-bloodedness" and "cold-bloodedness" are metabolic strategies suited to
particular anatomies and behaviors, and are not somehow superior or inferior.

We need to consider vertebrate physiology in a slightly more sophisticated
way. It turns out that many "cold-blooded" animals develop warm blood temperatures when they are operating at a physiologically optimum temperature.
Indeed, a snake sunning itself can have a higher temperature than a mammal.
R. Silverstein, writing in *Harpers*, called the terms "warm-blooded" and "cold-
blooded" "convenient ideational shorthand" for terms that describe vertebrate
temperature regulation more meaningfully: endothermy and ectothermy.

ENDOTHERMY AND ECTOTHERMY

Temperature Regulation among Vertebrates

All vertebrates regulate their temperature. Some, the **ectothermic** (*ecto* – outside; *therm* – heat) organisms, do so using external sources of heat, whereas others, the **endothermic** (*endo* – inside) organisms, regulate temperature internally. We can add a second dimension to our thinking about this by noting that in some organisms, called **poikilotherms** (*poikilo* – fluctuating), the temperature fluctuates, but in others, called **homeotherms** (*homeo* – same), the temperature remains constant. Humans, our own experiences tell us, are endothermic homeotherms: When we are unable to maintain our body temperature, we get very sick. Most lizards, on the other hand – ectothermic poikilotherms – are best transported from place to place by keeping them cool in food coolers: The animals are simply slowed down. Lizards and snakes are *functionally* homeothermic: They function optimally at constant, relatively elevated temperatures. The difference between them and an endotherm is that they can survive in coolers simply by allowing their metabolism to slow down in a way that an endotherm cannot. Ectotherms can *tolerate* decreases in core temperature, whereas endotherms must internally *regulate* their core temperatures.

On the other hand, perhaps more foreign to us are endothermic poikilotherms like bats and hummingbirds, which have different levels of activity associated with different core (body) temperatures. For example, a resting bat goes into a trancelike state called **torpor**, in which the temperature of the animal is truly considerably lower than when the bat is flying. The same kind of thing occurs in bears, in which the core temperature of a bear in hibernation is somewhat lower than when it is active. The hyrax, a small mammal, regularly sunbathes to raise its core temperature.

It is generally thought that the terms *ectotherm* and *endotherm* represent two

---

BOX 14.1

TO HAVE AND TO HAVE HOT

ALTHOUGH ENDOTHERMY is *characteristic* of birds and mammals, it is by no means restricted to these groups. Physiologists have known for some time of plants(!) that can regulate heat in a variety of ways, the most common being to decouple the metabolic cycle so that energy, instead of being stored or used in growth, is released as heat. Several snakes are known to generate heat while brooding eggs, although this is accomplished by muscle exertion. Certain sharks and tunas can retain heat from their core muscles by a kind of diffusion called counter-current circulation, and a variety of insects, including moths, beetles, dragonflies, and bees, are known to regulate their body temperatures. Interestingly, endothermy is not characteristic of these groups of organisms. Simply, it is known that some of them do maintain temperatures warmer than those of the medium (air or water) in which they are living. Indeed, it has been estimated that endothermy has evolved independently at least 13 different times.

This, of course, differs from the idea that endothermy is *diagnostic* of a particular group. Indeed, endothermy is characteristic of but two groups: birds and mammals. In these, all of the organisms belonging to the group possess the character.

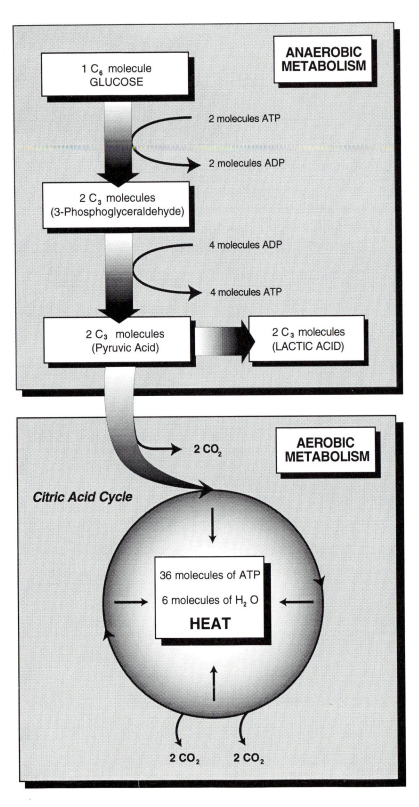

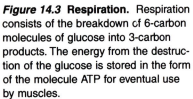

Figure 14.3 **Respiration.** Respiration consists of the breakdown of 6-carbon molecules of glucose into 3-carbon products. The energy from the destruction of the glucose is stored in the form of the molecule ATP for eventual use by muscles.

biochemically different methods of obtaining heat. The terms *poikilotherm* and *homeotherm*, however, are endpoints in a spectrum that runs from maintaining a constant temperature to having a fluctuating temperature. Although many animals do cluster at the familiar endpoints, many do not.

Endothermic and Ectothermic Metabolism

METABOLISM. Temperature control is not the only issue in endothermic and ectothermic tetrapods. Indeed, it is the very nature of the **metabolism** itself, that is, the sum of the chemical reactions in the cells of the organism, and the effect that the differences in endothermic and ectothermic metabolisms have on activity, that are significant.

To understand the differences between the metabolisms of endothermic and ectothermic tetrapods, we must review some of the basic biology of metabolism. In all organisms, energy is obtained from the breakdown of a molecule called ATP (adenosine triphosphate) to ADP (adenosine diphosphate) in the following way:

$$ATP \longrightarrow ADP + energy$$

The goal, therefore, is to store energy in the form of ATP, so that when energy is required – for example, during muscle movement – it can be released when a high-energy phosphate bond in ATP is broken to produce ADP.

The storage of energy in the form of ATP occurs through **cellular respiration**, which is the breakdown of carbohydrates through a series of oxidizing reactions. **Carbohydrates** are a family of molecules, including sugar, whose chemical bonds, if broken, release energy. **Oxidation** means that oxygen has bonded with whatever is being oxidized. In respiration, the chemical bonds in the carbohydrates are broken as oxidation takes place, and energy is released as heat, as well as stored in ATP. In fact, the oxidative breakdown of a single molecule of glucose (a simple carbohydrate) can produce 38 new molecules of ATP. The system, however, is not nearly 100% efficient: ATP production captures about 40% of the energy of the bonds of the carbohydrates. The remaining 60% is released as heat (Figure 14.3).

Organisms respire oxygen, then, to drive these complex oxidation reactions, so that energy may be stored in the form of ATP. As the energy output of the organism is increased, the amount of ATP needed is increased, and hence more oxygen is consumed and more heat is produced. This is why breathing and heart rates increase when we exercise: We are using more energy, requiring more ATP to be generated, and thus we need more oxygen. We get hot when we exercise in part because of the heat released as ATP is produced.

There is a point, however, at which the organism reaches its maximum rate of oxygen consumption. When this point is exceeded, lactic acid forms within the cells as a product of the oxidative breakdown of the carbohydrates. The presence of lactic acid in too great a concentration is detrimental to the organism, and thus when the respiration becomes **anaerobic** (without oxygen; that is, the need for oxygen exceeds the ability of the cells to respire it), the organism is fast approaching its maximum energy output.

METABOLISM IN ENDOTHERMS AND ECTOTHERMS. With regard to the onset of anaerobic metabolism, endotherms and ectotherms are very different. At body

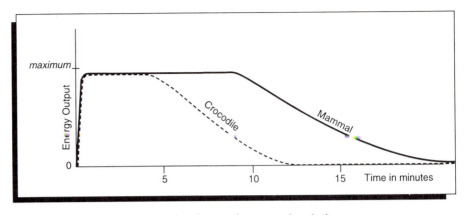

Figure 14.4 Energy output vs. time in ectotherms and endotherms.

temperatures equal to that of an endotherm, ectotherms can increase their oxygen consumption between 5- and 10-fold. It's very much like the throttle on an engine: When the throttle is opened, the speed of the engine increases, its power output increases, and its gas consumption increases. Ectotherms increase the amount of oxygen they consume to 5–10 times above their warmed resting rate before lactic acid begins to accumulate.

An endotherm, on the other hand, is also capable of increasing its oxygen consumption to 5–10 times above its resting rate. Its resting rate, however, is approximately where an ectotherm's point of lactic acid formation occurs. Thus, the oxygen consumption *and* energy output of an endotherm are considerably greater than those of an ectotherm.

This has major consequences for activity. The *amount* of energy required to move equivalently sized endotherms and ectotherms is similar; that is, it takes about as much energy for a 1-kg lizard to move as for a 1-kg mammal. However, an endotherm can produce far more energy than can an ectotherm before it (the endotherm) begins anaerobic metabolism (lactic acid production). This means that generally, *endothermic tetrapods are capable of higher levels of activity sustained over longer periods of time than are ectothermic tetrapods* (Figure 14.4).

Ectothermic tetrapods have compensated in a variety of significant ways. Most important, many have developed ways to sustain short bursts of anaerobic respiration. Such short bursts allow for about 2–5 minutes of relatively high activity and allow ectotherms a range of active behaviors, including guarding territory and young, hunting using speed and endurance, and actively defending themselves. The differences between endotherms and ectotherms, therefore, are even greater than simply the source of heat necessary to optimize the metabolic reactions carried on in the organism. Metabolic differences translate into some very real differences in physical capabilities. One final factor, however, must be figured into our understanding of these organisms: cost in energy.

It has been estimated that it costs 10–30 times as much energy to maintain an endothermic metabolism as to maintain an ectothermic metabolism. Much of the energy is expended on maintaining a constant body temperature. An analogy can again be made between endotherms and an engine with no idle setting. The machine is always revving at high speeds (because it has no idle, a lower adjustment) and using up large quantities of fuel (a form of energy).

In a world of finite resources, this means that a limited source of energy can support far fewer endotherms than ectotherms. Suppose, for example, that an organism had available to it all the energy resources on an acre of land. Because each endotherm requires more energy than each ectotherm, that acre of land – all other things being equal – is capable of supporting fewer endotherms than ectotherms. Thus, if the number of organisms is any measure of success, ectotherms will be more numerous than endotherms, because ectotherms are more economical: They simply do not cost as much to maintain.

We can return, therefore, to the question we asked earlier: If endothermy is superior to ectothermy, why are ectotherms *still* so numerous and diverse? A possible answer is that endothermy and ectothermy are simply two different metabolic strategies for optimizing use of resources. One depends upon admittedly costly sustained levels of activity, whereas the other substitutes for sustained activity short bursts of activity and greater economy.

DINOSAUR ENDOTHERMY: THE EVIDENCE

Reconstructing aspects of the metabolism of extinct vertebrates requires circumstantial evidence drawn from a wide variety of sources. But as is the case in the courtroom, evidence that is circumstantial may be all that is available to convict. Here, we categorize the evidence bearing upon fossil metabolism as anatomical, histological, ecological, zoogeographic, phylogenetic, and geochemical, borrowing the first five of those categories from Yale University's J. H. Ostrom, one of the pioneers in this field.

Anatomy

POSTURE. As we observed in Chapters 4 and 5, all dinosaurs are thought to have maintained a fully erect posture. Among living vertebrates, a fully erect posture occurs only in birds and mammals, both of which are endothermic. A number of paleontologists, therefore, have argued that the fully erect limb position in dinosaurs is certainly suggestive of endothermy.

The relationship between posture and endothermy did not at first seem to be one of cause and effect. Why would an organism need a fully erect posture to be an endotherm? Think of it the opposite way, however: An erect stance is inherently unstable. The fine neuromuscular control necessary to maintain a fully erect stance would be possible only within the temperature-controlled environment afforded by an endothermic metabolism.

Recent work by D. R. Carrier (see also Box 4.2) suggests another relationship between endothermy and posture. Carrier observed that when an animal with sprawling posture moves, the trunk of the organism flexes from side to side as the animal walks. Such flexion reduces the amount of air that can fill the lungs. For example, when the body twists to the right, the right lung is squeezed and the amount of air that it can accept is much less than the amount of air that it can accept when the body isn't twisted (Figure 14.5).

In the case of tetrapods with sprawling or semi-erect postures, Carrier noted an irony: Because the lungs are alternately being compressed during locomotion (through side-to-side flexion of the trunk), when the animal needs the most air,

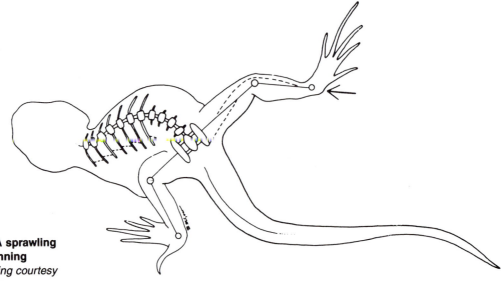

Figure 14.5 A sprawling vertebrate running quickly. *Drawing courtesy of D. R. Carrier, Paleobiology 13:327.*

it gets the least. Carrier suggested, therefore, that efficient locomotion became possible without flexion of the torso, and thus the lungs could fill to capacity.

His hypothesis suggests a causal relationship between posture and gait, on the one hand, and endothermy on the other. Considered in this way, a fully erect posture could be a prerequisite for an endothermic metabolism.

Two other simple correlations have been noted between anatomy and endothermy. The first, that long-leggedness is characteristic of living endotherms whereas living ectotherms possess relatively stubby limbs, was noted by Bakker. The second is the observation that among living tetrapods, the only bipeds are endotherms. Although bipedalism is not obviously a prerequisite for endothermy, if the living tetrapods are any guide, endothermy may be a prerequisite for *obligate* bipedalism.

TRACKWAYS, LIMB ANATOMY, AND INFERRED ACTIVITY LEVELS. A number of scientists have had the idea that metabolism could be reflected in behavioral differences, which in turn might be reflected in aspects of the skeletal anatomy. First and foremost in this regard is evidence from dinosaur trackways. Trackways show what was suspected from anatomy: Dinosaurs moved with a fully erect posture in which the feet were placed – as are our own – directly beneath the body (Figure 14.6). Moreover, in a variety of small bipedal dinosaurs such as dromaeosaurids and the smaller ornithopods, thin, elongate (gracile) bones (in which the thigh is short relative to the length of the calf) suggest high levels of running activity – behavior certainly not characteristic of modern ectotherms.

And what of the larger dinosaurs, especially those that were not bipedal? Here the issue becomes murkier. In 1987, Bakker published estimates of the locomotory speeds of dinosaurs. To calculate them, he used relationships pioneered by British functional morphologist R. McNeill Alexander. Alexander in 1976 determined that the walking speeds of all tetrapods can be calculated from a combination of footprint spacing (stride length) and the length of the hindlimb (see Box 14.2). Such estimates were of reasonably constant accuracy no matter which liv-

Figure 14.6 A Middle Jurassic theropod trackway from the Entrada Formation, Utah, U.S.A. *Photograph courtesy of Martin Lockley.*

ing tetrapod was tested. Alexander reasoned that if the relationship applies to living tetrapods, it could also apply to dinosaurs, where the hindlimb length and stride length (from trackways) were known. He concluded that those dinosaurs studied moved relatively slowly, at least while making tracks.

Bakker continued Alexander's work, using studies of modern animals and applying these to a variety of fossil tetrapods, many of which were dinosaurs. Among modern vertebrates, 95% of his estimations fell within 2.5 times the measured speed. This allowed rather much latitude in his estimates of speeds, and he unhesitatingly concluded that dinosaurs moved at speeds equal to, or greater than, those of equivalent-sized mammals. Bakker imagined a 5- or 10-ton *Triceratops* achieving a thundering "gallop" (Figure 14.7).[1]

Galloping notwithstanding, could large dinosaurs have moved at the speeds Bakker proposed? In all of Bakker's reconstructions, the *forelimbs* of quadrupedal

---

[1] The idea of galloping dinosaurs is problematical. The primitive gait in amniotes is rather different from this, as it involves a hindlimb-forelimb mixture (i.e., right hindlimb → right forelimb → left hindlimb → left forelimb). Galloping, on the other hand, is a distinctive means of locomotion that involves an arcuate bound from the hindlimbs to the forelimbs. R. A. Thulborn and C. Janis have suggested that the gallop possibly originated from the bounding motion of ancestral small mammals. As mammals became larger, the backbone stiffened considerably (presumably to support the increased weight), but the bounding motion was retained in the form of the gallop, which still involved the hindlimb-to-forelimb motion. Dinosaurs, as we have seen, have no history of a small, bounding, quadrupedal ancestor; they evolved from bipedal forms. Thus, for large quadrupedal varieties to evolve a true gallop would require a complex rewiring of their neuromuscular systems. For this reason, many paleontologists reject the idea of galloping dinosaurs. Recently, however, a crocodile was reported (and filmed) in a full gallop! Obviously, then, the gallop is not uniquely mammalian, even if it did evolve in mammals as Janis and Thulborn suggest.

dinosaurs are treated like those of mammals: They are fully erect and move in a predominantly fore–aft plane. Most quadrupedal dinosaurs have been reconstructed as having semi-erect forelimbs; early workers, blissfully unaware of the kinds of debates that their reconstructions would engender, reconstructed them that way because the bones of the forelimbs seem to suggest outward flexing at the elbow. Indeed, all paleontologists agree that there is some flexure at the elbow; the question is, does the elbow bow out or is the flexure facing rearward (for example, see Figure 9.2)? Bakker has marshaled a variety of arguments to support his contention that the forelimbs in ceratopsians are fully erect, including reconstructions of a well-formed elbow joint and a highly mobile shoulder blade. A detailed rebuttal of Bakker's forelimb reconstructions by R. E. Johnson of the

---

BOX 14.2

IN THE TRACKS OF THE DINOSAURS

TRACKWAYS are the most tangible record of locomotor behavior and provide the most direct indicator of gait and the only independent test of hypotheses of gait based upon anatomical reconstructions. When footprints are arranged into alternating left-right-left-right patterns, they demonstrate that all dinosaurs walked with a fully erect posture. But how can trackways also give us an indication of locomotor speed?

We begin with stride length – that is, the distance from the planting of a foot on the ground to its being planted again. When animals walk slowly, they take short strides, and when animals are walking quickly or running, they take considerably longer strides. This much is intuitive for anyone trying to catch a bus about to pull away from the curb. Now, consider the situation when you are being chased by something smaller than you. The creature chasing you must take long strides for its size, and more of them too, just to keep up. So there is clearly a size effect during walking and running, and this will likely be different for different kinds of animals under consideration.

How, then, to relate stride, body size, and locomotor speed? British functional morphologist R. McNeill Alexander provided an elegant solution to this problem by considering **dynamic similarity**. Dynamic similarity is a kind of conversion factor: It "pretends" that all animals are the same

size and that they are moving their limbs at the same rate. With these adjustments for size and footfall, it doesn't matter if you're a small or large human, a dog, or a dinosaur. All will be traveling with "dynamic similarity"; only speed will vary. That variable Alexander terms *dimensionless speed*.* It is dimensionless speed that has a direct relationship with relative stride length. Stride length, of course, can be measured from trackways, which in turns allows us for the first time to calculate locomotor speed in dinosaurs.

To see how all this works, let's use Alexander's example of the trackway of a large theropod from the Late Cretaceous of Queensland, Australia. The tracks are 64 cm long, which Alexander, from other equally sized theropods, estimated must have come from a theropod with a leg length of about 2.56 m (Alexander determined that in this group of animals, limb length approximately equals four times the foot length). The stride length of these tracks is 3.31 m, so the relative stride length (stride length/limb length) is 1.3. The dimensionless speed for a relative stride length of 1.3 is 0.4. And from all these measures, this Australian theropod must have been traveling reasonably quickly, at about 2.0 m/sec, or 7.2 km/hr.

---

*Dimensionless speed may appear oxymoronic, but it is defined as real speed divided by the square root of the product of leg length and gravitational acceleration (i.e., dimensionless).

Milwaukee Public Museum and Ostrom, using the forelimbs of the ceratopsian *Torosaurus,* suggests that the issue is far from resolved. They believe that the elbow was partially flexed and angled somewhat away from the side of the body.

Again, the differences between the ancestries of dinosaurs and mammals may play a telling role. In mammals, a fully erect posture evolved in a quadrupedal ancestor; however, in dinosaurs the fully erect posture evolved in a bipedal ancestor. Dinosaurs that have become secondarily quadrupedal (see Chapter 5) may therefore be very unlike mammals in design, and the quadrupedal adaptations of dinosaurs need not necessarily exactly match those of mammals.

ASSORTED ADAPTATIONS FOR PROCESSING HIGH VOLUMES OF FOOD.
Remembering that endotherms require more energy than ectotherms, if it could

---

As complicated as this approach appears, it represents the best method for estimating the actual speeds implied by trackways. But what about the fastest speeds a dinosaur might have been capable of?

In 1982, R. A. Thulborn (University of Queensland) put together the means by which absolute locomotor abilities could be calculated. Thulborn's work relied heavily upon Alexander's slightly earlier studies on speed estimates from footprints and the relationship between body size, stride length, and locomotor speed among living animals. Alexander determined that relative stride length has a direct relationship with locomotor speeds at different kinds of gaits (e.g., walking, running, trotting, galloping):

Locomotor Velocity =
$$0.25(\text{gravitational acceleration})^{0.5}$$
$$\times (\text{estimated stride length})^{1.67} \times (\text{hindlimb height})^{-1.17}$$

Thulborn used this equation to estimate a variety of running speeds for more than 60 dinosaur species. The first group of estimates were for the walk–run transition, where stride length is approximately two to three times the length of the hindlimb. A potentially more important estimate – especially for dinosaurs fleeing certain death or pursuing that all-important meal – is maximum speed, which Thulborn calculated using maximum relative stride lengths (which range from 3.0 to 4.0) and the rate of striding, called limb cadences (estimated at $3.0 \times$ hindlimb length$^{-0.63}$).

Although we report some of these speeds elsewhere in this book, it is of some value to summarize the overall disposition of Thulborn's estimates. For small bipedal dinosaurs – which would include certain theropods and ornithopods – all appear capable of running at speeds of up to 40 km/hr. Ornithomimids, the fastest of the fast, may have sprinted up to 60 km/hr. The large ornithopods and theropods were most commonly walkers or slow trotters, probably averaging no more than 20 km/hr. Thus the galloping, sprinting *Tyrannosaurus,* however attractive the image, did not impress Thulborn (or us) as likely.

Then there were the quadrupeds. Stegosaurs and ankylosaurs walked at no more than a pokey 6 to 8 km/hr. Sauropods probably moved at 12 to 17 km/hr. And ceratopsians – galloping along full throttle like enraged rhinos? Thulborn estimated that they were capable of trotting at up to 25 km/hr.

Are these estimates accepted uncritically by all? P. Dodson has argued that these calculations applied to humans would suggest that humans can run as quickly as 23 km/hr! So it is possible that these calculations overestimate the speeds at which dinosaurs could run. On the other hand, anatomically, dinosaurs are not humans; these calculations could be perfectly valid for dinosaurs but be unapplicable to humans. At a minimum, they do give some indication of the relative speeds of dinosaurs; for example, how quickly *T. rex* might have run relative to *Triceratops.*

Figure 14.7 **Gregory Paul's dynamic restoration of several galloping** *Triceratops.* *Illustration courtesy of G. S. Paul.*

be shown that all dinosaurs required large amounts of food to function, an endothermic metabolism for Dinosauria might be implied.

Most salient in this regard are the sophisticated chewing adaptations of ceratopsians and ornithopods, particularly hadrosaurids (see Chapters 9 and 10). Moreover, because they allow breathing and chewing to take place simultaneously, secondary palates are commonly associated with more efficient feeding. Ankylosaurs and hadrosaurids both bear well-developed secondary palates, suggesting that these animals had the benefit of being able to breathe and chew simultaneously.

The relationship to metabolism of these specializations for efficient food processing is by no means clear. Hadrosaurids and ceratopsians clearly had developed chewing mechanisms as efficient as, or more efficient than, those found in modern herbivorous mammals. Birds, however, which are fully endothermic, do not have secondary palates. These organisms little rely upon the food's being manipulated at the mouth; it is sent onward for treatment by the gizzard. Ankylosaurs, though possessed of a secondary palate, had very small teeth and none of the dental batteries characteristic of the ornithopods (Chapter 10). Indeed, secondary palates are known in modern turtles and crocodiles (as well as mammals), so their significance in terms of endothermy is not clear.

BRAINS. Most dinosaurs appear to have been considerably smarter than supposed in the first half of this century. In the late 1970s, the EQ concept (Box 14.3) was developed further by University of Chicago paleontologist J. A. Hopson. Hopson used **brain endocasts**, internal casts of the braincases of dinosaurs. This was done by painting latex on the inside of a fossilized braincase. When the latex dried and

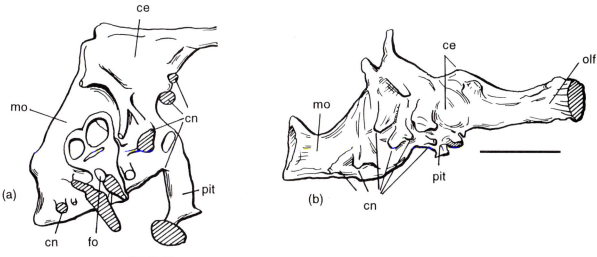

Figure 14.8 Brain endocasts of (a) *Plateosaurus* (see also Figure 4.11) and
(b) *Tyrannosaurus*. Note scale differences. Keep in mind that the skull of an adult *T. rex*
was 1.5–2.0 m long. Based upon brain size, neither of these dinosaurs was apparently a
spectacular thinker! Abbreviations: fo – fenestra ovalis (where vibrations from the eardrum
are sensed by the brain); ce – cerebrum; cn – cranial nerves; mo – medulla oblongata; pit –
pituitary; olf – olfactory region (sense of smell). Scale for (a) = 3 cm; scale for (b) = 6 cm.
Plateosaurus after Galton (1990); Tyrannosaurus *after Hopson (1980).*

was peeled off, it produced a cast of the inside of the braincase (Figure 14.8).

How good an indicator of the brain is the internal shape of the braincase? The braincases of living birds and mammals exactly conform to the shape of the brain. However, in the living lizards, snakes, and crocodiles, this is not the case: Their brains are considerably smaller than their braincases. Hopson, however, corrected for this situation in dinosaurs, producing a minimal brain size estimate, and thus a minimal EQ. This was then compared with the estimated body weight of various dinosaurs (Box 14.4). Hopson found that non-avian theropods in general tended to have the highest EQs, with some ornithopods coming a close second. Indeed, *Troodon* (a small theropod) had an EQ that is well within the modern bird or mammal range. Coelurosaurs in general tended to have higher EQs than the larger-bodied theropods. As we have seen, the lowest EQs were among the sauropods, ankylosaurs, and stegosaurs. Still, many dinosaurs were at least within the EQ range of crocodilians and other living "reptiles."

Hopson concluded that the EQs of dinosaurs generally fell somewhere between those of modern ectotherms and endotherms:

If the brain required by a vertebrate is primarily a function of its total level of activity and therefore reflects its total energy budget, then coelurosaurs appear to have been metabolically as active as living birds and mammals, carnosaurs and and large ornithopods appear to have been less active but nevertheless significantly more so than typical living reptiles, and other dinosaurs appear to have been comparable to living reptiles in their rates of metabolic activity. (J. A. Hopson, 1980, p. 309 in R. D. K. Thomas and E. C. Olson [eds.], *A Cold Look at the Warm-Blooded Dinosaurs*, American Association for the Advancement of Science Selected Symposium no. 28)

Histology

One of the truly exciting and original lines of evidence for dinosaur endothermy comes from the science of **bone histology**, or the study of bone tissues. Fossil bone may preserve fine anatomical details that are visible under a microscope. With a rock-cutting saw, one can slice across a slab of fossil dinosaur bone in much the way a butcher uses a band saw to cut across a leg of beef. To see the full details of the bone, a thin slice can be mounted on a glass slide and ground down to a thickness that allows light to be transmitted through it (about 30 microns). Thus the fine-scale structure becomes visible under a microscope (Figure 14.9).

BOX 14.3

DINOSAUR SMARTS

HOW CAN WE MEASURE the intelligence of dinosaurs?[*] The short answer is, not easily! However, it is clear that at a very crude level, there is a correlation between intelligence and brain–body weight ratios. Brain–body weight ratios are used because they allow the comparison of two differently sized animals (that is, brain–body weight ratios allow one to compare a chihuahua with a St. Bernard). The correlation suggests that in a general way, the larger the brain–body weight ratio, the smarter the organism. Indeed, mammals have higher brain–body weight ratios than fish and are generally considered more intelligent.

A modification of this idea was first proposed by psychologist H. J. Jerison, who in the early 1970s developed a measure called the "encephalization quotient" (EQ). The EQ measures how large an animal's cerebral hemisphere is in proportion to the rest of its brain. Jerison noted that, based upon EQ, living vertebrates cluster into two groups, endotherms and ectotherms. The endotherms show greater encephalization (higher EQs) and the ectotherms show lower encephalization (lower EQs). Thus, for Jerison, endotherms and ectotherms could be distinguished by brain size. Figure Box 14.3 shows the EQs of some dinosaur groups.

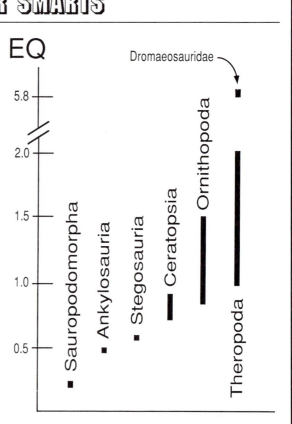

Figure Box 14.3 EQs of dinosaurs compared.
Modified from Hopson (1980).

---

[*]A more fundamental question is, What is intelligence? As applied here, intelligence refers to the ability to learn, and perhaps the capability for abstract reasoning. The measurement of "intelligence" in humans is a realm of inquiry freighted with a notoriously racist history and consequently much emotional baggage. A superb history of this is found in Stephen J. Gould's 1981 book, *The Mismeasure of Man*. Here, we are discussing intelligence in far cruder terms – at the level of the comparison of the intelligence of a crocodile and a dog.

Figure 14.9 **Ceratopsian bone under a microscope.** Compare with Figure 14.10; note presence of Haversian bone. Each cell is about 0.3 mm in diameter.

The idea of bone histology indicating rates of growth and aspects of metabolism in fossil reptiles is largely the work of A. J. de Ricqlès of the University of Paris (France). Ricqlès's insight was the observation of Haversian bone in birds, mammals, and dinosaurs.

HAVERSIAN BONE. Organisms grow, and their bones grow with them. Bones grow by **remodeling**, which involves the resorption (or dissolution) of bone laid down first – **primary bone** – and the redeposition of a kind of bone called **secondary bone**. Secondary bone is laid down in the form of a series of vascular canals called **Haversian canals**, and resorption and redeposition of secondary bone can occur repeatedly during remodeling. When this process is reiterated, a type of Haversian bone known as **dense Haversian bone** is formed. This bone has a distinctive look about it; the canals and their rims are closely packed (Figure 14.10).

Among living vertebrates, dense Haversian bone is found only in endotherms: in many mammals and birds. Among extinct vertebrates, dense Haversian bone has been observed in dinosaurs, pterosaurs, and the advanced synapsids (therapsids), including mammals. It was not too great a leap of faith to propose that dinosaurs, too, must have been endotherms.

THE SIGNIFICANCE OF HAVERSIAN BONE. Dense Haversian bone is due to a variety of factors, one of which may be endothermy. Haversian canals are known to be correlated with size and age; they are possibly correlated with the type of bone they replace, the amount of mechanical stress undergone by the bone in which they are found, and nutrient turnover (the interaction between soft tissue and developing bony tissue). Ricqlès writes:

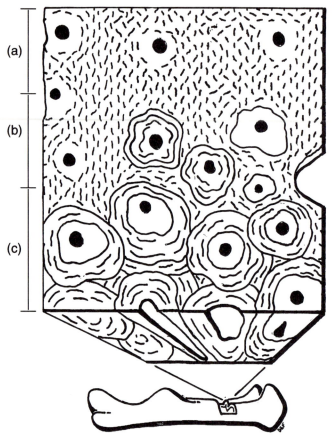
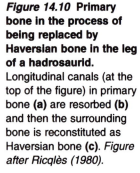

Figure 14.10 **Primary bone in the process of being replaced by Haversian bone in the leg of a hadrosaurid.** Longitudinal canals (at the top of the figure) in primary bone **(a)** are resorbed **(b)** and then the surrounding bone is reconstituted as Haversian bone **(c)**. *Figure after Ricqlès (1980).*

[A] high rate of metabolism cannot be inferred from the mere *occurrence* of remodeled [dense Haversian] bone, but is rather reflected by the rate and extent of bone replacement. For instance, a 50-year-old turtle and a 5-year-old dog might show the same amount of Haversian reconstruction, but the physiological significance of these tissues would be very different, particularly in regard to metabolic rate. (A. J. de Ricqlès, 1980, p. 125 in R. D. K. Thomas and E. C. Olson [eds.], *A Cold Look at the Warm-Blooded Dinosaurs,* American Association for the Advancement of Science Selected Symposium no. 28)

And what does this mean for endothermic dinosaurs? It means that, assuming that dense Haversian bone formed in dinosaurs at rates comparable to those in mammals, dinosaurs probably lived for life spans approximating those of Recent mammals, and dinosaurs could have had rates of bone growth similar to those found in mammals. If such rates really occurred, they would be in good agreement with conditions that might be expected to be found with an endothermic metabolism.

RATES OF BONE GROWTH AND LAGs.   What is really known about rates of dinosaur bone growth? Some of the early work in the field was done by

J. Peterson, at that time an undergraduate student of J. R. Horner's at Montana State University. As described in Horner's and J. Gorman's book, *Digging Dinosaurs*, Peterson correlated dense Haversian bone with overall limb development. In doing so, she was able to establish developmental stages of growth in the hadrosaurid *Maiasaura*; she could take a cross section of bone and tell from the development of the Haversian canals which growth *stage* the animal had reached. Using rates of bone growth obtained from living organisms, Peterson and Horner estimated that *Maiasaura* grew at an astounding 3 m/yr!

Quite different conclusions about different dinosaurs were reached in the late 1980s by British histologist R. E. H. Reid, and in the 1990s by his former student, A. Chinsamy, now at the South African Museum in Cape Town. These researchers recognized **lines of arrested growth** (**LAGs**), or bands in the bone, which they took to represent yearly increments. Chinsamy worked on the sauropodomorph *Massospondylus* and on an early theropod, the small, light-bodied *Syntarsus*, both from South Africa. From these saurischians she estimated *rates* of growth. She estimated that *Massospondylus* took 15 years to reach 250 kg (17 kg/year), whereas *Syntarsus* took 7 years to reach an estimated 20 kg (3 kg/year). These rates were somewhat slower than those that might have been predicted by Peterson for *Maiasaura* (Figure 14.11). The appearance of the secondary bone preserved in the thighs of these organisms more closely approximated that of modern birds than that of a crocodile. Chinsamy concluded that the sauropodomorph and the theropod she studied showed an apparent mix of features characteristic of endothermic and ectothermic metabolisms: Relatively slow rates of development are characteristic of an ectothermic metabolism, but the dense Haversian bone looked more like that of living endotherms (birds) than living ectotherms (crocodilians).

LAGs have come to play significant roles in the debate on dinosaur endothermy. LAGs occur when bone growth stops in modern ectotherms, usually as a result of temperature fluctuations caused by seasonality. Living endotherms – birds and mammals – grow independently of temperature and seasonality, and thus do not possess well-developed LAGs. Reid noted that LAGs occur in many different kinds of dinosaurs, suggesting to him that growth in dinosaurs was more susceptible to external climatic influences than had been predicted by Bakker. Reid, too, postulated a physiology somehow "intermediate" between the physiology of mammals and birds, on the one hand, and that of lizards and crocodiles on the other.

Chinsamy also studied LAGs in the Mesozoic bird *Patagopteryx*. *Patagopteryx* is an enantiornithine bird and thus was fully capable of flight (and would obviously have been feathered). Chinsamy found LAGs in *Patagopteryx*, leading her to the conclusion that the early bird's metabolism was subject to seasonal growth, even though it clearly bore feathers. Her conclusion was that the modern avian endothermic metabolism occurred *after* the development of feathers (see Chapter 13)!

D. J. Varricchio, from the Museum of the Rockies in Montana, recently studied cross sections of *Troodon* shins. In these he observed LAGs, which he interpreted as seasonal cessation of bone growth. Here again was a suggestion that the metabolism of these theropods was not truly like that of modern mammals and may have responded to seasonally induced temperature changes.

LAGs are not without their pitfalls. It is not at all clear that they represent *yearly* increments. Indeed, one hadrosaurid fossil is reported to have different numbers of LAGs for its arms and its thighs! Moreover, the appearance (or absence) of LAGs has never really been tightly correlated with climate. Like all other supposed indicators of metabolism, LAGs are not the last word.

In 1979, P. A. Johnston, a paleontologist now at the Royal Tyrrell Museum of Palaeontology in Alberta, Canada, observed concentric growth rings in the teeth of saurischian and ornithischian dinosaurs. Among modern tetrapods, such growth rings are typically found in ectotherms. The rings, like LAGs, are thought to represent seasonal cycles. Among endotherms such patterns are rare, because the relatively constant, elevated metabolic rates ensure growth at a constant rate. Johnston noted that the rings found in the dinosaur teeth were very much like those found in crocodile teeth from the same deposits. His conclusion, therefore, was that dinosaur teeth suggested an ectothermic metabolism for their owners.

BOX 14.4

WEIGHING IN A DINOSAUR

WHEN THE YOUNG CASSIUS CLAY (Muhammed Ali) weighed in for his title fight with Sonny Liston, he simply stood on a scale. On the other hand, to estimate the weight of a dinosaur is a more daunting proposition.

There are generally two ways that such weight estimations are approached. The first is based upon a relationship between limb cross-sectional area and weight. This relationship has some validity, because obviously as a terrestrial beast becomes larger, the size (including cross-sectional area) of its limbs must increase. The question is, does it increase in the same manner for all tetrapods? If so, a single equation could apply to all. It is clear, however, that it cannot. As noted by J. O. Farlow, weight is dependent in part upon muscle mass and muscle mass is really a consequence of behavior. Therefore, weight estimates of dinosaurs are in part dependent upon presumed behavior. For example, reconstructing the weight of a bear involves assumptions of muscle bulk and gut mass very different from those in reconstructing the weight of an elk (Figure Box 14.4a). Indeed, the cross-sectional area of their limb bones might be

identical, but they might weigh very different amounts. Moreover, our knowledge of dinosaurian muscles and muscle mass is rudimentary. This method, although convenient and used by a number of workers (including R. T. Bakker), has the potential for serious misestimations of dinosaur weights.

A second method, pioneered by American paleontologist E. H. Colbert in the early 1960s, involves the production of a scale model of the dinosaur and the calculation of its displacement in water (Figure Box 14.4b). That displacement could then be (1) multiplied by the size of the scale model (if the model was 1/32 of the original, the weight of the displaced water would have to be multiplied by 32), and (2) further modified by some amount to a number corresponding to the specific gravity of body tissues. But what is the specific gravity of body tissues of an extinct tetrapod? Based upon studies with a baby crocodile, Colbert determined that baby crocodiles, at least, have a specific gravity of 0.89. Unfortunately, there is no uniform specific gravity shared by all tetrapods. Studies with a large lizard (*Heloderma*)

The idea was met with a welter of strong criticism. Much of the criticism suggested that the rings might not be solely the result of yearly fluctuations in temperature and/or moisture. Other critics observed that growth rings can occur in mammals such as desert bighorn sheep and African buffalo.

Ecology

OF PREDATORS AND THEIR PREY. Perhaps the most tantalizing – and frustrating – evidence bearing upon dinosaur endothermy comes from **paleoecology**, the study of ancient interactions among organisms. The idea – again from Bakker – was rooted in the fact that endotherms require more energy than ectotherms. Because he was interested in *energy* (and not simply numbers of organisms), all of his calculations involved masses of organisms. Total weight of the organisms involved, that is, their **biomass**, was assumed by Bakker to directly reflect energy. If predators were endothermic, they should require more energy than if they were

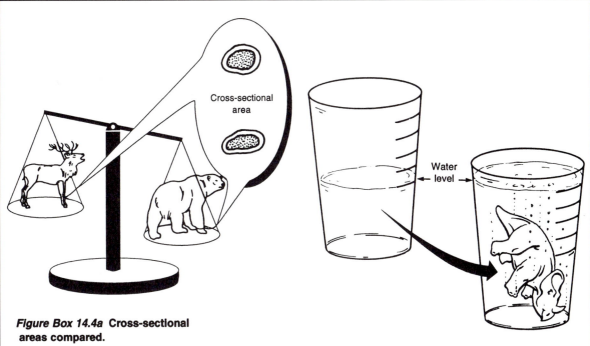

Figure Box 14.4a Cross-sectional areas compared.

Figure Box 14.4b Weighing a dinosaur.

showed that the specific gravity of that lizard was 0.81. Among mammals, it would not be surprising to find the specific gravity of a whale differing from that of a cheetah, which might in turn differ from that of a gazelle. In short, although the displacement method is perhaps a bit more accurate than limb cross-sectional calculations, it is still dependent upon inferred muscle mass (and thus behavior).

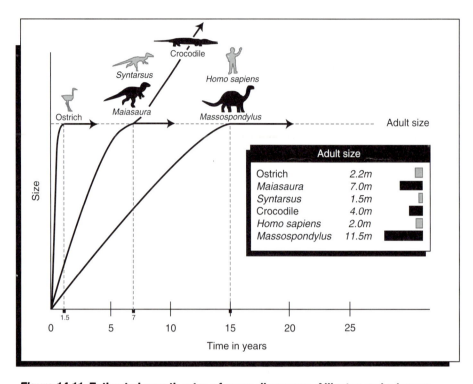

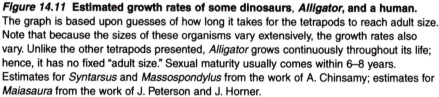

Figure 14.11 Estimated growth rates of some dinosaurs, *Alligator*, and a human.
The graph is based upon guesses of how long it takes for the tetrapods to reach adult size.
Note that because the sizes of these organisms vary extensively, the growth rates also
vary. Unlike the other tetrapods presented, *Alligator* grows continuously throughout its life;
hence, it has no fixed "adult size." Sexual maturity usually comes within 6–8 years.
Estimates for *Syntarsus* and *Massospondylus* from the work of A. Chinsamy; estimates for
Maiasaura from the work of J. Peterson and J. Horner.

ectothermic, and this should in turn be reflected in the weight proportions in the
community of predator and prey. Bakker wrote:

> [A] meter of heat production in extinct vertebrates is the predator-prey ratio:
> the relation of the "standing crop" of a predatory animal to that of its prey.
> The ratio is a constant that is a characteristic of the metabolism of the preda-
> tor, regardless of the size of the animals of the predator-prey system. The rea-
> soning is as follows: The energy budget of an endothermic population is an
> order of magnitude larger than that of an ectothermic population of the same
> size and adult weight, but . . . the yield of prey tissue available to predators . . .
> is about the same for both an endothermic and an ectothermic population.
> . . . The maximum energy value of all the carcasses a steady-state population
> of lizards can provide is about the same as that provided by a prey population
> of birds or mammals of about the same numbers and adult body size.
> Therefore a given prey population, either ectotherms or endotherms, can
> support an order of an order of magnitude greater biomass of ectothermic
> predators than of endothermic predators, because of the endotherms' higher
> energy needs. (R. T. Bakker, "Dinosaur Renaissance," *Scientific American*,
> April 1975, pp. 61–62)

With that premise, Bakker obtained modern predator–prey biomass[2] ratios, with the intent to compare them with the ancient ones. The modern ones were relatively accessible from published accounts. It turned out that size played a role; a community of large herbivores might yield 25% of its total biomass as prey, whereas a community of small herbivores might produce 600% of its total biomass as prey.[3] How then to factor in the issue of size, which would obviously affect the final ratio? Bakker proposed that endpoints in the size spectrum of predators and prey effectively cancel each other out, and the result is that uniform predator–prey biomass ratios exist for endotherms, regardless of size. Bakker calculated that the predator–prey biomass ratio for ectothermic organisms is 40%, and the predator–prey biomass ratio for endothermic organisms is 1%–3%. Here then was an order of magnitude difference in the biomass ratios that ought to be recognizable in ancient populations.

Now he was armed with a tool from modern ecosystems that he believed could reveal the energy requirements of ancient ecosystems. By counting specimens of predators and presumed prey in major museums and by estimating the specimens' living weights, he attempted reconstructions of the biomasses of fossil assemblages. His results seemed unequivocal: Among the dinosaurs, the predator–prey biomass ratios were very low, ranging from 2% to 4%. He interpreted this low number as indicative of endothermic dinosaurian predators (Figure 14.12).

Bakker's study, for all its creativity and originality, raised a howl of protest. Bakker never made his exact methods or numerical results widely available to other scientists through his publications. Moreover, he never acknowledged the number of assumptions embodied in his conclusions. J. O. Farlow, reviewing Bakker's conclusions, noted that even the initial premise – that there is an order of magnitude difference between ectothermic predator–prey biomass ratios and endothermic predator–prey biomass ratios – may not hold in all cases. Bakker's assumption that prey are approximately the same size as the predators is clearly not correct (consider a bear eating a salmon) and has drastic effects on the resultant ratio. For example, in the case of frogs and lizards eating insects, predator–prey biomass ratios range from 7% to 60%. Indeed, we really don't know who the prey of many carnivorous dinosaurs were. If we assume that such prey were about the same size as the carnivore, we can find suitable candidates. On the other hand, it is reasonable to suppose that carnivorous dinosaurs consumed prey from a range of sizes.

The predator–prey calculation assumes that the number of predators is limited by the amount of food (in this case, prey) available. But it isn't at all clear that this is the only factor involved; a variety of spatial, social, behavioral, and health factors also constrain predator populations.

The predator–prey biomass calculation also assumes that all deaths are the result of predation – that there can be no mortality due to other causes. However, when the predatory–prey biomass ratio was calculated by Farlow to incorporate deaths by predation only, the *maximum* ectothermic predator–prey biomass ratio was 20%. In reality, of course, because death comes to populations by means other

---

[2] He called them "predator–prey ratios," but in fact they were estimated predator biomass–estimated prey biomass ratios. Here, for convenience, we'll refer to them as predator–prey biomass ratios.

[3] How could a community produce six times as much biomass as there was community in the first place? The answer is time-averaging. Bakker noted, "[A] herd of zebra yields from about a fourth to a third of its weight in prey carcasses a year, but a 'herd' of mice can produce up to six times its weight because of its rapid turnover, reflected in a short life span and high metabolism per unit weight" (R. T. Bakker, "Dinosaur Renaissance," *Scientific American*, April 1975, p. 62).

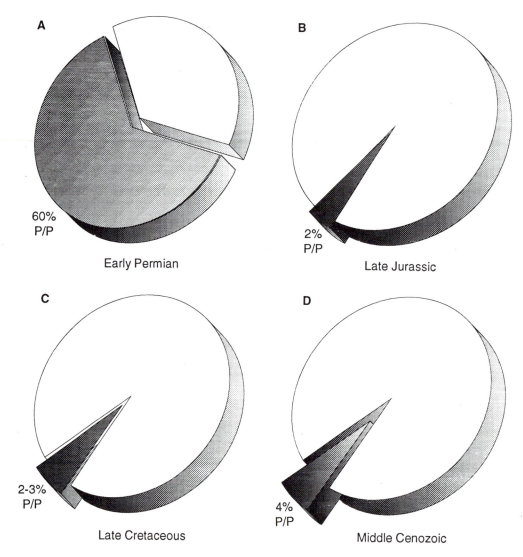

A

60%
P/P

Early Permian

B

2%
P/P

Late Jurassic

C

2-3%
P/P

Late Cretaceous

D

4%
P/P

Middle Cenozoic

Figure 14.12 **The proportions of predators to prey in selected faunas in the history of life, as reconstructed by R.T. Bakker. (A)** Early Permian of New Mexico; **(B)** Late Jurassic of North America; **(C)** Late Cretaceous of North America; **(D)** Middle Cenozoic of North America. Predators are shaded; prey are white.

than predators, this number could be smaller. And as noted by Farlow, such ratios could, and do, vary significantly when the turnover rate – that is, the rate at which a population produces a new generation as the older generation dies off – of the predators differs substantially from that of the prey.

There are serious and legitimate problems with using fossils for such calculations. Most obvious are difficulties in estimating the true weight of dinosaurs. Several estimates are available (see Box 14.4), but even discounting great differences in these values, small differences in weight estimates can substantially alter the predator–prey biomass ratios.

Another problem is that preservation of dinosaur faunas can be very mislead-ing. In the field, dinosaur material is found with a variety of biases in preservation. How can one ever be sure that the proportions of the living community are rep-resented? Because we can't, paleontologists commonly talk about fossil assem-blages. The word *assemblage* reminds us that the proportions of organisms that we find in a particular assemblage may not be the same as the proportions of the same animals in the ecosystem in which those animals lived.

Bakker obtained data by counting specimens in museums. But this is going out of the frying pan and into the fire: Dinosaur fossils are commonly collected by museums to obtain rare or particularly well-preserved specimens, not to attempt to reconstruct ancient communities. It is unlikely that museum collec-tions truly represent the correct proportions of the living community.

Bakker's idea is a brilliant study in creative thinking; however, so much uncer-tainty is introduced through his method that his results can hardly be conclusive.

Zoogeography

As noted earlier in this chapter, modern large ectotherms are confined within a limited range of about 20° north and south of the equator. In the Northern Hemisphere, Late Cretaceous sites have been found in Alaska, the Yukon, and Spitsbergen. In the Southern Hemisphere, Early Cretaceous dinosaur sites are known from the southern part of Victoria, Australia. Both of these are above the Late Cretaceous Arctic and Antarctic circles.

What were these high latitude localities like? In North America, evidence derived from theoretical considerations and from fossil plants suggests that the region experienced a mean annual temperature of 8°C, and a mean cold month temperature of 2°–4°C. By comparison, based upon stable isotopes, the Australian Early Cretaceous site had an estimated mean annual temperature of 5°C, with tem-peratures descending to as low as –6°C. Overall, the climate of the Australian local-ity is thought to have been humid, temperate, and strongly seasonal. Certainly sites at both the northern latitudes and the southern latitudes experienced extended periods of darkness, and we can be reasonably sure that, at least occasionally, tem-peratures in winter descended below freezing (recall that there is still uncertainty about whether or not there was permanent polar ice; see Chapter 2).

The assemblage from North America includes hadrosaurids, tyrannosaurids, and troodontids. The dinosaur assemblage from Australia includes a large thero-pod (possibly *Allosaurus*) and some apparently large-brained hypsilophodontid ornithopods (including many juveniles) with highly developed vision. Along with these dinosaurs are fish, turtles, pterosaurs, plesiosaurs, birds (known solely from feathers), and, incredibly, an amphibian that apparently had gone extinct during the Early Jurassic everywhere else in the world.

This mix of animals is not particularly easy to interpret in terms of endothermy or ectothermy. P. V. Rich and T. H. Rich from Monash University, Victoria, Australia, and their colleagues studied the southern faunas and remarked on the large brains and well-developed vision of the ornithopods, which they sug-gested were helpful during periods of extended darkness. Other members of the southern assemblage, however, were not so equipped. The Riches and colleagues noted that all of the animals except the species of *Allosaurus* could potentially have survived winter by burrowing. The dinosaurs in the assemblage could have survived by migration; however, where they might have migrated to is an open question, and migration was certainly not an option for the turtles and labyrinthodonts, and possibly not for the plesiosaurs.

In the case of the North American faunas, only *Troodon* is of a size that could make burrowing feasible. Migration was potentially a solution to inclement winter weather, although the dinosaurs would have had to migrate for tremendous distances before temperatures warmed sufficiently.

In 1987, G. S. Paul published an evocatively illustrated argument for the idea that dinosaurs living in the high latitudes must have been endothermic (Figure 14.13). He noted that a large ectotherm, the Late Cretaceous crocodilian *Phobosuchus* (regarded by many as *Deinosuchus*, comparable in size to the Alaskan and Canadian hadrosaurids and tyrannosaurids) is found in temperate latitudes

Figure 14.13 Gregory Paul's vision of a tyrannosaurid drinking at a nearly frozen Arctic pond at night. *Drawing courtesy of G. S. Paul.*

and yet is absent in the northernmost deposits. He suggested, too, that those smaller, ectothermic members of the assemblage who could burrow, did. Finally, he concluded that ectothermic dinosaurs made sluggish by the cold would be easy prey for organisms – mammalian or not – with more active metabolisms.

Burrowing was obviously unfeasible for the hadrosaurids and tyrannosaurids in the assemblage (because of their tremendous size). As for migration, based upon theoretical calculations of energy requirements, Paul rejected it as a viable option for the dinosaurs in the northern assemblage. With migration energetically untenable, and ectothermy in an Arctic climate putting potential ectotherms at a strong selective disadvantage, Paul concluded that the large dinosaurs in the Arctic assemblage had to be some type of endotherm. Moreover, he suggested that the Arctic dinosaurs may have maintained some insulatory covering (such as fur or feathers) at least seasonally.

There are some inherent weaknesses in this line of reasoning. Even where it is best known, *Phobosuchus* is a rare animal, and thus its absence at high latitudes hardly comes as a surprise. Moreover, there has been no evidence for burrowing of any sort reported. Finally, the obvious (nonmigrating) ectotherms from southern high latitudes suggest that the absence of *Phobosuchus* in the Arctic may have been due to factors other than the cold.

In 1991, a publication by biologist J. R. Spotila and colleagues reassessed theoretically the energy requirements for migration. Spotila, a veteran of studies of dinosaur endothermy and behavior, concluded on theoretical grounds that migration was a possibility and even a likelihood for medium-sized to large dinosaurs. Spotila's view was based on energy calculations for a large ectotherm, the seagoing leatherback turtle (*Dermochelys*). Water conducts heat at a rate far *greater* than does air, which means that heat loss would be even more severe for *Dermochelys* than it would have been for a dinosaur. In fact, for very large (voluminous) dinosaurs such as sauropods, the issue might be the release of heat to avert overheating, rather than the retention of heat!

Phylogeny

Birds are dinosaurs (Chapter 13) and modern birds are endothermic. Our question must be, therefore, at what point during dinosaur evolution did endothermy evolve? If it evolved very early in the evolution of theropods, the ancestor of dinosaurs (Chapter 5) may have been endothermic and thus all dinosaurs would have inherited the metabolism. On the other hand, if it evolved with Aves, then perhaps birds – and birds alone – are the only endotherms in Dinosauria.

An important clue comes with insulation. All small- to medium-sized modern endotherms are insulated with fur or feathers.[4] Indeed, pterosaurs are suspected endotherms, in part because they are known to have been covered with a fur-like coat. There is a certain sense to this; if an ectotherm depends upon external sources for heat, why develop a layer of protection (insulation) from that external source? *Archaeopteryx*, with its plumage, is therefore usually considered to have been endothermic (but see Chapter 13). Because skin impressions are not known from non-avian coelurosaurs, it is not clear whether they might have borne feathers. These theropods certainly are good candidates for endothermy; with their avian skeletal morphology and histology and with high EQs, they are as likely to have been endothermic as any dinosaur. If so, they might have been covered with a coating of downy feathers for insulation, rather than with primary and secondary feathers for flight.

While in this speculative vein, it is interesting to note – as did R. S. Seymour in 1976 – that all living endotherms possess four-chambered hearts. Seymour argued that the four-chambered heart system, in which the oxygenated blood is completely separated from the deoxygenated blood, is probably a prerequisite for endothermy. The argument is based upon the idea that endothermy requires relatively high blood pressures in order to constantly perfuse complex, delicate organs such as the brain with a constant supply of oxygenated blood. Such high

---

[4]Although all mammals are fur bearing, the large mammals in tropical climates have very sparse coats that do not insulate. Here, a combination between temperature and the size of the organism is assuring heat retention.

blood pressures, however, would "blow out" the alveoli in the lungs. For this reason, mammals separate their blood into two distinct circulatory systems: the blood for the lungs (pulmonary circuit) and the blood for the body (systemic circuit). The two separate circuits require the four-chambered heart – a pump that can completely separate the circuits. Part of Seymour's argument was based upon the inference that in sauropods like *Brachiosaurus*, tremendous blood pressures would be required to bring the oxygenated blood to the brain (see Chapter 11). Thus the heart would *have* to be a double-pump (one for oxygenated and one for unoxygenated blood) four-chambered organ.

Would such a heart be possible in dinosaurs? The nearest living relatives of dinosaurs – crocodiles and birds – both possess four-chambered hearts. The distribution of birds and crocodiles on the cladogram suggests that it is likely that a heart with a double-pressure pumping system was present in basal Archosauria.

Geochemistry

Another tantalizing clue about dinosaur metabolism was very recently (1994) proposed by W. Showers and R. E. Barrick, stable isotope geochemists at the University of Southern California. Recall from Chapter 2 that the proportion of the rare stable isotope of oxygen (^{18}O) varies as temperature varies; therefore, if a substance contains oxygen, one can learn something about the temperature at which that substance formed by the ratio $^{18}O/^{16}O$. In the case of bone, the oxygen is contained in phosphate (PO_4) that forms part of the mineral matter in the bone. Thus, knowing the oxygen isotopic composition of the bone, one can learn something of the temperature at which the bone formed (Figure 14.14).

Barrick and his co-workers analyzed samples all around the skeleton of *Tyrannosaurus*. Their idea was that if the animal was ectothermic, there should be a large temperature difference between parts of the skeleton located deep within the animal (i.e., ribs and the trunk backbone) and those located toward the exterior the animal (i.e., limbs and tails). If, however, the animal was endothermic, then because the body would be maintaining its fluids at a constant temperature, there should be little temperature difference between bones deep within the animal and those more external. The difference – or lack of difference – in temperatures should be reflected in the proportions of $^{18}O/^{16}O$.

The studies showed that the change in temperature between the core and the extremities in *T. rex* was 4°C or less. Barrick and Showers's conclusion was, therefore, that *Tyrannosaurus* was endothermic.

This idea has been no freer of criticism than others that preceded it. If the animal were a mass homeotherm (and not a true endotherm), critics argued, temperatures of the extremities would be like those of the core. Also, some critics believe that the bones have recrystallized. Such recrystallization can take place on a very fine scale and can be difficult to recognize, even under a microscope.

Barrick and some co-workers also studied skeletons of several different dinosaurs. Interestingly, although the differences between exterior and interior bones showed little isotopic variation, the isotopic data from different skeletons showed, overall, more variability than might be expected in an endotherm. Barrick and his co-workers interpreted this to mean that the animal may – as an adult – have experienced a shift to lower metabolic rates (thereby increasing the overall isotopic variability).

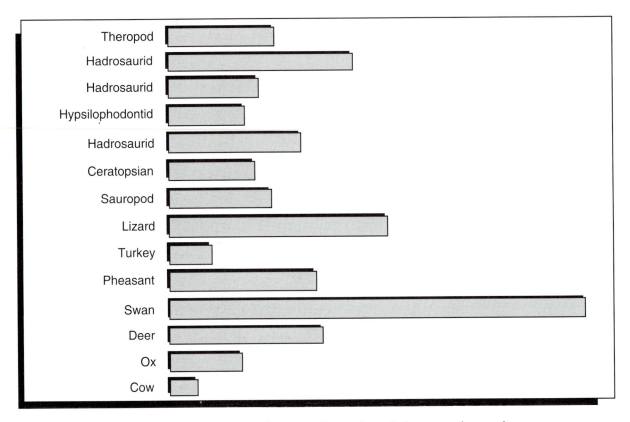

Amount of isotopic variation ——▶ increasing

Figure 14.14 **Stable isotopic variation within and between living and extinct organisms.** Three hadrosaurids are represented (*Corythosaurus, Edmontosaurus, Hypacrosaurus*), one theropod (*Tyrannosaurus*), one hypsilophodontid (*Orodremus*), a sauropod (*Camarasaurus*), and a juvenile ceratopsian. *Modified from a figure courtesy of R. Barrick.*

The Nose Knows

The latest news about dinosaur thermoregulation comes from their snouts! Very recent studies by W. J. Hillenius (University of California at Los Angeles) and J. Ruben (Oregon State University) argue that endothermy – which we already know is metabolically very expensive – requires the lungs to replenish their air (i.e., ventilate) at a high rate. And high ventilation rates automatically increase loss of body water, unless something is done to prevent it. What modern mammals and birds do is build complex bony projections, called **respiratory turbinates**, at the front end of their nasal cavities; the surface area of these mucous-membrane–covered turbinates pulls a great deal of moisture out of the air before it leaves the nose. This is an excellent trick, solving as it does the problem of water loss associated with increased lung ventilation. Moreover, it might be yet another clue to physiology in extinct animals.

So what about dinosaurs – did they have have respiratory turbinates?

Although a number appear to have had olfactory turbinates (indicative of a well-developed sense of smell), apparently none – as far as we presently know – had respiratory turbinates to allow them to recoup their bodily moisture. On this basis, Ruben argues that non-avian dinosaurs were probably not endothermic.

TOWARD A CONSENSUS?

The debate over dinosaur endothermy climaxed with the 1980 publication of a special American Association for the Advancement of Science volume called *A Cold Look at the Warm-Blooded Dinosaurs*. In general, the contributions in the book struck a compromise between Bakker's concept of dinosaurs having bird/mammalian levels of endothermy and the view of dinosaurs as oversized, uninspired crocodiles.

The most conservative viewpoint was that dinosaurs were **mass homeotherms**; that is, their very bulk precluded fluctuations in their core temperatures. Though the idea might apply to the very largest dinosaurs, many were – excluding tails – relatively small creatures. Thus, because of their mass, the largest dinosaurs must have retained heat. Smaller dinosaurs, however, had high surface-to-volume ratios and would have cooled rapidly.

The consensus of opinion has come to be that as regards endothermy among dinosaurs, one size does *not* fit all. Clearly, smaller dinosaurs must have been metabolically very different from larger dinosaurs. Spotila recently applied the concept of **gigantothermy** to large dinosaurs. Gigantothermy is a kind of modified mass homeothermy that mixes large size with relatively low metabolic rates with control of blood circulation in peripheral tissues. Spotila and his colleagues noted that, among endotherms, size is inversely correlated with metabolic rate. That is, the larger the endotherm, the slower its relative metabolic rate. This makes a lot of sense in the context of mass homeothermy: Because large animals (with low surface-to-volume ratios) retain heat, high levels of metabolic activity (and associated heat generation) would heat the animal beyond its ability to cool off. In other words, it would suffer the biological equivalent of a meltdown.

To avert biological meltdown, Spotila and colleagues argued, large dinosaurs might have resorted to gigantothermy. Large size would help retain heat, and control of circulation to the extremities would modify heat dissipation. Gigantothermy in combination with migration, they argued, would be as viable for these organisms as for a typical large migrating mammalian endotherm.

And what of the smaller dinosaurs? Spotila and his colleagues argued that if they maintained mammalian metabolic rates, they would have had to maintain an insulatory coating to prevent too rapid a loss of heat (as a modern bird or mammal does). If they did not maintain a mammal/avian level of endothermy, Spotila and colleagues suggested, perhaps they burrowed during cold months. Intermediate-sized dinosaurs, Spotila and colleagues suggested, enjoyed the best of two worlds: When very young, they maintained mammalian/avian levels of endothermy, but as they got older and larger, they could have switched over to a more gigantotherm-type of metabolism. Still, as we have seen, a gigantothermic metabolism, though viable in a marine organism (the leatherback turtle), may not be fully compatible with large terrestrial size and with mammalian activity levels.

EPILOGUE. When the data are ambiguous or apparently conflicting (as they are in the case of the metabolism of dinosaurs), we rely upon intuition and predisposition. Ironically, we tend to be most strident and dogmatic when our understanding is weakest. Strongly worded, oversimplified treatments of the problem occasionally still pop up in the popular press, but most specialists have come to recognize that dinosaur metabolism was likely nonuniform across the group, and unique to dinosaurs. Here, we'll let Spotila and his colleagues have the last word, their viewpoint on dinosaur endothermy being much in accord with our own:

> The concept of dinosaurs as avian- or mammalian-style endotherms has intuitive appeal to the layperson, and is infinitely preferable to the image of dinosaurs as the cultural icons of stupidity, sluggishness, extinction, and above all, failure.... An eager public has been extremely receptive to popularizers willing to offer a readily comprehensible account of dinosaurs as highly active, racing, dancing endotherms.... However, the view of dinosaurs that the public has embraced so eagerly is partisan and flawed.... [This view] has derived from a philosophically ... suspect concept (that endothermy is superior to ectothermy ...), and from doctrinaire [views] that insist that all dinosaurs were endothermic that have failed to keep abreast with relevant developments in biophysics. (J. R. Spotila et al., 1991, *Modern Geology* 16:223–224)

Modern analysis has vastly increased the amount of information that can be brought to bear on the question of dinosaur endothermy, but with the increase in understanding comes an increase in our appreciation of the complexity of the problem. The view of dinosaurs as ancient archosaurian versions of mammals, complete with poses and metabolisms reminiscent of mammals, is hard to reconcile with the diverse array of anatomical, histological and geochemical data – much of it outlined in this chapter – that continues to suggest that they were something else: something neither mammalian, nor avian, nor reptilian.

Important Readings

Andrews, R. C. 1953. All About Dinosaurs. Random House, New York, 146 pp.

Bakker, R. T. 1975. Dinosaur Renaissance. Scientific American, April 1975:58–78.

Bakker, R. T. 1986. The Dinosaur Heresies. William Morrow and Company, New York, 481 pp.

Bakker, R. T. 1987. Return of the dancing dinosaur; pp. 38–69 *in* S. J. Czerkas and E. C. Olsen (eds.), Dinosaurs Past and Present, Vol. I. Natural History Museum of Los Angeles County, Los Angeles.

Bakker, R. T., and P. M. Galton. 1974. Dinosaur monophyly and a new class of Vertebrates. Nature 248:168–172.

Barrick, R. E., and W. J. Showers. 1994. Thermophysiology of *Tyrannosaurus rex*: evidence from oxygen isotopes. Science 265:222–224.

Chinsamy, A. 1993. Bone histology and growth trajectory of the prosauropod dinosaur *Massospondylus carinatus* Owen. Modern Geology 18:319–329.

Chinsamy, A. 1993. Image analysis and the physiological implications of the vascularization of femora in archosaurs. Modern Geology 19:101–108.

Chinsamy, A., L. M. Chiappe, and P. Dodson. 1994. Growth rings in Mesozoic birds. Nature 368:196–197.

Colbert, E. H., R. B. Cowles, and C. M. Bogert. 1947. Rates of temperature increase in the dinosaurs. Copeia 1947:141–142.

Farlow, J. O. 1990. Dinosaur energetics and thermal biology; pp. 43–55 *in* D. B. Weishampel, P. Dodson, and H. Osmólska (eds.), The Dinosauria. University of California Press, Berkeley.

Feduccia, A. 1972. Dinosaurs as reptiles. Evolution 27:166–169.

Desmond, A. 1975. The Hot-Blooded Dinosaurs. Dial Press, New York, 238 pp.

Hillenius, W. J. 1994. Turbinates in therapsids: evidence for Late Permian origins of mammalian endothermy. Evolution 48:207–299.

Hopson, J. A. 1977. Relative brains size and behavior in archosaurian reptiles. Annual Reviews of Ecology and Systematics 8:429–448.

Hopson, J. A. 1980. Relative brain size in dinosaurs – implications for dinosaurian endothermy; pp. 287–310 *in* R. D. K. Thomas and E. C. Olson (eds.), A Cold Look at the Warm-Blooded Dinosaurs. American Association for the Advancement of Science Selected Symposium no 28.

Jerison, H. J. 1973. Evolution of the Brain and Intelligence. Academic Press, New York, 482 pp.

Johnston, P. 1979. Growth rings in dinosaur teeth. Nature 278:635–636.

Ostrom, J. H. 1969. Terrestrial vertebrates as indicators of Mesozoic climates. Proceedings of the North American Paleontological Convention, pp. 347–376.

Ostrom, J .H. 1969. Osteology of *Deinonychus anthirropus*, an unusual theropod from the Lower Cretaceous of Montana. Bulletin of the Peabody Museum of Natural History – Yale 30:1–165.

Ostrom, J. H. 1974. Reply to "Dinosaurs as reptiles." Evolution 28:491–493.

Ostrom, J .H. 1976. *Archaeopteryx* and the origin of birds. Biological Journal of the Linnaean Society of London 8:91–182.

Paul, G. S. 1988. Physiological, migratorial, climatological, geophysical, survival, and evolutionary implications of Cretaceous polar dinosaurs. Journal of Paleontology 62:640–652.

Reid, R. E. H. 1987. Bone and dinosaur endothermy. Modern Geology 11:133–154.

Reid, R. E. H. 1990. Zonal "growth rings" in dinosaurs. Modern Geology 15:19–48.

Rich, P. V., T. H. Rich, B. E. Wagstaff, J. McEwen Mason, C. B. Douthitt, R. T. Gregory, and E. A. Felton. 1988. Evidence for low temperatures and biologic diversity in Cretaceous high latitudes of Australia. Science 242:1403–1406.

Ricqlès, A. J. de. 1980. Tissue structures of dinosaur bone – functional significance and possible relation to dinosaur physiology; pp. 103–109 *in* R. D. K. Thomas and E. C. Olson (eds.), A Cold Look at the Warm-Blooded Dinosaurs. American Association for the Advancement of Science Selected Symposium no 28.

Ruben, J. 1991. Reptilian physiology and the flight capacity of *Archaeopteryx*. Evolution 45:1–17.

Ruben, J. 1995. The evolution of endothermy in mammals and birds: from physiology to fossils. Annual Review of Physiology 57:69–95.

Russell, L. S. 1965. Body temperature of dinosaurs and its relationships to their extinction. Journal of Paleontology 242:497–501.

Spotila, J. R., M. P. O'Connor, P. Dodson, and F. V. Paladino. 1991. Hot and cold running dinosaurs: body size, metabolism, and migration. Modern Geology 16:203–227.

Thomas, R. D. K., and E. C. Olson, eds. 1980. A Cold Look at the Warm-Blooded Dinosaurs. American Association for the Advancement of Science Selected Symposium no. 28, 514 pp.

Wieland, G. R. 1942. Too hot for the dinosaur! Science 96:359.

CHAPTER 15

DINOSAURS IN SPACE AND TIME

AMONG DINOSAURIA, what is really known about who lived and died with whom? We have found a spectacular fossil of *Protoceratops* apparently in a *Velociraptor* death grip (Figure 15.1). Here is a documented interaction; the two fossils were found together in a pose that suggests that they were alive when they came together. Other documented interactions are Late Jurassic theropods and sauropods: Tooth marks of a large carnivorous theropod (or theropods) have been found on sauropod bones. Where fossils end, our imaginations take over: The potential confrontation between *Tyrannosaurus* and *Triceratops* has fired imaginations since the remains of both beasts were first discovered, but of course no fossil has ever been found of the two locked in mortal (or any other sort of) combat.

Put another way, could a duckbill ever have been scarfed down by *Dilophosaurus*? (No, they lived 120 million years apart.) How about *Triceratops* goring *Tarbosaurus*? (No, they lived on different continents.) How about *Staurikosaurus* snacking on *Protoceratops*? (No, they lived about 150 million years apart *and* on different continents!) This last pairing is no more likely than seeing cavemen and *T. rex* together; in fact, even less likely: The oldest "caveman" is separated from the likes of *T. rex* by a mere 64 (or so) million years, but the earliest dinosaur is separated in time from *T. rex* by about 160 million years! Understanding who the dinosaurs are means understanding where they reside in space (that is, their **geographic distribution**) and their distribution in time (or **temporal distribution**).

Figure 15.1 **"Fighting dinosaurs" from Tugrikin-shireh, Mongolia.** The fossils were found locked in a death embrace in an ancient desert dune field. It is possible that they were overcome in a sandstorm. To the left is the skeleton to *Protoceratops*; to the right, *Velociraptor*, with its claws hooked on the frill and in the stomach cavity of *Protoceratops*.

GEOGRAPHIC DISTRIBUTION

As can be seen from Table 15.1, despite uneven records of preservation (Chapter 2), dinosaurs have been found – at one time or another – on all continents, and most of the time on most continents.

A consequence of continental connections (see Chapter 2) is similarities in the faunas. If the faunas appear very similar to each other, then there is evidence of some land connection to allow the fauna of one continent to disperse to the other continent. In contrast, faunas that have developed in relative isolation tend to be distinct, with their own characteristics. This type of distinctness is called **endemism**. A fauna that is closely allied with a particular geographic region (large or small) is called **indigenous** or **endemic**. A region that is populated by distinct faunas unique to it is said to show *high endemism*. High endemism can be caused by widely separated continents, because there is no opportunity for faunal interchange. During the time of Pangaea, when continents were coalesced and the opportunities for faunal interchange were high, the earth was characterized by

TABLE 15.1 Distribution of Dinosaurs on Continents during the Mesozoic Era

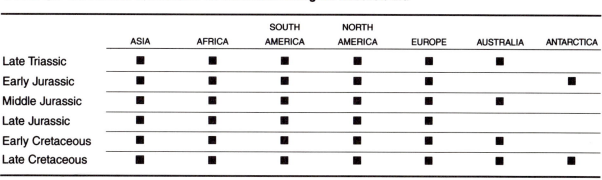

	ASIA	AFRICA	SOUTH AMERICA	NORTH AMERICA	EUROPE	AUSTRALIA	ANTARCTICA
Late Triassic	■	■	■	■	■	■	
Early Jurassic	■	■	■	■	■		■
Middle Jurassic	■	■	■	■	■	■	
Late Jurassic	■	■	■	■	■		
Early Cretaceous	■	■	■	■	■	■	
Late Cretaceous	■	■	■	■	■	■	■

unusually *low endemism*. Here, then, is an example of geology driving and modifying large-scale patterns of evolution.

PRESERVATION. Table 15.1 shows the distribution of dinosaurs among the continents through time. Uncritically studied, it suggests that during the Mesozoic, dinosaurs migrated into and emigrated from Australia and Antarctica. The paucity of dinosaur remains from both of these continents, however, is surely more a question of preservation and inhospitable conditions for collecting fossils than a question of where dinosaurs actually lived. Methods of estimating the completeness of fossil preservation have been developed (Box 15.1), but these have not been applied to most dinosaur faunas.

TEMPORAL DISTRIBUTION. The University of Pennsylvania's P. Dodson has estimated that many dinosaur genera lived 4–8 million years. This means that every 4–8 million years, one genus died out and another took its place. This rate of taxonomic turnover (4–8 million years) may have been accelerated by mass extinction events (see Chapter 16) that periodically left the earth depleted and vacant for colonization by new dinosaurian faunas.[1] The sequence of dinosaurs through time and space is like a grand pageant through earth history, in which each interval of time has a characteristic fauna that takes on a characteristic quality.

The manner in which these faunas are handled here *condenses* many millions of years and large geographic regions. It is presented in this way to summarize the 160 or so million years of global dinosaur evolution. Considered on that scale, it is a reasonable approach. But recall that dinosaur genera turn over at rates of 4–8

---

[1] The rate of dinosaur species turnover has interesting consequences. Obviously, fossils are found within sedimentary rocks. Sedimentary rocks are commonly grouped into bodies called formations, which are simply rocks that can be mapped and which share similar features. Most terrestrial formations represent several million years of deposition and particular environments, which means that a dinosaur species found in one formation will probably not occur in the one above or below it: Too much time would have elapsed, and usually an environmental change is represented. Thus successive formations contain different species; there are very few instances of a particular dinosaur species found in one formation as well as in the one above or below it. It is therefore very difficult to track the fates of individual species; terrestrial deposition being episodic as it is, species associated with one formation are long gone before the time of the next formation.

BOX 15.1

COUNTING DINOSAURS

THERE ARE SEVERAL WAYS that the true diversity of dinosaurs can be estimated. In this box we introduce one of these ways and exemplify it with Ceratopsia.

This book has been emphasizing the evolutionary relationships of the different dinosaur groups using cladistics and cladograms. These cladograms portray the closeness of relationship of the different dinosaur taxa. Cladograms can also be seen as the relative sequence of evolutionary events. For animals with a fossil record – such as dinosaurs – this relative sequence can be compared with the real sequence of appearance in the stratigraphic record. If we are lucky enough to have a dense, unbiased fossil record or one that happened to capture critical evolutionary events, then the relative, phylogenetic sequence ought to mirror the stratigraphic sequence.

In addition, because the history of life is continuous, the combination of phylogeny and stratigraphy has a lot to say about the presence, and meaning, of gaps in the fossil record. Suppose Dinosaur X and Dinosaur Y were each other's closest relative; thus they share a unique common ancestor. Suppose also that Dinosaur X is known from rocks dated at 100 million years ago and Dinosaur Y came from 125-million-year-old rocks.

The ancestor of X and Y has to be *at least* 125 million years old. And if this is true, then there must be some not-yet-sampled history between this ancestor and Dinosaur X – to the tune of 25 million years. That 25-million-year gap can be called a **minimal divergence time** (**MDT**); it can be calculated for any two taxa provided their phylogenetic relationships and stratigraphic occurrences are known.

Finally, this sort of work can be used to calculate taxonomic diversity during times of both good fossil samples and bad. Lineages that have left no known physical record (through fossils) of their existence are called **ghost lineages** by New York systematists M. Norell and M. Novacek. MDTs are measures of the duration of ghost lineages (Figure Box 15.1).

Ghost lineages help us estimate diversity. When combined with rough counts of actual critters, ghost lineages embellish our estimates of diversity by putting in the "ghosts" that have yet to be discovered.

Ceratopsia Counted

It has sometimes been claimed that ceratopsians have one of the best fossil records among all dinosaur groups. Here we consider how such a

---

million years. If this is so, when we present, for example, the huge fauna from the last 30 million years of the Cretaceous, we are combining in time, and possibly in space, animals that certainly did *not* live together. So, considered on a global scale of tens of millions of years, we can recognize large-scale evolutionary patterns from the lists in this chapter. But considered on a smaller scale, the relationships of the faunas to each other may be illusory.

The fact that we are North American scientists, in combination with the fact that European and North American fossil localities have been studied longer than those elsewhere *and* the fact that those studies are more accessible to us than those elsewhere, gives this chapter a Western slant. This is unfortunate. We have said this before and we will undoubtedly say this again: The answers to many of the questions posed in this book certainly reside in the localities and fossils of Asia, South America, Antarctica, Australia, and Africa.

A caveat is necessary: Our summary cannot be fully comprehensive; here we

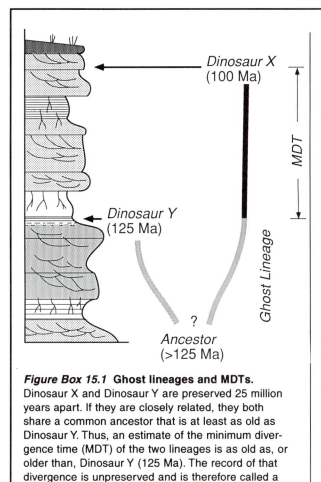

Figure Box 15.1 Ghost lineages and MDTs.
Dinosaur X and Dinosaur Y are preserved 25 million years apart. If they are closely related, they both share a common ancestor that is at least as old as Dinosaur Y. Thus, an estimate of the minimum divergence time (MDT) of the two lineages is as old as, or older than, Dinosaur Y (125 Ma). The record of that divergence is unpreserved and is therefore called a ghost lineage (in gray on the figure).

statement might be examined. We begin with some very raw data and choose species, a choice that reflects the more general importance of species in evolutionary biology. In all, there are at least 32 ceratopsian species, fewer than are found in theropods, sauropodomorphs, and ornithopods, yet more than in ankylosaurs, stegosaurs, and pachycephalosaurs.

If we average the raw data over the duration of each higher taxon, ceratopsians apparently produced new species at the rate of 1 every 1.9 million years, compared to a high of 1 new species per 1.4 million years for sauropodomorphs and a low of 1 per 5.6 million years for stegosaurs. How can we get at the question of the quality of the fossil record? MDTs can be used as estimates of the relative completeness of a group of organisms whose phylogeny and stratigraphic distribution are known.

To estimate overall diversity of a particular group, we must include ghost lineages as well as those that are known. For ceratopsians, MDT values range from 0 to nearly 30 million years, with an average just over 5 million years. These are among the smallest MDTs for all Dinosauria, suggesting that the fossil record of this group is comparatively pretty well represented. Furthermore, actual ceratopsian species counts are nearly 70% of the total ceratopsians after ghost lineages have been added to our diversity calculations. On these measures, ceratopsians do indeed have one of the best records of all of the major dinosaur groups.

will highlight only the most famous faunas and localities, and the environments that they represent. Many places have produced a wealth of indeterminate material, as well as eggshells and ichnotaxa. We have not attempted to summarize these.

TRIASSIC–JURASSIC

Late Triassic
(231–208 Ma)

Recall that the Late Triassic was the time of Pangaea: All the continents were coalesced into a single supercontinent (Figure 15.2). We saw in Chapter 2 that the increased landmass exerted a strong influence on climate and that, generally, conditions were strongly seasonal, possibly with greater heat and aridity than are found today.

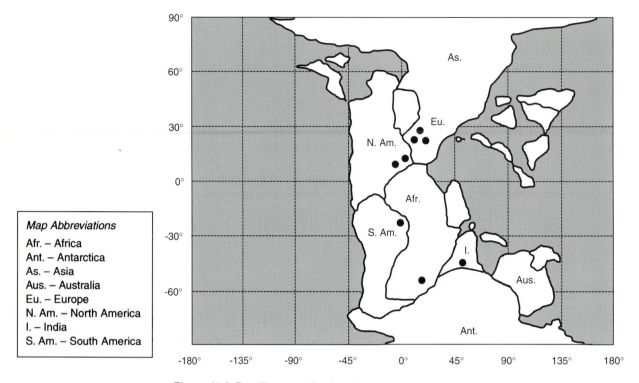

Map Abbreviations
Afr. – Africa
Ant. – Antarctica
As. – Asia
Aus. – Australia
Eu. – Europe
N. Am. – North America
I. – India
S. Am. – South America

Figure 15.2 **Rectilinear projection of continents and major dinosaur-bearing localities during the Late Triassic.** *Reconstruction by Paleogeographic Information System, M.
I. Ross and C. R. Scotese.*

Studies by a variety of scientists have begun to suggest that the Late Triassic
was a time of low vertebrate faunal endemism. This suggests that the land connections to all continents must have greatly enhanced the potential for faunal
interchange over what it is today. As we have seen (Chapters 2, 5, and this chapter), the fauna was not yet dinosaur dominated; rather, it was a unique mixture.
Members of this fauna from the therapsid clade included the squat, beaked, tusked
herbivorous dicynodonts (see Plate IA), and a variety of carnivorous and herbivorous animals that must have acted and looked very like mammals. Finally, of
course, the earliest mammals – tiny, shrewlike, insectivorous creatures (Figure
14.1) – were present.

But the Late Triassic fauna was not only therapsid dominated. Also present
were a variety of archosaurs. Among the most common members of the fauna
were the crocodile-like **phytosaurs** (long-snouted, aquatic fish-eaters; Figure
15.3). Then there were a cloud of carnivorous forms, some large, quadrupedal,
and heavily armored, and some small, lightly built, and unarmored. The earliest
turtles appeared during the Late Triassic.

Our reckoning of Late Triassic terrestrial vertebrates would not be complete
without some mention of pterosaurs. The very first pterosaurs appeared during
Late Triassic time, at about the same time as dinosaurs (see the following).
Already by this time they were highly specialized for flight, suggesting that significant evolution occurred before the Late Triassic.

Finally, toward the end of of the Late Triassic, there appeared the earliest dinosaurs.

THE EARLIEST DINOSAURS. The *precise* antiquity of the earliest known representatives of Dinosauria is difficult to determine; intercontinental correlations of non-marine sediments (particularly between South America and North America), an absence of reliable radiometric dates, and difficulties intercalating the terrestrial sediments into a time scale largely based upon marine rocks, hinder precision. It is agreed by most paleontologists, however, that the dinosaur fossil record unambiguously extends back to the older of two subdivisions of the Late Triassic, the **Carnian**, which denotes the time interval from 231 Ma to 223 Ma. University of Chicago sedimentologist R. Rogers has obtained a date to 228 Ma for sediments containing *Eoraptor*, the first known dinosaur (see Chapters 5 and 12). Probably dinosaurs appeared on earth slightly before this (Middle Triassic dinosaur footprints have been claimed, but it is hard to diagnose true dinosaurs from footprints), but their record is generally said to extend from Carnian time.

ASIA. Dinosaurs from the Asian Late Triassic are poorly represented. Material is known from only India and the People's Republic of China (China). Theropods and prosauropods are the only material preserved.

AFRICA. Africa, particularly South Africa, has proven to be one of the truly important sources of information about the Late Triassic. There the famous rocks of the Stormberg Series – the remnant of a number of great ancient meandering river systems – preserve some of the most spectacular fossils known. Among the impressive finds from South Africa are a host of beautifully preserved therapsids, so close to mammals that the line between these and true mammals is very hard to pinpoint. Among dinosaurs, the prosauropod *Melanorosaurus* is best represented.

EUROPE. Late Triassic dinosaurs are known from Germany, France, the United Kingdom, and Belgium. Again, prosauropods and theropods dominate, including

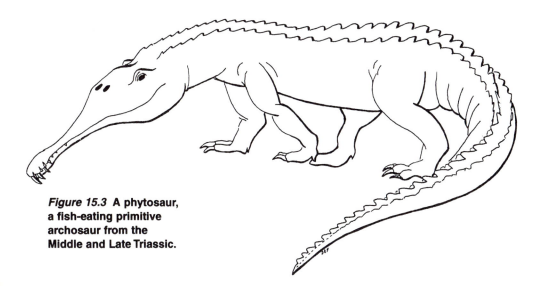

Figure 15.3 **A phytosaur, a fish-eating primitive archosaur from the Middle and Late Triassic.**

such famous prosauropods as *Plateosaurus* and *Thecodontosaurus*. Among theropods, most of the finds are called *Megalosaurus* – that ill-defined genus to which so much European theropod material has been referred – but this almost certainly reflects how poorly defined *Megalosaurus* is.

SOUTH AMERICA.   South America has only recently been recognized as an extremely important source of our knowledge about the earliest dinosaur faunas. The beautiful gray-white badlands of the Ischigualasto Formation of Argentina have produced forms like the primitive ornithischian *Pisanosaurus*, melanorosaurid prosauropods, and the putative basal theropods *Staurikosaurus*, *Herrerasaurus*, and most recently, *Eoraptor* (Plate IA). As we observed in Chapters 5 and 12, these probable theropods may be not only fundamental to who we think theropods are, but also to who we think dinosaurs are.

NORTH AMERICA.   Thick, fossiliferous Late Triassic sequences are preserved on the east coast and from the southwestern part of the United States. In the eastern part of the United States, running from Georgia north to Nova Scotia (Canada) is a series of rocks spanning from the late Middle Triassic to the Early Jurassic that were deposited in fluvial and lacustrine (lake) environments as the dismemberment of Pangaea was initiated (see Chapter 2). These rocks preserve a variety of theropod and ornithischian footprint faunas, first interpreted in the middle 1800s by Amherst professor E. Hitchcock as birds. Subsequent finds, as well as restudy of Hitchcock's collection by a variety of different researchers, have revealed the non-avian identities of these dinosaur prints.

For many years, the American southwest was one of the important sources of knowledge about Late Triassic dinosaurs. There, the banded mudstones and sandstones of the fluvial and lacustrine Chinle Formation of Arizona (beautifully exposed in Petrified Forest National Park; Plate IB), New Mexico, and Utah, provided collectors with a variety of footprint faunas referred to Theropoda and Ornithischia, and the theropod *Coelophysis*. The Chinle of New Mexico also produced one of the most famous dinosaur bonebeds known: the Ghost Ranch *Coelophysis* quarry. There in the mottled purple, red, and gray mudstones, hundreds of complete, articulated skeletons of *Coelophysis* are preserved. Finally, in western Texas and eastern New Mexico, the Late Triassic Dockum Group has produced light-bodied theropods referred to *Coelophysis*, as well as some ornithischian material.

AUSTRALIA.   Scattered footprints referred to theropods and prosauropods are known from Queensland.

ANTARCTICA.   No Late Triassic dinosaurs are known from Antarctica.

Early Jurassic
(208–178 Ma)

The Early Jurassic was the first time on earth when dinosaurs truly began to dominate terrestrial vertebrate faunas (Figure 15.4). Although they shared the limelight with some relict anamniotes (Chapter 4), a few therapsids (including the

COLOR PLATES

Paintings by Brian Regal

LATE TRIASSIC

Plate IA A hot, humid, lazy day in the Late Triassic of Argentina, represented by the Ishigualasto Formation. A large, herbivorous dicynodont (a kind of therapsid) suns itself on a log.

Plate 1B The Late Triassic of Arizona, U.S.A. A small theropod, *Coelophysis*, drinks from one of the rivers that deposited the sediments now called the Chinle Formation.

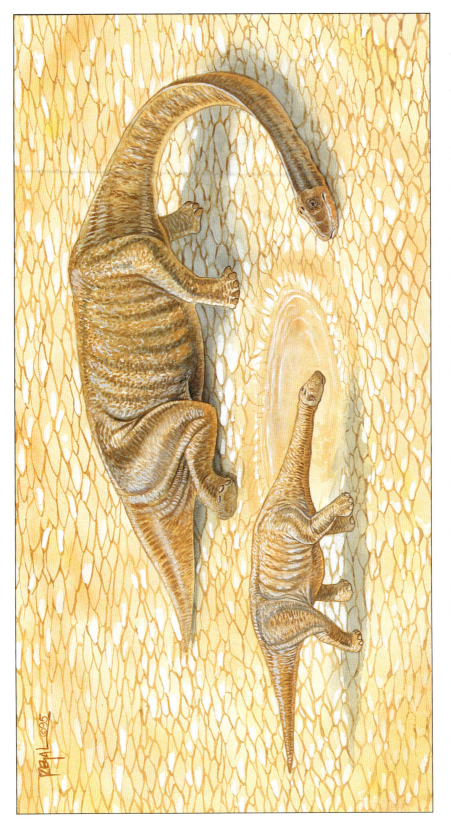

Plate IIA The Kota Formation of India. Intensive wet-dry climate cycles of the Early Jurassic may have produced the scene shown here: thirsty barapasaurs desiccating in a drying lake bed.

Plate IIB *Dilophosaurus*, a 6-m theropod, walks along a river bank in the Early Jurassic of Arizona, U.S.A. The Kayenta Formation preserves the bones of this carnivorous dinosaur, as well as sediments from the river systems in which it lived.

MIDDLE JURASSIC

Plate IIIA The Middle Jurassic of Sichuan Province, China, provides the best insights into this poorly known time interval. Here, the stegosaur *Huayangosaurus* delivers a stinging swat to the theropod *Sinraptor*.

Plate IIIB Volcanoes, mudslides, and dinosaurs characterized the Middle Jurassic La Boca Formation of NE Mexico. A small ornithopod waits out a volcanic eruption and ash fallout in this reconstruction.

Plate IVA Sichuan Province, China, also produces a very rich Late Jurassic fauna. Here, a herd of the sauropod *Mamenchisaurus* is accompanied across a river by an unnamed pterosaur.

Plate IVB The Late Jurassic Morrison Formation of the western interior of the U.S.A. has produced one of the richest dinosaur faunas of all time. In this Morrison scene, a marauding *Ceratosaurus* bullies a pair of *Camptosaurus*.

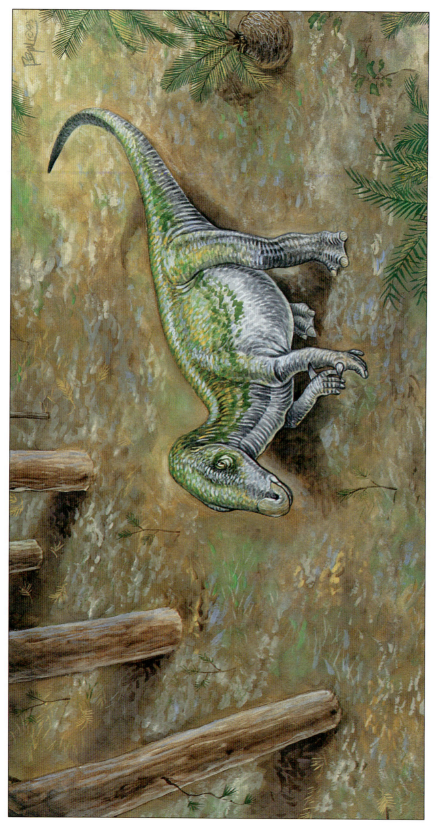

Plate VA The Early Cretaceous Wealden Formation of the United Kingdom. In a richly forested environment, *Iguanodon* pioneers a new use for its extraordinary thumb spike: scratching its arm.

Plate VB The Early Cretaceous Cloverly Formation of Wyoming, U.S.A., suggests a grim game of death: A pack of lightly built *Deinonychus* takes on a curled, hunkered-down, defensive nodosaurid ankylosaur.

LATE CRETACEOUS

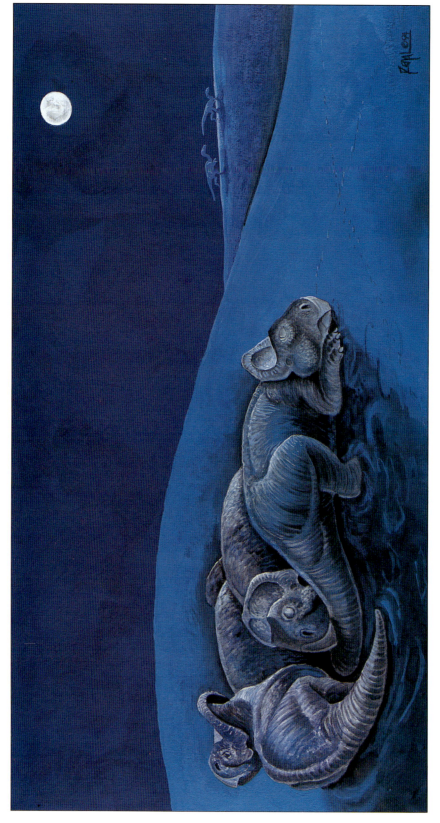

Plate VIA Midnight at the oasis. A moon rises over *Protoceratops* sleeping in an erg, now preserved as the Djacochta Formation at Tugrikin-shireh, Gobi Desert, Mongolia.

Color Plates 12

Plate VIB *Parasaurolophus* in the forested fluvial environments of the Dinosaur Park (Oldman) Formation, Late Cretaceous, Alberta, Canada.

Plate VIIA Twilight in Transylvania. The Hateg Basin of Romania contains the remains of ornithopods such as *Telmatosaurus* and *Rhabdodon*. These latest Cretaceous dinosaurs apparently lived on volcanic islands in the midst of an archipelago.

Plate VIIB The latest Cretaceous, fluvial Hell Creek Formation preserves some of the last, and best-known, dinosaurs. Here a mother *Triceratops* pauses over her dead baby in a forested thicket of the Hell Creek floodplain.

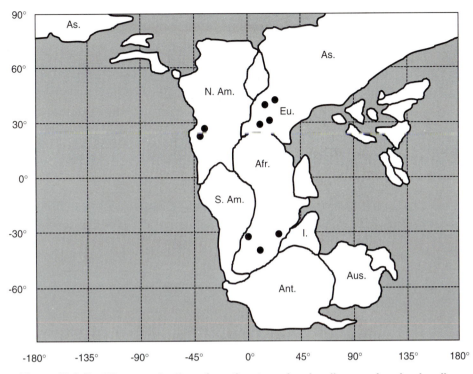

Figure 15.4 Rectilinear projection of continents and major dinosaur-bearing localities during the Early Jurassic. Map abbreviations as in Figure 15.2. *Reconstruction by Paleogeographic Information System, M. I. Ross and C. R. Scotese.*

rather diminutive contemporaneous mammals), turtles, and, of course, pterosaurs and the newly evolved crocodilians, many of the players in the game were by now dinosaurs. Interestingly, the Early Jurassic faunas inherited some of the quality of low endemism that characterized the Late Triassic world. The unzipping of Pangaea was in its very earliest stages, and it had not gone on so long that the fragmentation of the continents was yet reflected through increased global endemism.

By the Early Jurassic, dinosaurs were well represented. Ornithischians had now made their appearance. These ornithischians were rather primitive, but already we can recognize two major groups of ornithischians (Cerapoda and Thyreophora). Although "prosauropods" were still abundant during the Early Jurassic, true sauropods made their appearance during this time interval.

ASIA. The Asian Early Jurassic depends primarily upon China. In the southern province of Yunnan, the lowest beds of the extremely fossiliferous dark red Lufeng Series have produced a variety of dinosaurs, including the "prosauropods" *Lufengosaurus* and *Yunnanosaurus;* remains claimed to be heterodontosaurid (*Dianchungosaurus*); and a creature thought to be a primitive thyreophoran (*Tatisaurus*). Material referred to the ceratosaur *Dilophosaurus* has been found wrapped around a *Yunnanosaurus*, prompting speculation by the Chinese discoverers that *Dilophosaurus* was dining on the "prosauropod." Fossiliferous outcrops from Sichuan, China (more famous to most Westerners for spicy, delicious cuisine than for dinosaurs) include *Lufengosaurus* and material referred to primitive sauropods (*Zizhongosaurus*). Other significant finds in Asia include theropod material and *Barapasaurus*, the basal sauropod from the Indian subcontinent (Plate IIA).

AFRICA. Like its Late Triassic wealth, southern Africa has a rich and fossiliferous Early Jurassic. Orange Free State has boasted prosauropods *Massospondylus* and *Thecodontosaurus*, as well as the heterodontosaurid ornithopod *Lanasaurus* and some eggshell fragments. Early Jurassic finds in Cape Province include *Massospondylus* and the small ornithopod *Heterodontosaurus*. In Lesotho, the same "prosauropods" have been found, as well as the primitive ornithischian *Lesothosaurus*. Footprints thought to belong to theropods have been recorded in all these areas.

Zimbabwe has produced a variety of footprint faunas (including theropods and sauropodomorphs) and is renowned for the primitive sauropod *Vulcanodon*, recovered from a group of fossiliferous, sedimentary rocks now called the *Vulcanodon* beds. The theropod *Syntarsus* (see section on the Early Jurassic of North America) has also been recovered, as well as the "prosauropods" *Massospondylus* and *Euskelosaurus*.

In northern Africa, Morocco has produced a variety of footprint faunas, referred to theropods, sauropods, and (questionably) even stegosaurs.

EUROPE. The Early Jurassic of Europe is poorly represented. A variety of theropod fragments not assignable to particular genera are known, however; from England come the ceratosaur *Sarcosaurus* and the basal thyreophorans *Emausaurus* and *Scelidosaurus*. Early Jurassic footprints have been recorded in France and Germany.

SOUTH AMERICA. Early Jurassic footprints referred to theropods have been found in three localities in Brazil.

NORTH AMERICA. The Early Jurassic of North America is again divided between the east coast and the southwest. In eastern North America, the "prosauropods" *Ammosaurus* and *Anchisaurus* are known from Nova Scotia, Canada, and from the Newark Supergroup in the eastern United States. A variety of footprints referred to theropods and ornithischians are known from river and lake sediments that filled rift valleys that developed along the eastern coast of the continent as North America commenced its separation from Africa; representatives of these ichnofaunas are known all along the eastern edge of the North American continent.

In the southwest, the banded, fluvial Glen Canyon Group of Arizona, southern Utah, and New Mexico has produced a variety of important dinosaurs. Glen Canyon sedimentation records the development of a large meandering river system that was eventually overrun by desert sands. This complex suite of sedimentary environments has produced a variety of footprint faunas, as well as fossils of the theropods *Dilophosaurus*, *Syntarsus*, and *Segisaurus* (Plate IIB). Remains of *Ammosaurus* and *Massospondylus* have also been recovered. Finally, the gracile, basal thyreophoran *Scutellosaurus* has also been recovered from these sediments.

AUSTRALIA. No Early Jurassic dinosaurs are known from Australia.

ANTARCTICA. *Crylophosaurus*, a crested theropod, is now known from the Late Jurassic of Antarctica. The find also included other theropod material and remains of a large "prosauropod."

Middle Jurassic
(*178–157 Ma*)

The Middle Jurassic has historically been an enigmatic time in the history of terrestrial vertebrates. As noted in Chapter 2, Middle Jurassic terrestrial sediments are quite uncommon (Figure 15.5). When we look at the total diversity of tetrapods through time (see Box 15.2), the curve all but bottoms out during the Middle Jurassic. Did vertebrates undergo massive extinctions at the end of the Early Jurassic? Probably not. More likely, the curve is simply reflecting the serendipitous absence of terrestrial Middle Jurassic sediments on earth. Without a good sedimentary record to preserve them, we can have little knowledge of the faunas that came and went during that time interval.

Regardless, the Middle Jurassic must have been an important time in the history of dinosaurs. We know that the dismemberment of Pangaea was underway by this time. Moreover, by the Late Jurassic, dinosaurs had diversified tremendously, and endemism was again on the rise. Many of the diverse faunal elements that characterized earlier faunas – nonmammalian therapsids, for example – were

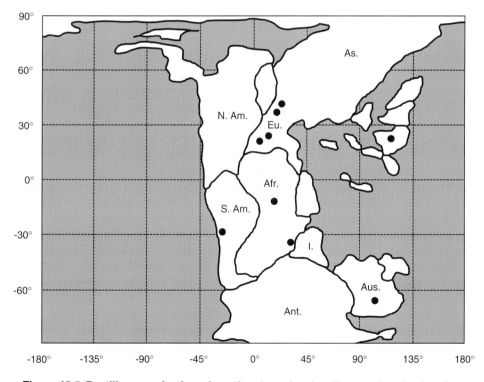

Figure 15.5 Rectilinear projection of continents and major dinosaur-bearing localities during the Middle Jurassic. Map abbreviations as in Figure 15.2. *Reconstruction by Paleogeographic Information System, M. I. Ross and C. R. Scotese.*

BOX 15.2

THE SHAPE OF TETRAPOD DIVERSITY

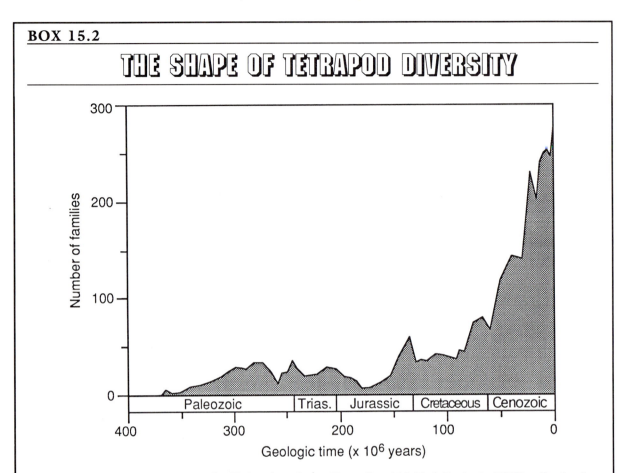

OVER THE PAST 10 OR SO YEARS, the University of Bristol's M. J. Benton has been compiling a comprehensive list of the fates of tetrapod families through time (Figure Box 15.2). We see several interesting features of the curve that results from this compilation. Note the drop in families during Middle Jurassic time. This, as we have seen, is an **artifact**, that is, a specious result. This particular one comes from the lack of localities, more than from a true lack of families during the Middle Jurassic. Then, notice the huge rise of families during the Tertiary. Some of this may be real, and perhaps attributable in part to Tertiary birds and mammals (both of whom are tremendously diverse groups), but some of it might be another artifact: the inescapable fact that as we get closer and closer to the Recent, fossil biotas become better and better known. This is because more sediments are preserved as we get closer and closer to the Recent, and

Figure Box 15.2 **M. J. Benton's (1986) estimate of vertebrate diversity through time.** On the x-axis is time; on the y-axis is diversity as measured in numbers of families.

a greater amount of sedimentary rocks preserved means more fossils. The big spike at the end of the Jurassic is the Morrison Formation, a unit that preserved an extraordinary wealth of fossils.

So a curve like Benton's requires skill to understand and to factor out the artifacts. Nonetheless, we can see that generally, dinosaurian diversity increased throughout dinosaurs' stay on earth, and as they progressed through the Cretaceous, dinosaurs continued to diversify. The increase in diversity shown in Benton's diagram may be a reflection of the increasing global endemism of the terrestrial biota, itself driven by the increasing isolation of the continental plates.

largely out of the picture. The insignificant exception to this, of course, was the mammals, who were still around, but hanging on by the skin of their multicusped, tightly occluding teeth. Thus by the Late Jurassic, dinosaurs had truly consolidated their hold on terrestrial ecosystems. This probably commenced sometime during the Middle Jurassic, which must therefore have been a kind of pivot point in the history of dinosaurs. It's a shame that we cannot know more of this crucial time.

ASIA. The Middle Jurassic of Asia, particularly of China, is clearly the best terrestrial Middle Jurassic in existence. There, from the centrally located province of Sichuan, a variety of dinosaurs have been recovered, including the theropods *Gasosaurus* and *Kaijiangosaurus*, the sauropods *Shunosaurus* and *Datousaurus*, the hypsilophodontid ornithopod *Yandusaurus*, and a stegosaur, *Huayangosaurus* (Plate IIIA). New finds from the Sino-Canadian expedition include the big theropods *Monolophosaurus* and *Sinraptor*. The diversity is simply unmatched on other continents.

AFRICA. Morocco and Algeria have produced Middle Jurassic fossils. Specimens referred to *Cetiosaurus* have been obtained, as has some indeterminate theropod material. Two brachiosaurid sauropods have been recovered from the island of Madagascar: *Bothriospondylus* and *Lapparentosaurus*.

EUROPE. A century and a half of study have yielded modest results, particularly in England. There, a variety of footprint faunas have been recovered. Beyond this, some sauropod material has been recovered, as well as material from the theropod *Megalosaurus*. Material from *Lexovisaurus*, a stegosaur, has also been found, along with the jaw of a nodosaurid ankylosaur called *Sarcolestes*. By and large, the finds are incomplete, and a number of theropods are known that are difficult to refer to a particular genus. From France, more material referred to *Megalosaurus* has been recovered, as well as the theropods *Piveteausaurus* and *Poekilopleuron*. More *Lexovisaurus* material has been recovered from France. A locality in Portugal has produced some theropod material.

SOUTH AMERICA. Chile and Argentina both have Middle Jurassic localities. These have produced the sauropods *Amygdalodon* and *Volkheimeria*, and a theropod, *Piatnitzkysaurus*.

NORTH AMERICA. The only known dinosaur-producing Middle Jurassic localities in North America are from Mexico. One is a footprint locality, and the other is Huizachal Canyon in northeastern Mexico, a locality that has produced ornithischian and theropod teeth (Plate IIIB). This locality also has a pterosaur, crocodilians, and therapsids (including true mammals), suggesting that here, at least, some of the low endemism that characterized previous terrestrial ecosystems was still in effect.

AUSTRALIA. Queensland has produced a footprint fauna and the sauropod *Rhoetosaurus*.

ANTARCTICA. No Middle Jurassic dinosaurs are known from Antarctica.

Late Jurassic
(157–145 Ma)

Recall from Chapter 2 that by the time of the Late Jurassic, global climates had stabilized and were generally warmer and more equable (less seasonal) than they now
are. Polar ice, if present, was reduced. Sea levels were higher than today. The fact
that the earth's plates had by this time migrated to more disjunct positions than
they had in the Late Triassic suggests that we can expect to find greater endemism
in the terrestrial faunas of the Late Jurassic, as indeed we do (Figure 15.6).

For North America and arguably for the world, the Late Jurassic could be
called the Golden Age of Dinosaurs. Many of the dinosaurs that we know and love
were Late Jurassic in age. Somehow that special Late Jurassic blend of equable climates, small brains, and massive sizes epitomizes dinosaur stereotypes and exerts
a fascination on humans.

In fact, the Late Jurassic produced a remarkable variety of forms. Many *were*
large: gigantic sauropods (*Brachiosaurus*, *Diplodocus*, *Camarasaurus*, among others) as well as theropods that, if not *T. rex*–sized, still reached upwards of 16 m.
Of course, not all theropods were large; small, gracile ones abounded, and as we
have seen (Chapter 13), it was during the Late Jurassic that the first unambiguous birds appeared. Moreover, this was the time of stegosaurs, hypsilophodontids,
and primitive iguanodontians.

Who else was out there? Turtles had by this time explored the terrestrial *and*

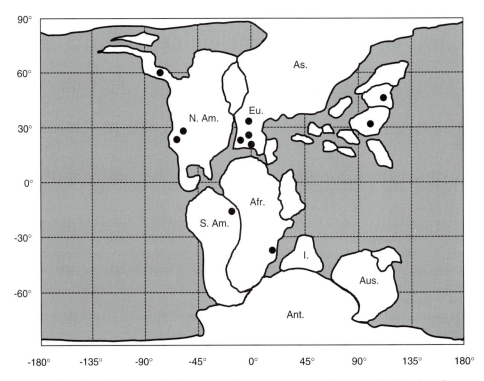

**Figure 15.6 Rectilinear projection of continents and major dinosaur-bearing locali-
ties during the Late Jurassic.** Map abbreviations as in Figure 15.2. *Reconstruction by
Paleogeographic Information System, M. I. Ross and C. R. Scotese.*

marine realms. Pterosaurs had undergone much evolution, and although primitive forms with tails continued through the Late Jurassic, more derived tailless forms (called pterodactyloids) had evolved. Crocodilians underwent a remarkable radiation; in addition to other more conventional types, one Late Jurassic model with fins was fully seagoing. Some of the early representatives of lizards appeared during this time. By Late Jurassic time, dinosaurs had consolidated their dominance of terrestrial vertebrate faunas. A suggestion of continental connections in the Late Jurassic remains in the faunas: Recall that *Brachiosaurus* and *Dryosaurus* have been found in both North America and Africa (Tendaguru).

ASIA. The Asian Late Jurassic, although not yet as rich as the North American records, is proving to be an up-and-coming contender. Most of the material comes from Sichuan, China, although some material has been found in India and Thailand. Among the large theropods, *Szechuanosaurus* and *Yangchuanosaurus* are most memorable. An impressive, diverse group of sauropods has been found, including *Mamenchisaurus* (Plate IVA), *Tienshanosaurus*, *Omeisaurus*, and *Euhelopus*. At least three genera of stegosaurs are known, including the beautifully preserved *Tuojiangosaurus*, *Chialingosaurus*, and *Chungkingosaurus*.

AFRICA. The Late Jurassic of Africa is almost (but not quite) synonymous with the Tendaguru beds of Tanzania (see Box 11.1). Excluding Tendaguru, footprints are known from Niger and Morocco, and sauropods are known from Malawi and Zimbabwe. South Africa produced a stegosaur (*Paranthodon*), some sauropod material, and some ornithopod material referred questionably to the Early Cretaceous *Iguanodon*. But it was Tendaguru that produced the spectacular sauropod fauna (*Dicraeosaurus*, *Barosaurus*, *Brachiosaurus*, *Tornieria*), a small ornithopod (*Dryosaurus*), a primitive stegosaur (*Kentrosaurus*), and a variety of theropods (referred to *Ceratosaurus*, *Megalosaurus*, *Elaphrosaurus*, and *Allosaurus*, among others) whose precise affinities remain uncertain.

EUROPE. The Late Jurassic of Europe is no less rich, though much more fragmentary. A number of finds have been (questionably?) referred to *Megalosaurus*, that charter member of Sir Richard Owen's Dinosauria. Other Late Jurassic European theropods include *Metriacanthosaurus* and a variety of finds difficult to assign within Theropoda. No discussion of this time interval in Europe would be complete without mentioning Solnhofen (Chapter 13), the ancient, brackish-water lagoon deposit that preserved beautiful specimens of the light-bodied theropods *Compsognathus* and *Archaeopteryx*.

A variety of indeterminate sauropod fragments have been found throughout Europe, including brachiosaurids, diplodocids, and camarasaurids. Beyond these, stegosaurs are well represented, including *Lexovisaurus* and *Dacentrurus*. The ornithopod *Hypsilophodon* has been described, as has the nodosaurid ankylosaur *Dracopelta*. In summary, the European fauna is a reasonably diverse, important fauna, still showing strong links to the African fauna and, as we shall see, to the North American fauna.

SOUTH AMERICA. The South American Late Jurassic to date consists of trackways from Chile and Argentina, and some undescribed dinosaur remains (including reported sauropod material) from Colombia and Chile.

NORTH AMERICA. It is North America that has produced the Late Jurassic fauna that has so caught the public imagination. That fauna comes almost exclusively from the Western Interior of North America, from a sedimentary unit called the Morrison Formation (Plate IVB). The Morrison is a thick wedge of sediment that formed as material was shed eastward off the ancestor of the modern Rocky Mountains. It has the distinction of being the largest terrestrial formation in North America, being present in Arizona, New Mexico, Utah, Colorado, and Wyoming. It represents a variety of different ancient environments, including rivers (and their floodplains), lakes, and even some coastal barrier-bar sequences. It was the Morrison to which paleontologists scurried to collect dinosaurs in the Great Dinosaur Rush, it was the Morrison that produced famous Wyoming localities like Como Bluff and Bone Cabin Quarry (in which the ground was so littered with fossil bone that a cabin was made out of bone fragments), and it was the Morrison that yielded up mass dinosaur graves like the Cleveland-Lloyd Quarry and what has now become Dinosaur National Monument (Chapter 12). In short, the Morrison is a fabulously rich unit that continues to produce significant Late Jurassic fossils.

The roster of dinosaurs from the Morrison is impressive. Among theropods, *Allosaurus*, *Ceratosaurus*, *Marshosaurus*, *Stokesosaurus*, *Coelurus*, *Ornitholestes*, and *Torvosaurus* are best known. Among sauropods, the Morrison fauna includes *Diplodocus*, *Camarasaurus*, *Dystylosaurus*, *Haplocanthosaurus*, *Apatosaurus*, *Brachiosaurus*, *Ultrasauros*, *Supersaurus*, *Seismosaurus*, and *Barosaurus*. Then there are the ornithopods *Camptosaurus*, *Dryosaurus*, *Drinker*, and *Othnielia*, and the stegosaur *Stegosaurus*. Plus eggshells. Plus trackways. Plus who knows what in the future?

AUSTRALIA. No Late Jurassic dinosaurs are known from Australia.

ANTARCTICA. No Late Jurassic dinosaurs are known from Antarctica.

CRETACEOUS

Early Cretaceous
(145–97 Ma)

As we have seen, the first part of the Cretaceous was a time of enhanced global tectonic activity (Figure 15.7). With this came increased continental separation, as well as greater amounts of CO_2 in the atmosphere. This in turn probably produced warmer climates through the "greenhouse effect." Climates from the Early through mid-Cretaceous (to about 96 Ma) were therefore warmer and more equable than today.

The Early Cretaceous marks the rise of the largest representatives of Ornithopoda. Ankylosaurs also became a significant presence among herbivores of the Early Cretaceous times. Finally, the earliest known representative of Ceratopsia, a group that was to be very significant in the Late Cretaceous, made its appearance in mid-Cretaceous time, toward the end of the Early Cretaceous.

Moreover, the balance of the faunas seems to have changed. During the Late

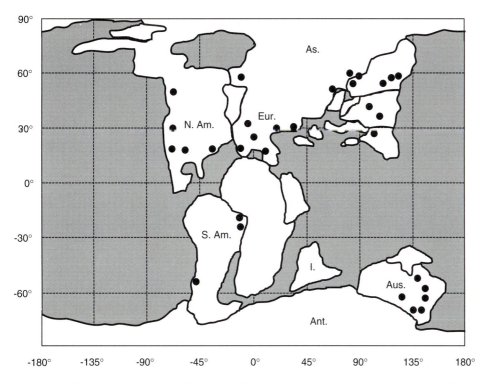

Figure 15.7 **Rectilinear projection of continents and major dinosaur-bearing locali-ties during the Early Cretaceous.** Map abbreviations as in Figure 15.2. *Reconstruction by Paleogeographic Information System, M. I. Ross and C. R. Scotese.*

Jurassic, sauropods and stegosaurs were the major large herbivores, with ornithopods being represented primarily by smaller members of the group. Now, in the Early Cretaceous (and in fact, throughout the Cretaceous), ornithopods became the dominant large herbivores. Sauropods and stegosaurs were still present, but the significance of these groups seems to have been greatly reduced. Was the spectacular Cretaceous ascendancy of Ornithopoda due to the feeding innovations developed by the group? Perhaps; however, as we have seen, such a question is not easily answered by the fossil record. The parallel success of Ceratopsia in Late Cretaceous time and the independent invention by that group of similar feeding innovations suggest that sophisticated oral food processing probably didn't hurt!

Lastly, the Early Cretaceous was a time of increased endemism. The days of truly global faunas were coming to a close by the end of the Early Cretaceous, and instead, distinctive regional faunas were beginning to appear, such as the little ceratopsian *Psittacosaurus*. This pattern was most marked by the Late Cretaceous, but was well underway by the Early Cretaceous.

ASIA. The Early Cretaceous record is really quite extensive, involving some countries commonly associated with dinosaurs (China and Mongolia) and some not (South Korea, Thailand, and Japan). North-central and central Asia (including Mongolia) have produced a variety of significant material, including the ankylosaur *Shamosaurus*, the theropod *Harpymimus*, a dromaeosaurid, and the enigmatic ornithopod *"Iguanodon" orientalis*. Also from north and central Asia come several species of the earliest representative of Ceratopsia, *Psittacosaurus*.

From China comes a large amount of material, and unlike the material from north and central Asia, much of it is reminiscent of the Late Jurassic. The theropods *Kelmayisaurus* and *Chilantaisaurus* have been described, as has the sauropod *Mongolosaurus*. A stegosaur, *Wuerhosaurus*, was also described. More "modern" faunas include an ornithomimid, the ornithopod *Probactrosaurus*, and, of course, *Psittacosaurus*.

From South Korea a good deal of sauropod material has been recorded, including some bones called *Ultrasaurus* (not to be confused with the sauropod *Ultrasauros* from North America). Thailand and Japan have produced theropod and ornithopod skeletal material and footprints.

AFRICA. The Early Cretaceous of Africa is primarily known from the northern part of the continent and from a remarkable fauna preserved in Niger. From Morocco, Algeria, and Tunisia come the theropod *Carcharodontosaurus* and the sauropods *Rebbachisaurus* and *Brachiosaurus*. *Elaphrosaurus*, a possible ceratosaur, has been recorded from Morocco and from Libya. Tunisia produced the interesting theropod *Spinosaurus*, whose unusually elongate neural arches are strangely echoed in the important primitive ornithopod from Niger, *Ouranosaurus*. *Bahariasaurus* and *Afrovenator*, both theropods; a hypsilophodontid, *Valdosaurus*; and the sauropods *Rebbachisaurus*, *Aegyptosaurus*, and *Astrodon* were also recovered from Niger.

EUROPE. The European Early Cretaceous is tremendously rich and was very significant in establishing the early ideas about Dinosauria in the minds of Victorian and Edwardian scientists such as Owen, Mantell, Buckland, and Seeley. Most famous are the Wealden fossil beds, a suite of fluvial rocks in southern England, which produced a distinguished roster including the theropods *Megalosaurus*, *Altispinax*, and most recently, *Baryonyx*; the sauropods *Cetiosaurus*, *Pelorosaurus*, *Chondrosteosaurus*, *Macrurosaurus*, *Pleurocoelus*, and *Titanosaurus*; the ornithopods *Iguanodon*, *Hypsilophodon*, and *Valdosaurus*; the stegosaur *Craterosaurus*;[2] and the the ankylosaurs *Acanthopholis* and *Hylaeosaurus* (Plate VA).

France and Belgium are almost as rich, with France producing a theropod (*Erectopus*) and a sauropod (*Aepysaurus*) and Belgium's being home to the spectacular find of 31 complete, mostly articulated *Iguanodon* specimens (see Chapter 10). Spain and Germany have also yielded Early Cretaceous faunas referable to many of these European genera, including *Hypsilophodon*, *Valdosaurus*, *Cetiosaurus*, and the enigmatic, headless, basal marginocephalian, *Stenopelix*.

SOUTH AMERICA. To date, the Early Cretaceous of South America has consisted largely of footprint faunas, with the exception of Argentina, from which the extraordinary ceratosaurian theropod *Carnotaurus* and a brachiosaur, *Chubutisaurus*, were recovered.

---

[2] This fossil was probably reworked from rocks older than the Wealden.

NORTH AMERICA. A great deal of Early Cretaceous is known from North America, particularly from the western and central parts of the United States, as well as from the eastern part of the country. From the Western Interior of the continent are a variety of theropods, such as dromaeosaurids (including *Deinonychus* and the newly discovered, fearsome *Utahraptor*), and troodontids (Plate VB). Some sauropodan material has been recorded, but sauropods do not seem to have been the presence in the Early Cretaceous of North America that they were in the Late Jurassic. Although sauropod and stegosaur material has been recorded, the predominant herbivores seem to have been iguanodontans like *Tenontosaurus*, ankylosaurs like *Silvisaurus*, *Sauropelta*, and *Nodosaurus*, and smaller herbivores such as hypsilophodontids (*Zephyrosaurus*). Clearly, sauropods and ornithopods were not ecologically interchangeable; the long-necked sauropods can be expected to have been eating food located in very different places from the location of the food eaten by ornithopods. Yet, some kind of ecological succession does appear to have been in the process of taking place during this time in North America.

A couple of other selected odds and ends from the Early Cretaceous deserve mention. Sauropod trackways from the Paluxy Formation and from Glen Rose, both in Texas, have been the subject of much debate and have been used as evidence to show (1) that sauropods provided protection for their young in the center of herds; (2) that sauropods were aquatic; and (3) that humans lived at the same time as the dinosaurs. None of these ideas is now considered valid.

Finally, the Arundel Formation of Maryland, then and now on the east coast of the United States, has proven to be productive. Material from there consists of theropods (including an ornithomimid), the sauropod *Astrodon*, and *Tenontosaurus*.

AUSTRALIA. By Early Cretaceous time, Australia was still tightly connected to the rest of Gondwana. It is not surprising, therefore, that relatively recent finds from Australia have shown that it, too, possessed a dinosaurian fauna typical of its time. Several theropods are known: *Kakuru*, *Rapator*, and a creature referred to *Allosaurus*. There is some sauropod material, including *Austrosaurus*. Among ornithischians, however, there is an impressive fauna: a nodosaurid ankylosaur, *Minmi*; an iguanodontan, *Muttaburrasaurus*; and a variety of other, smaller ornithopods: *Leaellynasaura*, *Atlascopcosaurus*, and *Fulgurotherium*. More time and effort will undoubtedly harvest further riches Down Under.

ANTARCTICA. No Early Cretaceous dinosaurs are known from Antarctica.

Late Cretaceous
(97–65 Ma)

The Late Cretaceous! Never before in their history were dinosaurs so diverse, so numerous, and so incredible. The Late Cretaceous boasted the largest terrestrial carnivores in the history of the world (tyrannosaurids), a host of sickle-clawed brainy (*and* brawny) killers worthy of any nightmare (or Hollywood movie), a marvelous diversity of duck-billed and horned herbivores, armored dinosaurs, and dome-heads.

Climate seems not to have affected diversity. In fact, although diversity increased, Late Cretaceous climates deteriorated from former, equable mid-Cretaceous time onward. As we have seen (Chapter 2), this occurred concomitantly with a marine regression, which undoubtedly played a role in the destabilization of climate. Conditions were not dreadful; they certainly were no worse than they are today, and possibly considerably better. What is significant for our story, however, is that the deterioration was not a sudden drop near the end (or beginning) of the Late Cretaceous. It was a steady deterioration that marked the entire 30 or so million years of the time interval. At the very end of the Mesozoic, there is no evidence for a sudden drop in temperatures or for any significant modification of climate.

ENDEMISM. Endemism among dinosaurs reached an all-time high during the Late Cretaceous (Figure 15.8). Southern continents tended to maintain the veteran Old Guard: a large variety of sauropods, some ornithopods, ornithomimosaurs, and in South America, the unusual abelisaurid theropods. In northern continents, however, new faunas appeared that were very different from previous ones. Among herbivores, sauropods were still present, although not as common as before. Stegosaurs left the world entirely (except one final record in India). But in their place, new creatures roamed, including pachycephalosaurs, ceratopsians, and hadrosaurids. As we have seen, many among these new inhabitants of the earth may have traveled in herds (flocks?), and large-scale migrations have been proposed. Among carnivores, large bipeds simply got larger, culminating in the extremely short-armed tyrannosaurids. Although carnivores of nearly equivalent size had existed previously, nothing shaped quite like tyrannosaurids had existed previously or has existed since. Likewise, a northern Late Cretaceous panorama would show herds (flocks) of ornithomimids roaming – a sight not common in previous times.

The Late Cretaceous record of dinosaurs is largely divided up between North America and Asia. This is not to suggest that the other continents are not represented; as Table 15.1 indicates, they clearly are. It simply means that we have a much better understanding of the northern faunas than of the southern ones and that our story could change dramatically as more and better dinosaur material from southern continents is unearthed. North America and Asia, however, share not only a rich record, but an extremely strong similarity between their respective faunas. These two continents share (almost exclusively) hadrosaurids, pachycephalosaurs, ceratopsians, tyrannosaurids, and ornithomimosaurs. This faunal similarity has been recognized for some years; however, in the late 1980s and early 1990s, P. J. Currie of Canada's Royal Tyrrell Museum, D. A. Russell of the Royal Ontario Museum, and the University of Chicago's P. C. Sereno used cladograms and a detailed analysis of faunal distributions to suggest that many of the North American faunas had their origins in Mongolia. They proposed that there may have been multiple migrations across a Bering land bridge (much as humans are thought to have migrated to North America some 64 million years later!). Once again, we have an example of geological forces (tectonics) modifying evolution.

FLOWERS. A final note about Late Cretaceous landscapes: It was during mid-Cretaceous times that flowering plants, the **angiosperms**, underwent a tremendous evolutionary burst.

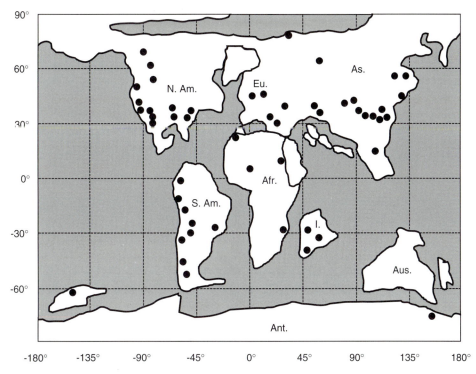

Figure 15.8 **Rectilinear projection of continents and major dinosaur-bearing localities during the Late Cretaceous.** Map abbreviations as in Figure 15.2. *Reconstruction by
Paleogeographic Information System, M. I. Ross and C. R. Scotese.*

A number of workers have pondered the relationship between dinosaurs and
the plants upon which they fed. All recognize that **co-evolution** must have
occurred, that is, that plants and herbivorous dinosaurs must have evolved in
response to one another. A claim by R. T. Bakker, that dinosaurs "invented"
angiosperms, is a strong statement of the co-evolutionary response between plants
and herbivorous dinosaurs (Box 15.3). A more measured view of the evolutionary interactions between Mesozoic and Tertiary plants and herbivores was conducted by S. Wing of the National Museum of Natural History (Smithsonian
Institution) and B. Tiffney at the University of California at Santa Barbara. They
described the probable diets of dinosaurian and other herbivores and the ways in
which contemporary plants reacted to this kind of "predation." They found that
during the middle of the Mesozoic, few vertebrates fed very selectively upon the
relatively slow growing conifers, cycads, and ginkgoes that formed the majority of
terrestrial floras.

Even with the origin and spread of angiosperms during the early part of the
Cretaceous Period, feeding consisted of low browsing and rather generalized grazing (much like a lawn mower "grazes" over whatever is in its path). Because many
of these Mesozoic herbivores were also very large and may have lived in large herds,
they probably also cleared expansive areas, trampling, mangling, uprooting, and
otherwise disturbing areas that otherwise might be colonized by plants.

Such low-level, generalized feeding and such disturbances of habitats tended to emphasize fast growth in plants, but it is thought that the animals did not contribute significantly to seed dispersal. Thus, the picture of Mesozoic plant–herbivore interactions appears to be one in which (1) plants produced vast quantities of offspring to ensure the survival of the family line into the next generation and (2) herbivores took advantage of the rapidly and abundantly reproducing resource base to maintain their large populations of large individuals. Plant–herbivore coevolution during the Mesozoic appears to have been based on habitat disturbance, generalized feeding, and rapid growth and turnover among plants.

With the extinction of large herbivores at the end of the Cretaceous and the collapse of the plant–herbivore interactions that resulted, another suite of interactions developed. These emphasized more selective feeding by the much smaller mammalian herbivores, the development of seed dispersal complexes, and the evolution of fleshy fruits and seeds.

ASIA.   As we stated previously, Asia has a spectacularly rich Late Cretaceous record. The roster of Late Cretaceous dinosaurs from northern Asia (excluding Mongolia and China) includes theropods of tyrannosaurid and therizinosaurid affinities, some sauropod (titanosaur) material, a variety of ornithopod material (including the endemic hadrosaurids *Aralosaurus*, *Nipponosaurus*, and *Jaxartosaurus*), and ankylosaur material. From India were recovered a variety of theropods (including *Indosuchus*) and a large amount of sauropod material (note India's geographic location during the Late Cretaceous), mainly referable to *Titanosaurus*. A final find of note from India is *Dravidosaurus*, the latest surviving stegosaur.

It is the Mongolian faunas, however, that have captured the world's imaginations since their early discovery by R. C. Andrews in the 1920s (see Box 7.1). From a series of disconnected localities popping up throughout the Gobi Desert (the southern part of Mongolia), dinosaurian riches have been (and continue to be) recovered (Plate VIA). Among Theropoda, there are the tyrannosaurid *Tarbosaurus*; the ornithomimosaurs *Anserimimus*, *Garudimimus*, and *Gallimimus*; the small-bodied maniraptorans *Oviraptor*, *Ingenia*, *Conchoraptor*, *Saurornithoides*, *Adasaurus*, *Elmisaurus*, *Borogovia*, *Velociraptor*, and *Hulsanpes*. Then there are the strange things: *Avimimus*, *Deinocheirus*, *Mononykus*, and the segnosaurs *Therizinosaurus*, *Segnosaurus*, and *Erlikosaurus*. Mongolia has sauropods as well: *Opisthocoelicaudia*, *Nemegtosaurus*, *Quaesitosaurus*. The ceratopsian fauna is composed of *Protoceratops*, *Microceratops*, and *Bagaceratops*; there are an ornithopod fauna consisting of hadrosaurids (*Saurolophus* and *Barsboldia*) and a rich ankylosaur fauna: *Tarchia*, *Saichania*, *Pinacosaurus*, *Amtosaurus*, *Talarurus*, and *Maleevus*. Finally, new faces on the scene include the pachycephalosaurs *Prenocephale*, *Tylocephale*, *Homalocephale*, and *Goyocephale*.

China has been likewise productive. Large theropod material referred to *Tarbosaurus*, *Chingkankousaurus*, *Alectrosaurus*, and *Shanshanosaurus* has been found, along with the smaller theropods *Saurornithoides*, *Archaeornithomimus*, and *Velociraptor*. Segnosaur finds include *Alxasaurus*, *Segnosaurus*, and *Enigmosaurus*. Among herbivores, China has produced some sauropods, but the hadrosaurid fauna is remarkable: *Tsintaosaurus*, *Tanius*, *Shantungosaurus*, *Bactrosaurus*, and *Gilmoreosaurus*. China has ankylosaurs (*Pinacosaurus*) as well as pachycephalosaurs (*Wannanosaurus*, *Micropachycephalosaurus*). Finally, ceratopsians are represented by *Microceratops*.

Eastern Asia also has a Late Cretaceous: Japan and Laos have hadrosaurid and sauropod representatives.

AFRICA.   Africa has a modest Late Cretaceous dinosaur record, known from Morocco, Egypt, Madagascar, Algeria, Niger, Kenya, and South Africa. This fauna includes theropods (*Bahariasaurus*, *Carcharodontosaurus*, *Spinosaurus*, *Elaphrosaurus*, *Majungasaurus*), titanosaur sauropods, and some pachycephalosaur material (*Majungatholus*).

EUROPE.   Recall from Chapter 2 that Europe during the Late Cretaceous was a series of disconnected small islands and waterways. Because of this, it is not surprising that thick piles of sediment did not have the opportunity to accumulate to represent this time interval in this region. Europe, therefore, does not have an abundant Late Cretaceous; however, some areas of Spain, France, Belgium, Austria, and Romania have proven productive (Plate VIIA). Indeterminate theropod material referred to *Megalosaurus*, dromaeosaurids, and troodontids has been discovered; however, sauropods have proven more abundant: The titanosaurs *Titanosaurus*, *Hypselosaurus*, and *Magyarosaurus* have all be found. In addition to these have been found the enigmatic ornithopod *Rhabdodon*, as well as *Telmatosaurus* (a hadrosaurid) and nodosaurid ankylosaur material including *Struthiosaurus*.

SOUTH AMERICA.   The Late Cretaceous of South America is becoming better known, and with this is emerging a fauna that is rather different from the North America–Asia fauna. The Late Cretaceous of South America contains a rich sauropod fauna, including *Titanosaurus*, *Antarctosaurus*, *Saltasaurus*, *Laplatasaurus*, and *Argyrosaurus*. Some theropods are known, including the abelisaurids

---

BOX 15.3

DINOSAURS INVENT FLOWERING PLANTS

IN HIS POPULAR BOOK *Dinosaur Heresies*, R. T. Bakker insists that dinosaurs "invented" flowering plants. The germ behind Bakker's hypothesis is that Late Jurassic herbivores, epitomized by sauropods, were essentially high browsers, whereas Cretaceous herbivores, epitomized by ornithopods, ankylosaurs, and ceratopsians, were largely low browsers. Bakker argued that Cretaceous low browsers put tremendous selective pressures on existing plants, so that survival could occur only in those plants that could disseminate quickly, grow quickly, and reproduce quickly. Angiosperms, he argued, are uniquely equipped with those capabilities. In his scenario, Bakker has Cretaceous low-browsing dinosaurs eating virtually all the low shrubbery, and plants responding

by developing a means by which animals simply couldn't keep up with the growth, reproduction, and dissemination of the plants.

How likely is this? We are not sure. Troubling, of course, is the strongly North American and Asian cast of this hypothesis; southern latitude faunas seem to have had just the faunal compositions that Bakker claims would *not* have put intense selective pressure on contemporary low-growth floras. Yet, angiosperms were radiating worldwide by Late Cretaceous time. Still, what is significant about Bakker's hypothesis is that in it, he, as well as others such as S. Wing and B. Tiffney, are clearly recognizing and attempting to define the co-evolution between dinosaurs and plants.

Xenotarsosaurus and *Abelisaurus*, some material referred to ornithomimids, and the small theropod *Noasaurus*. There are records of hadrosaurid ornithopods from Argentina.

NORTH AMERICA. The Late Cretaceous of North America includes the greatest abundance and diversity of dinosaurs of all time. This is partly because Canada and the United States have been prospected by more or less well-funded field parties for nearly 150 years, but it is also simply the luck of the draw: Extensive Late Cretaceous sediments are preserved on the east and south coastal plain of the United States, as well as in the Western Interior of the continent, in a gigantic swath of sedimentary rocks running from the south (Mexico) all the way through much of the western part of the United States, to the central and even northern parts of the Canadian provinces of British Columbia, Alberta, and Saskatchewan, all the way up to the north slope of Alaska. These rocks were deposited as a thick, east-facing wedge of sediment when the ancestral Rocky Mountains developed (Chapter 2) and were drained by east- and southeast-flowing river systems. The gigantic wedge of sediment produced by those streams includes some of the most famous Late Cretaceous rocks of all time. In Canada, the fluvial Milk River, Dinosaur Park (Oldman), Judith River, Two Medicine, and St. Mary's River formations, as well as the deltaic Scollard and Horseshoe Canyon formations, have all produced fabulous collections (Plate VIB). In the Western Interior of the United States, the fluvial Judith River, Two Medicine, Laramie, Lance, and Hell Creek formations, as well as the San Juan Basin region of Texas and New Mexico, have produced many of the dinosaur faunas (Plate VIIB).

Who are these dinosaurs? A list would include tyrannosaurids (*Tyrannosaurus, Albertosaurus, Aublysodon, Daspletosaurus*), ornithomimids (*Struthiomimus, Dromiceiomimus, Ornithomimus*), elmisaurids, troodontids (*Troodon*), dromaeosaurids (*Dromaeosaurus, Saurornitholestes*), *Chirostenotes*, hypsilophodontids (*Thescelosaurus, Parksosaurus, Orodromeus*), hadrosaurids (*Brachysaurolophus, Corythosaurus, Edmontosaurus, Gryposaurus, Kritosaurus, Lambeosaurus, Maiasaura, Prosaurolophus, Parasaurolophus*), pachycephalosaurs (*Stegoceras, Gravitholus, Pachycephalosaurus, Ornatotholus, Stygimoloch*), ceratopsians (*Anchiceratops, Arrhinoceratops, Avaceratops, Brachyceratops, Centrosaurus, Leptoceratops, Pachyrhinosaurus, Pentaceratops, Chasmosaurus, Styracosaurus, Montanoceratops, Torosaurus, Triceratops*), and ankylosaurs (*Ankylosaurus, Edmontonia, Denversaurus, Panoplosaurus, Euoplocephalus*). In all of this, a few sauropods have been recovered, including some diplodocid material and the Texas titanosaurid *Alamosaurus*.

A few finds have been made in the central and eastern parts of the United States, including the theropod *Dryptosaurus*; an ornithomimid; and the hadrosaurid *Lophorhothon* and *Hadrosaurus*. From Mexico come the theropod *Labocania*, as well as a variety of other theropod, ankylosaur, hadrosaurid, and ceratopsian material referred to many of the genera described previously.

What does this all mean? Factoring out more rocks and longer, better-funded collecting, were dinosaurs in North America (and Asia) more diverse (and abundant) than in the rest of the world? Possibly – indeed, some authors have suggested that the Northern Hemisphere was simply where most of the progressive dinosaurs hung out.

Uncondensed, how diverse was the Late Cretaceous North American fauna?

Studies at a detailed level, in which a real biota from a single slice of time is reconstructed, are few and far between (remember that "Late Cretaceous," for example, condenses 32 million years into a single point!). One such study, however, was published in 1987 by T. M. Lehman of Texas Tech University. Lehman attempted to reconstruct who might be living where during a typical "day" of the Late Cretaceous in North America.

To do this, Lehman tried to identify **biogeographic** zones in the latest Cretaceous – that is, regions in which latest Cretaceous faunas characterized by particular dinosaurs lived. Lehman identified three biogeographic zones in the latest Cretaceous of North America: (1) a southern-central fauna, which is dominated by the sauropod *Alamosaurus* (but contains hadrosaurids, ankylosaurs, ceratopsians, and theropods as well), (2) a northern fauna dominated by *Triceratops* (but containing also theropods, ankylosaurs, duckbills, hypsilophodontids, and pachycephalosaurs), and (3) a central fauna dominated by *Triceratops* and *Leptoceratops* (but which contains theropods, ankylosaurs, and hypsilophodontids). Each of these faunas was tied not only to a *geographic* region, but also to a *paleoenvironment*. The *Triceratops* fauna lived in a low, broad, humid, coastal plain setting adjacent to the Western Interior seaway that bisected North America during this time; the *Leptoceratops* fauna lived adjacent to the uplifting Rocky Mountains, where temperatures are thought to have been cooler; and the *Alamosaurus* fauna lived in a series of intermontane basins (as the ancestral Rocky Mountains underwent their latest Cretaceous uplift) and was therefore subject to seasonal wetting and drying.

In the case of the *Leptoceratops* fauna, Lehman's sample is not quite as extensive as would be ideal, but it is certainly a first step toward understanding the population dynamics of dinosaurs during a particular slice of time. And, the Western Interior of the North American continent is paleontologically rich enough, and has been studied in enough geological detail, to make this kind of analysis possible.

The rich faunas of the Late Cretaceous of North America are only equaled in richness by those preserved in China and Mongolia, and comparison between the two reveals interesting similarities and contrasts. First, those North American faunas that are known, lived along a broad swath of coastal plain. Water appears to have been very abundant, and dinosaur remains are preserved in the rivers, floodplains, and deltas that were deposited as the ancestral Rocky Mountains to the west were slowly eroded. By contrast, the environments where all dinosaurs in China and Mongolia appear to have lived (and died) were quite different. The early Late Cretaceous Asian dinosaurs lived in semiarid or arid internally draining basins, much like the present-day Great Basin of Utah and Nevada. Water was not abundant, as it was in North America at the same time, and was mainly restricted to playa lakes and low areas that existed between regions of sand dunes. In the later part of the Late Cretaceous, at least in Mongolia, this situation changed, and dinosaur material is found in deposits representing the remains of ancient rivers. How the transition from dry environments to wetter environments modified Asian dinosaur behavior or morphology – or diversity – is unclear. What is clear is that relative to North American counterparts living during the early– to middle–Late Cretaceous, Asian dinosaurs generally endured somewhat drier conditions.

We have noted that the North American Late Cretaceous fauna is progressive:

It contains primarily rather advanced representatives of the various dinosaur clades. Exceptions to this are known (for example, *Thescelosaurus*, a primitive-appearing but large hypsilophodontid from the latest Cretaceous of North America), but by and large, the fauna is quite highly derived and the last-appearing dinosaurs are the most derived members of their clades. Such is not the case for the Asian faunas, and as noted by Canadian paleontologists T. Jerzykiewicz and Russell, the Late Cretaceous of China and Mongolia contains a curious mixture of quite advanced forms and ones judged to be more primitive. For example, the variety of troodontids and dromaeosaurids preserved in this region is impressive; these are considered to be relatively advanced dinosaurs. On the other hand, however, more primitive forms, such as segnosaurs, are also preserved.

Faunal compositions differed as well. The North American faunas are dominated by ornithopods (hadrosaurids, in this case) and ceratopsians. As we have seen, titanosaurids are known from the Late Cretaceous of North America (New Mexico and Colorado), but they are never common and do not extend particularly far north on the continent. Theropods never come close to achieving the numbers, and probably not the diversity, of their ornithischian counterparts. Here, then, the predator–prey ratios (Chapter 14) developed by Bakker seem to be best confirmed. In Asia, on the other hand, ornithischians are certainly well known during the Late Cretaceous, but there one is also struck by the variety of other dinosaurs: segnosaurs, sauropods, and an immense array of unusual theropods, such as ornithomimosaurs and oviraptorosaurs. Perhaps the most striking aspect of the Asian faunas in comparison to their North American counterparts is the absence of the extremely diverse ceratopsian fauna that characterizes North America. Likewise, there appears to be far less diversity among hadrosaurids in Asia than in North America. Do all these differences reflect the differences in paleoenvironments, or is some other factor indicated?

Many authors have noted strong similarities between the faunas of North America and Asia. Jerzykiewicz and Russell have estimated that 55% of the families present in Asia are also present in North America. The other 44% are endemic to Asia.[3] There is good evidence that at some point (or points) during the mid-Cretaceous, at least protoceratopsids, tyrannosaurids, ankylosaurids, and possibly hadrosaurids migrated to North America from Asia across the Bering Straits (a migration that is thought to have also been accomplished, about 64 million years later, by a group of mammals called *Homo sapiens*). Subsequent radiation of ceratopsians and hadrosaurids in North America could account for many (but certainly not all) of the faunal differences between North America and Asia.

AUSTRALIA. A single Late Cretaceous locality, producing a bonanza of theropod, sauropod, and ornithopod footprints, has been discovered in Queensland.

ANTARCTICA. The Santa Marta Formation of Antarctica has produced apparent ankylosaur fragments, and A. Milner of the British Museum (Natural History) discovered an as yet unnamed ornithopod.

---

[3]The families that occur in both North America and Asia are Dromaeosauridae, Troodontidae, Elmisauridae, Ornithomimidae, Tyrannosauridae, Hadrosauridae, Pachycephalosauridae, Protoceratopsidae, and Ankylosauridae; the families that are endemic to Asia are Oviraptoridae, Deinocheiridae, Avimimidae, Diplodocidae, Homalocephalidae, and Therizinosauridae. Not all of these taxa, however, are monophyletic.

THE TERTIARY:
After the Ball Is Over

With the end of the Cretaceous, non-avian dinosaurs disappeared from earth forever. Ironically, dinosaurs were at the peak of their diversity when they went extinct. How sudden or fast the extinction occurred remains contentious (see Chapter 17). Never again would the world see archosaur-dominated faunas; mammals, once they had become well entrenched as the dominant terrestrial vertebrates, would be no more likely to give up their place in Tertiary ecosystems to dinosaurs than dinosaurs, throughout the 160 million years of dinosaur incumbency, were likely to give up *their* place to mammals! Everybody knows that dinosaurs disappeared at the end of the Cretaceous. Or did they?

In 1987, the University of Notre Dame's J. K. Rigby, Jr., and his colleagues reported "Tertiary dinosaurs" – that is, a few relict specimens from eastern Montana that they claimed persisted past the end of the Cretaceous. The report was instantly controversial. Critics charged that Rigby and his colleagues didn't know exactly where the end of the Cretaceous was in the sediments that they were studying; after all – it was pointed out – the end of the Cretaceous was commonly recognized as the place where the last (youngest) dinosaur was preserved. It was also noted that even if Rigby's dinosaurs were preserved in sediments *younger* than Cretaceous in age (i.e., Tertiary), the bones were obtained as fragments in Tertiary river channels and thus were probably reworked from Cretaceous sediments by Tertiary-aged rivers.

Rigby and his colleagues adhere to their claim of Tertiary dinosaurs. Their view has not received wide acceptance. The issue was summed up nicely by University of Pennsylvania's Dodson at a meeting in 1991. Suppose it were true, Dodson concluded, and a few stragglers survived into the Tertiary. What difference would this make? "It's like driving along the freeway and seeing a Model T (Ford)," he said. "There may still be one or two relicts left driving, but that doesn't mean that anybody is still making the Model T!" In other words, for all practical purposes, dinosaurs (well, *non-avian* dinosaurs!) disappeared from the face of the earth at the end of the Cretaceous.

Important Readings

Bakker, R. T. 1986. The Dinosaur Heresies. William Morrow and Company, New York, 481 pp.

Benton, M. J. 1986. The Late Triassic tetrapod extinction events; pp. 303–320 *in* K. Padian (ed.), The Beginning of the Age of Dinosaurs, Cambridge University Press, New York.

Dodson, P. 1990. Counting dinosaurs: how many kinds were there? Proceedings of the National Academy of Sciences 87:7608–7612.

Fastovsky, D. E. 1989. Dinosaurs in space and time: the geological setting; pp. 22–33 *in* K. Padian and D. J. Chure (eds.), The Age of Dinosaurs. The Paleontological Society, Short Courses in Paleontology no. 2.

Haq, B. U., J. Hardenbol, and P. R. Vail. 1987. Chronology of fluctuating sea levels since the Triassic. Science 235:1156–1166.

Harland, W. B., R. L. Armstrong, A. V. Cox, L. E. Craig, A. G. Smith, and D. G.

Smith.   1989.   A Geological Time Scale. Cambridge University Press, Cambridge, U.K., 262 pp.

Jerzykiewicz, T., and D. A. Russell. 1991. Late Mesozoic stratigraphy and vertebrates of the Gobi Basin. Cretaceous Research 12:345–377.

Lehman, T. M. 1987. Late Maastrichtian paleoenvironments and dinosaur biogeography in the western interior of North America. Palaeogeography, Palaeoclimatology, Palaeoecology 60:189–217.

Norell, M. A., and M. J. Novacek. 1993. Congruence between suprapositional and phylogenetic patterns: comparing cladistic patterns with the fossil record. Cladistics 8:319–337.

Rigby, J. K., Jr. 1987. The last of the North American dinosaurs; pp. 119–134 *in* S. J. Czerkas and E. C. Olson (eds.), Dinosaurs Past and Present, Vol. II. Natural History Museum of Los Angeles County and University of Washington Press, Seattle, Wash..

Ross, M. I. 1992. Paleogeographic Information System/Mac Version 1.3. Paleomap Project Progress Report no. 9, University of Texas at Arlington, 32 pp.

Russell, D. A. 1993. The role of Central Asia in dinosaurian biogeography. Canadian Journal of Earth Sciences 30:2002–2012.

Wing, S. L., and B. H. Tiffney. 1987. The reciprocal interaction of angiosperm evolution and tetrapod herbivory. Review of Palaeobotany and Palynology 50:179–210.

Weishampel, D. B. 1990. Dinosaurian distribution; pp. 63–139 *in* D. B. Weishampel, P. Dodson, and H. Osmólska (eds.), The Dinosauria. University of California Press, Berkeley.

CHAPTER 16

RECONSTRUCTING EXTINCTIONS:
The Art of Science

WHAT HAPPENED TO THE DINOSAURS? It is a straightforward question for which paleontology ought to be able to provide a straightforward answer. But, as is so often the case, the question is deceptively simple.

Consider some of the problems with finding a "straightforward answer." Suppose you think, as people have suggested, that dinosaurs died because the mammals ate their eggs. How do you prove this? You are not likely to find fossils of vicious little mammals with dinosaur embryos in their jaws. Even if you were fortunate enough to find such a fossil, would it prove that dinosaurs died because the mammals ate their eggs? Where is the *test* of the hypothesis?

Or maybe you like the idea that an asteroid killed the dinosaurs. You might be right, but you won't find even a common Late Cretaceous dinosaur like *Triceratops* beneath an asteroid fragment, flattened like road pizza.

It is not enough to just identify potential causes for the extinction. That is like looking at all potential murderers in the world to solve a particular murder case. Some relationship between the hypothetical cause and the extinction has to be established. For example, it is now pretty clear (as we shall see) that a large asteroid struck the earth 65 Ma. So be it, but our question must be, what – if anything – does the asteroid have to do with this extinction? Before we can blame the extinction on an asteroid – or on anything else – we need to find some kind of connection between the presumed culprit and the extinction event itself. Indeed, in extinctions, as in a court of law, we affirm innocence until guilt is proven (or at least until the hypothesis is falsified!).

EXTINCTIONS

We all know that organisms have gone extinct. Certainly the present-day biota is not the same as one that existed 100 Ka (or 10 Ka), which in turn is not the same as one that existed 100 Ma, and so forth. The biota has changed, and the old many times has given way to the new.

Extinction itself is straightforwardly defined: When the birth rate fails to keep up with the death rate, we get extinction. But in a sense, that definition begs the question. It tells us nothing about the nature or causes of the extinction, questions which turn out to be far more challenging and interesting.

Paleontologists generally divide extinctions into two types, for which different causes are postulated. The first is the so-called **background extinctions**, isolated extinctions of species that occur in an ongoing fashion. The second type is called **mass extinctions**.

BACKGROUND EXTINCTIONS. Although background extinctions are less glamorous than mass extinctions, we know that they can be important: University of Tennessee paleobiologist M. L. McKinney has estimated that as many as 95% of all extinctions can be accounted for by background extinctions. Isolated species disappear due to a variety of causes, including out-competition, depletion of resources in a habitat, changes in climate, the development or destruction of a mountain range, river channel migration, the eruption of a volcano, the drying of a lake, or the destruction of a forest, grassland, or wetland habitat. Although some dinosaur extinctions coincided with mass extinction events (such as the Triassic–Jurassic boundary, or the Cretaceous–Tertiary boundary), by far the majority of favorite and famous dinosaurs – *Maiasaura, Dilophosaurus, Protoceratops, Deinocheirus, Styracosaurus, Cetiosaurus, Iguanodon, Ouranosaurus, Allosaurus*, to name a tiny fraction – were probably victims of background extinctions.

MASS EXTINCTIONS. Mass extinctions are thought to be qualitatively and quantitatively different from background extinctions. Four components are involved: *Large numbers* of species go extinct; *many types* of species go extinct; the effects must be *global*; and the effects must occur in a *geologically short* period of time. None of these has a truly precise definition, because there turn out to be no fixed rules for mass extinctions. For example, consider the criterion of a geologically short period of time. Suppose that a mass extinction takes place over 2 million years. If it took place 245 Ma, when viewed from a modern vantage point it appears geologically instantaneous. If, however, it took place today, an event spanning 2 million years' duration would seem extremely gradual. There is no precise moment in time when 2 million years magically transforms itself from gradual to instantaneous.

How, then, do we know that there even were mass extinction "events" and how can we recognize them? In 1983, University of Chicago invertebrate paleontologists D. Raup and J. Sepkoski published a compilation of invertebrate extinctions through time (Figure 16.1). The compilation shows that although extinctions characterize all periods (and it is these that are termed background extinctions), there are intervals of time in which extinction levels are signifi-

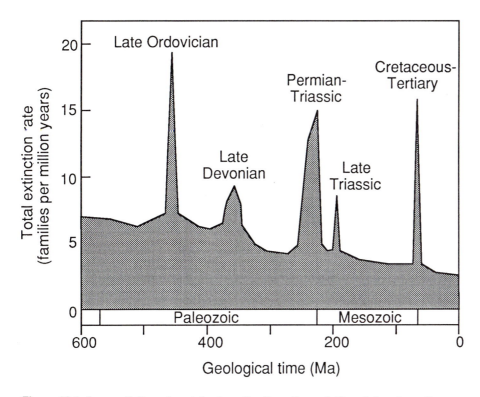

Figure 16.1 A compilation of vertebrate extinctions through time, taken from the work of D. M. Raup and J. J. Sepkoski. The five most significant were the Late Ordovician (438 Ma), the Late Devonian (380 Ma), the Permian–Triassic (245 Ma), the Late Triassic (208 Ma), and the Cretaceous–Tertiary (65 Ma).

cantly elevated above background levels. Such intervals are said to contain the mass extinctions. Raup and Sepkoski recognized 15 intervals of mass extinction, of which 5 clearly towered above the others. These 5 are an early Paleozoic mass extinction (Late Ordovician, 438 Ma); a middle Paleozoic mass extinction (Late Devonian, 380 Ma); a mass extinction that occurred right at the Paleozoic–Mesozoic boundary (Permian–Triassic boundary, 245 Ma); a Late Triassic mass extinction (208 Ma); and the Cretaceous–Tertiary extinction (65 Ma).

Raup and Sepkoski divided the 15 extinctions into "minor," "intermediate," and "major" mass extinctions, based upon the extent of the extinction above background. In the entire history of life, only one extinction qualifies as major: that is the Permian–Triassic (commonly called "Permo–Triassic") extinction. The remaining four of the Big Five – including the Cretaceous–Tertiary extinction, which involved the dinosaurs and thus will be the focus of our attention – were all classified by Raup and Sepkoski as intermediate. The rest of the 15 extinctions were considered minor, although probably not to the organisms that succumbed during them!

Raup and Sepkoski's approach is worthy of note. They looked at the *pattern* of extinctions and did not concern themselves with cause – at least not at first. The idea was to first establish the pattern, and, once this was understood, to see if the pattern of extinctions is in agreement with one or more potential causes. Here, we will consider the pattern of extinction and survival (or **survivorship**, as it is commonly called) and entertain hypothetical causes that are in agreement with the pattern. Establishing the pattern of the extinction, however, can be quite difficult. The Permo–Triassic extinction exemplifies this nicely.

In the Permo–Triassic extinction, as many as 96% of all species died, and as many as 52% of all families disappeared from the earth. Clearly, the magnitude of this extinction was extraordinary, measured by counting either the disappearances of species or the disappearances of families. But 52% and 96% are very different numbers. Which most accurately characterizes what actually occurred 245 Ma? Was the world 50% empty, 90% empty, or neither? And which is relevant in terms of cause?

Moreover, estimates of the duration of the Permo–Triassic extinction vary from essentially overnight to about 2 million years. Both are effectively geologically instantaneous, but the two are rather different in terms of possible causes. For example, one hypothesis has a decrease in continental shelf area causing the Permo–Triassic extinction. According to this idea, when the continents came together to form Pangaea, the world's shelf area – presumably where most of the victims lived – so decreased that many animals died out for lack of living space. Is this hypothesis plausible? Certainly so, if the extinction took 2 million years. Two million years is a time interval that is in agreement with the speed that the earth's plates move, and it is long enough for tectonic forces to assemble Pangaea and reduce significantly the total area of the earth's continental shelves. On the other hand, if the extinction really took place overnight, than the relatively slow-moving forces of the earth's tectonism probably cannot be held responsible, because they operate on time scales very different from overnight.

RESOLVING THE PAST

Reconstructing the *pattern* of an extinction (let alone hypothesizing a cause for it), however, is an enterprise fraught with room for error. Here, we address some of the problems inherent in this kind of analysis, so that reconstructed Cretaceous–Tertiary events are understood in context.

Fundamental to all problems with understanding the pattern of mass extinctions is the incompleteness of the fossil record, which simply does not preserve all the creatures that lived (and died). This means that, essentially, we must estimate the blanks and fill them in. Other difficulties associated with mass extinction reconstruction come from techniques that are necessarily used by paleontologists but, unfortunately, cause certain problems (even as they solve others). These include artificial range extensions and the somewhat complex issue of incongruent taxa. In the following, we explain each of these issues and discuss their ramifications.

Incompleteness of the Fossil Record: Resolution

The incompleteness of the fossil record is of two types: Rocks do not record all of the time that has elapsed, and fossils of every organism that has ever lived are not preserved. The issue is one of **resolution**. In the study of optics, resolution is the degree to which one can distinguish the detail of things; the more one can distinguish detail, the better the resolution is said to be.[1] So it is in geology and paleontology. The better we can distinguish all of the components of an event that has taken place in the past and that is preserved in the rock record, the better our *resolution* is said to be. Obviously, the business of understanding what took place in a mass extinction becomes in part the business of maximizing our resolution of the event.

TIME RESOLUTION. We saw in Chapter 2 that rocks preserve the only tangible record of the time that has elapsed on earth. And we observed that the record of time is rather incomplete (see Figure 2.1). Estimates of the percentage of total earth time not preserved in the rock record vary between 50% and 90%. This means that globally, as much as 9 out of every 10 hours simply have no rocks present to record the time that elapsed. Regardless of whether the rocks recording the time were eroded, or simply were never deposited in the first place, the resolution is poor.

With resolution so poor, the places where we are able to study a particular event, even if it was originally global, become very restricted. We need to go wherever the time interval is present that preserves the event. In the case of the Cretaceous–Tertiary boundary, many people – even professionals – are very surprised to discover that there are only about 20 localities, most of which are in North America, that preserve the last days of the dinosaurs. With the amount of Cretaceous–Tertiary rock that contains dinosaurs so restricted, care must be exercised in reaching conclusions about the patterns.

PALEONTOLOGICAL RESOLUTION. Simply because something once lived does not guarantee that it will be preserved. Organisms with hard parts, such as bone, tend to leave better records than those without hard parts (the latter including the vast majority of the biota), but even so, it is known that most fossil localities only yield but a small fraction of the biota that once lived in the environment represented by the locality. In part, we know this because of accidents of preservation: extremely rare localities where tremendous numbers and many types of organisms are preserved. By comparison with biotic diversity and abundance in modern environments, these special localities are probably preserving close to all of what lived in that place at that time. But, such marvels of preservation are very uncommon. By and large, localities preserve a very modest proportion of the original diversity. How modest? Paleobiologist McKinney makes the following suggestion:

> There are an estimated 5 to 50 million species of life on Earth today. Yet, since life first originated over 3.5 *billion* years ago, an estimated 1 to 3 billion species have come and gone. This means that *over 99% of all species that ever existed are extinct*. They will be found in the fossil record, if they are to be

---

[1]Televisions, computer monitors, comic books, and magazine pictures all rely upon the fact that human visual resolution is limited. If you had extremely well-developed eye sight, you would resolve a collection of dots instead of a picture. This was the underlying principle of the French impressionist Georges Seurat's *pointillisme* technique. Likewise, a magazine picture studied under a magnifying glass reveals the dots of color that our eye blends into an image. Recognizing art in the dots themselves, twentieth-century American artist Roy A. Lichtenstein painted on a scale in which the dots and the images both were easily visible to the unaided eye.

found at all. The question is, how complete is this record? . . . Of the 1 to 3 billion species that have existed, only one to a few million, or *less than 1%* have probably been fossilized. Of those fossilized, only about 10% have so far been discovered and described. (M. L. McKinney, 1993, *Evolution of Life*, Prentice Hall, Englewood Cliffs, N.J., pp. 123–4)

RESOLUTION AND THE RECONSTRUCTION OF EXTINCTION EVENTS.   Many authors have likened the fossil record to a series of snapshots of the history of life. Another analogy might be a series of "stills," taken at irregular intervals, from a motion picture film.

Consider a well-known extinction in terms of stills from a motion picture. Imagine a picture taken in eastern Montana, along Rosebud Creek, June 24, 1876. You'd see the cocky Lieutenant Colonel George Armstrong Custer leading 250 well-equipped bluecoats of the Seventh Cavalry. If we were lucky, the next frame might be as soon as the next day, but obviously the picture looks somewhat different; distributed across a west-facing slope just east of the Little Bighorn River would be the bodies of Custer and his thoroughly annihilated troops. The event was a landmark in American history – an important victory for Native Americans, a tragedy in light of the backlash that followed; and yet even this singular event in U.S. history – only 120 years old – is poorly understood. What happened? Was the Seventh Cavalry destroyed in 15 minutes or 5 hours? Did the Native American warriors prevail little by little, or was it virtually instantaneous? Where was Custer during the battle? Sitting Bull? Gall? Many aspects of the battle – including its length – remain poorly understood; as with any extinction, the lack of resolution hinders precise understanding of the events that took place.

In the fossil record, *the poorer the resolution, the more abrupt any extinction event looks*. In one fossil locality (in our cinematic analogy, the last pre-extinction "still"), all the known members of the fauna are present; in the next-younger fossil locality (the first postextinction "still"), many of them are gone. Did they die out suddenly or did they die out over a long period of time? Considered uncritically, evidence from the localities shows an apparently dramatic extinction. But if we had more resolution and could see steps representing the time between the localities, we could better assess exactly how abrupt or gradual the extinction was.

SIGNOR-LIPPS EFFECT.   The difficulties in accurately characterizing the pattern of an extinction are increased by sampling problems, themselves related to the distribution of fossils in the rocks. The idea, first articulated in 1982 by paleobiologists P. Signor and J. Lipps (at that time both at the University of California at Davis), has profound implications in the reconstruction of mass extinctions. Suppose that on the average, fossils are distributed at some regular interval through the rocks[2] (whose thickness, you recall, represents some amount of time). Suppose, too, that we specify a moment in time after which there will be no fossils: We will call that an **extinction boundary**. As one approaches that boundary – that is, as one begins in older sediments and works upward (therefore through younger and

---

[2]In fact, terrestrial fossils such as dinosaurs are almost never distributed at regular intervals through sediments. Rather, the environment represented by the kind of rock that entombs them controls their distribution. Here, however, we assume a regular distribution to more easily explain the Signor-Lipps effect.

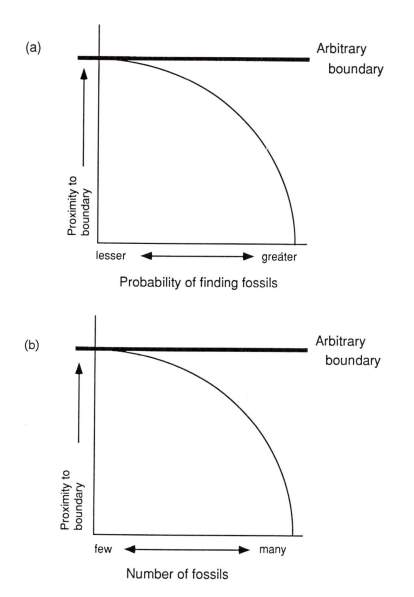

(a)

Arbitrary
boundary

Proximity to
boundary

lesser ⟷ greater

Probability of finding fossils

(b)

Arbitrary
boundary

Proximity to
boundary

few ⟷ many

Number of fossils

Figure 16.2 **The Signor-Lipps effect.** The fossils are randomly distributed throughout the
rock sequence; however, as any arbitrary boundary is approached, the probability **(a)** of
finding fossils becomes smaller and smaller as the search area decreases. Because of this,
the number of fossils found decreases **(b)**, with the result that the extinction appears gradual, even if the distribution of fossils is in fact constant throughout the rock sequence.

younger sediments) to the boundary – the chances of finding a particular fossil
become smaller in the remaining interval. This is because the volume of rock in
which a fossil could potentially be found is decreasing. Indeed, chances are that
there will be no fossils at exactly the moment of extinction and that the "last" one
(also the highest one) will occur somewhat before (or below) the boundary
(Figure 16.2). Luis Alvarez, a Nobel prize–winning physicist, likened it to

attempting to estimate the United States–Canada border by the home of the northernmost member of the United States Congress. Because these homes are randomly distributed with respect to the border, the border must be estimated to be somewhat north of the northernmost member's house.

Here is another way to consider this problem: Think of a handful of different coins that are tossed on the floor toward one wall of a room. As you get closer and closer to the wall (and the amount of area being searched becomes smaller and smaller), the probability of finding a particular type of coin—say a nickel—in the remaining area becomes smaller and smaller. Indeed, as the area being searched becomes smaller and smaller (as one approaches the wall), the probability of finding *any* coin becomes smaller and smaller. So it is with fossils. As a boundary is approached, the probability of finding fossils is diminished because the area being searched becomes smaller.

Suppose you are truly omniscient and you know that an extinction occurred suddenly, at a particular point in time. What would be the pattern of fossil distribution in the rocks? Because of the Signor-Lipps effect, as the place in the rock where the moment of extinction occurred (that is, the boundary) is approached, the number of fossils found (of all types) would gradually drop off. This is because the search area below the boundary is becoming smaller and smaller (as we get closer and closer to the zone of the extinction). The last occurrences of all fossils, therefore, might be some distance (and thus some amount of time) before the extinction boundary, because the rock is not simply a solid mass of fossils. With all such last occurrences at various distances from the boundary, the extinction itself would not appear to have occurred instantaneously.

Range Extensions

The ideas embodied in the Signor-Lipps effect had been known for some years before they were formally stated by Signor and Lipps. Thus, biostratigraphers have reasoned for a long time that where they found the last (most recent) occurrence of a fossil would almost certainly not represent the exact moment that the creature went extinct on earth, but probably a time somewhat before its ultimate extinction. This being true, it has been the standard practice for biostratigraphers to *extend* the known time ranges of organisms to the nearest biostratigraphic boundary (Figure 16.3).

This habit, although quite reasonable in light of what we have seen of fossil distributions in the geological record, is self-defeating when it comes to trying to reconstruct a mass extinction. Because the time ranges of the fossils have all been extended to common biostratigraphic boundaries, all disappearances have been orchestrated, as it were, to look synchronous. A paleontologist attempting to understand the dynamics of an extinction event from the published literature would learn that many extinctions took place simultaneously. We could easily conclude that a large, instantaneous, mass extinction had taken place (because everything appears to disappear suddenly)!

Would we be misled? Maybe yes, and maybe no. Because the last occurrence of a fossil is undoubtedly slightly older than the actual time that the species it represents ultimately disappeared from the face of the earth, extending the range is realistic. Extending it to the nearest biostratigraphic boundary, however, is highly

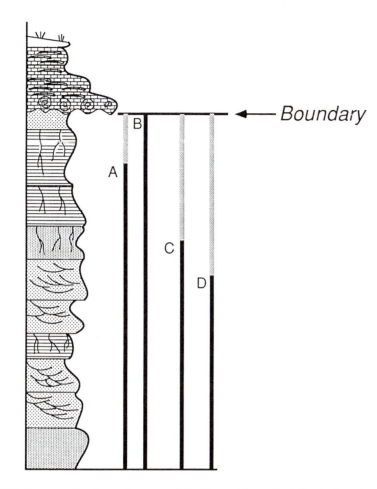

Figure 16.3 Range extensions. Because of the Signor-Lipps effect, biostratigraphers commonly extend the ranges of taxa to the nearest boundary. In this hypothetical case, the solid black lines represent the actual ranges of lineages A, B, C, and D. Only B goes to the boundary. The actual ranges of A, C, and D fall short of the boundary, but would be extended by biostratigraphers to the boundary (represented by the gray lines). Such extensions make the extinction appear catastrophic, even if it is not.

arbitrary and potentially misleading. Extending the range any other amount, however, would be equally arbitrary, and just as misleading. The convention of range extensions came about for a purpose, but uncritically accepting published stratigraphic ranges would lead one to the conclusion that a given mass extinction was far more sudden than it possibly was.

Choice of Taxonomic Level and the Problem of Incongruent Taxa

How we measure the magnitude of an extinction can have everything to do with how we perceive it. We have already seen that the Permo–Triassic extinction involved a 96% extinction at the species level but "only" a 52% extinction at the family level (52% still seems rather extreme to us!). Obviously, both of these measurements are correct; the implication is that much of the species-level diversity is bound up within a relatively few families.

On the other hand, confronted with an extinction whose severity we wish to measure, the choice of taxonomic level deserves some careful consideration. One might at first think that the most refined (lowest) taxonomic level would provide the most accurate reflection of an extinction. Indeed, it tells you in a detailed way who went extinct and who did not, but it can lead to problems. Simply put, many fossil finds cannot be identified down to the species level. We have a choice, therefore, of either not using a potential piece of data (the fossil) or using it at a higher taxonomic level.

For example, suppose you are interested in proportions of fossils in an assemblage (such as, for example, R. T. Bakker was when he formulated the predator–prey ratios; see Chapter 14). Unfortunately, preservation being what it is, you could not identify to species level every theropod fossil you found; indeed, suppose that you found 25 theropod fossils, of which you could identify only 2 below the level of Theropoda. Would it make sense to ignore the information about assemblage composition implied by 23 specimens and consider only 2 specimens? Clearly not; you could include *all* the specimens in your database by simply designating them as "theropods." On the other hand, the significance of your data will be more than slightly blurred by the use of higher taxonomic categories.

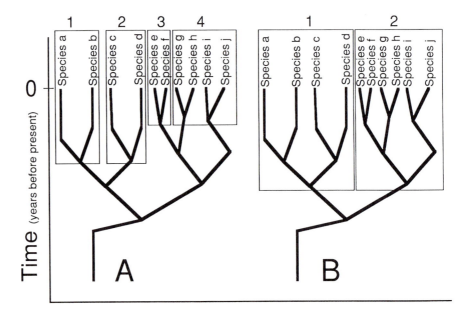

***Figure 16.4* Two hypothetical trees showing all the speciation events leading to 10 species (of course, we can never really know all the speciation events that have occurred through time).** In interpretation (A), the taxonomist has recognized four families. In interpretation (B), two families have been recognized. Because there is no absolute definition of a "family," neither interpretation is wrong; however, the taxonomic interpretation has made (A) appear more diverse than (B), even though both are interpretations of the same tree. If these families went extinct, should we say that two, or that four, families went extinct?

The use of higher taxonomic categories, however, leads to a much more complex problem. Because such higher categories – that is, categories above the species level – may not be directly comparable, it is very likely that in using higher categories for an analysis, one might be comparing apples and oranges. Such taxa may be **incongruent**: Even though they are both designated as families (or genera, or orders, or classes, or whatever), they may not truly be evolutionarily equivalent. There is no exact definition for what is meant by the the word "family" (or any other higher taxonomic category); who is to say that the groups that were used in calculating the 52% "familial" extinction at the Permo–Triassic boundary were really comparable (so that the extinction really was 52% at the *familial* level)?

We can analogize the problem this way: Suppose we are city planners and a developer wants to change local zoning from residential to commercial. We would like to know how dire the effects will be. One way to look at the problem is that 23 people will be made homeless. But if those 23 people really consist of four 5-person families (each including 3 children, and each family in a separate apartment) and 3 grandparents (each also in a separate apartment), it could be said that seven independent households will be made homeless. Suppose, however, that the fathers in two of the families are brothers, and the mothers in the other two families are sisters. The developer tells the town council, "Look, only two families will be affected by this zoning change." She is deliberately referring to *extended* families. Here, and in biology, because the term "family" (or any other taxonomic designation above the species) is not precisely defined, it takes on a variety of meanings that change its significance depending upon who the speaker is and what she hopes to convey.

The issue is really whether the higher taxa under discussion have equivalent branch points in the phylogenetic history of the organism (Figure 16.4). Because evolution consists (as we have seen) of repeated branchings of lineages, how can we be sure that the groups that we are comparing really represent exactly equivalent branch points? Because higher taxonomic categories are so poorly defined, many evolutionary biologists do not believe that taxonomic categories above the species level are useful in evolutionary reconstructions.

The argument has been made that as long as the categories are monophyletic, they can still be useful (because they represent groups that are of evolutionary significance). This may be true, but it does not negate the problems we have raised. It suggests that great care must be exercised in the way in which diversity in a mass extinction is measured, and that the assumptions that are embodied in the use of higher taxa must be articulated and well understood.

Important Readings

Donovan, S. K., ed. 1989. Mass Extinctions: Processes and Evidence. Columbia University Press, New York, 266 pp.

Gould, S. J., ed. 1993. The Book of Life. Ebury-Hutchinson, London, 256 pp.

McKinney, M. L. 1993. Evolution of Life. Prentice-Hall, Englewood Cliffs, N. J., 415 pp.

Norell, M. A., and M. J. Novacek. 1992. The fossil record and evolution: comparing cladistic and paleontologic evidence for vertebrate history. Science 255:1690–1993.

Raup, D. M. 1991. Extinction: Bad Genes or Bad Luck? W. W. Norton and Company, New York, 210 pp.

Raup, D. M., and J. J. Sepkoski. 1984. Periodicity of extinctions in the geological past. Proceedings of the National Academy of Sciences 81:801–805.

Sepkoski, J. J. 1986. Phanerozoic overview of mass extinction; pp. 277–295 *in* D. M. Raup and D. Jablonski (eds.), Patterns and Processes in the History of Life. Springer-Verlag, Berlin.

Signor, P. W., and J. H. Lipps. 1982. Sampling bias, gradual extinction patterns and catastrophes in the fossil record; pp. 291–296 *in* L. T. Silver and P. H. Schultz (eds.), Geological implications of impacts of large asteroids and comets on the Earth. Geological Society of America Special Paper no. 190.

CHAPTER 17

THE CRETACEOUS–TERTIARY EXTINCTION:
The Frill Is Gone

ONE OF THE MOST SIGNIFICANT ASPECTS of the **Cretaceous–Tertiary boundary** (commonly abbreviated **K/T**[1]) – and one that commonly surprises people when they hear of it for the first time – is that many things occurred whose magnitude and significance *far* transcended the mere extinction of some dinosaurs. Indeed, the earth was redistributing its continents into a form very much as we find them today and, apparently, a large extraterrestrial body – an asteroid – collided with the earth. Moreover, the biota underwent tremendous changes whose nature is completely beyond our own experience. Indeed, the evidence is very strong that for a period of time, the world's oceans were virtually "dead" – that the great cycles of nutrients that form the complex food webs in the oceans temporarily collapsed.

Our understanding of this extinction must incorporate all the changes that took place during this very dramatic time in earth history; to focus on a single event or group (such as dinosaurs) is missing the point. Our treatment of this extinction, therefore, will address all that is known of K/T boundary events, and any hypotheses that we propose must account for all that is known of what took place.

---

[1]The *T* in K/T obviously stands for the Tertiary Period. The *K* stands for the German word *kreidezeit,* which means "chalk time." The latest Cretaceous was first recognized at the well-known white chalk cliffs of Dover, England.

GEOLOGICAL RECORD OF THE LATEST CRETACEOUS

Tectonic Activity

The Late Cretaceous was a time of major tectonic activity, particularly in what is now (and what was then) the Pacific Basin. Such unusual activity is suggested by Late Cretaceous mountain building and volcanism. The Rocky Mountains were already undergoing their second orogenic pulse and would begin a third, and extremely important period, of growth before the Cretaceous was over. In South America, the Andean Mountain Range had begun to emerge. In Europe, the Alps were undergoing an orogenic pulse.

The driving force behind all this mountain building is thought to have been extensive sea-floor spreading in the Pacific. With spreading rates significantly enhanced over previous times, the edges of the Pacific Basin became zones of subducted oceanic crust. As the crust was thrust beneath the continent, volcanic arcs formed above subduction zones. Presently, we find the Pacific rimmed with large amounts of granite of Cretaceous age, representing the roots of the volcanos that must have been associated with the Cretaceous subduction complexes.

Along with the arc volcanism from the subducting lithosphere, intense basaltic volcanism occurred between 65 and 60 Ma (from the very end of the Cretaceous into the early Tertiary) in western and central India. This volcanic episode, called the **Deccan traps**, consisted of a series of volcanic lava flows that ultimately covered an area of $500,000$ km^2. These were not the typical kind of explosive Krakatoa/Mount Saint Helens/Pinatubo volcanism that blasts clouds of gas and glass high into the atmosphere, making eerie red sunsets visible around the globe. Rather, these were episodic flows called **flood basalts**, in which tremendous volumes of lava flowed from deep fissures in the earth's crust, forming a large lava plateau region.[2] These were the most extensive continental flood basalts of the past 200 Ma. Apparently the episode occurred over several million years; sediments interlayered between the volcanic episodes contain vertebrate and plant fossils. Some geologists have surmised that during this time interval of extended episodic volcanism, tremendous volumes of volatile gases – carbon dioxide, sulfur oxides, and possibly nitric oxides, among the most prevalent – were emitted into the atmosphere, possibly affecting global temperatures and damaging the ozone layer. Some workers have even speculated that this interval of volcanism, in combination with the generally increased explosive volcanism of the Late Cretaceous, produced an episode of acid rain caused by sulfuric, hydrochloric, and/or nitric acids. We will return to this idea again when we discuss the asteroid impact with the earth and will touch on its potential effects on the K/T extinction.

By Late Cretaceous time, the continents had assumed something close to their present distributions (see Chapters 2 and 15). Most significantly (for the distributions of dinosaurs), it appears that a land bridge connected North America and Asia, causing a remarkable similarity in the dinosaur faunas of the two regions. Africa, it appears, was an isolated continent; however, South America was connected to the Antarctica–Australia continental mass by a thin

---

[2] That windsurfing paradise, the Columbia Gorge, is formed from the Columbia River's cutting through the Columbia Plateau, a smaller, younger, North American version of the Deccan traps.

401

GEOLOGICAL RECORD OF THE LATEST CRETACEOUS

splint at the southern tip of the continent. The remnant of this splint is present-day Tierra del Fuego.

Oceans

As we have seen, the Cretaceous was a time of advancing (transgressing) and retreating (regressing) sea levels. In general, however, Cretaceous sea levels were much higher than sea levels are today. Because of this, throughout much of the Cretaceous, North America was partially or wholly bisected by the Western Interior seaway. Europe consisted of a complex series of islands and waterways. The other continental masses are less well known, but presumably all had higher sea level stands than one now finds (see Chapter 2).

The end of the Cretaceous was marked by a significant regression. It is thought that at this time, there was a temporary cessation in activity at the mid-oceanic spreading centers. When this cessation in activity occurred, the mid-oceanic ridges, which are prominent submarine topographic features during active spreading, cooled and sank, lowering their topographic elevation. The lowered topography is thought to have had the effect of increasing the volume of the ocean basins, which in turn lowered sea level.

Exactly when this regression occurred is a matter of some conjecture. Many paleontologists have claimed that it reached its maximum at 65 Ma and was therefore coincident with the extinctions. Others feel that there is good evidence that the regression actually occurred very slightly (several tens of thousands of years) before the K/T boundary. Regardless of exactly when the regression occurred, it is clear that by the end of the Cretaceous, more of the continents were exposed than had been in the previous 60 or so million years.

Climate

The latter half of the Cretaceous seems to have been a time of gentle cooling from the highs reached in the mid-Cretaceous. Although temperatures remained warm well into the Cenozoic, it is thought that by the Late Cretaceous, global mean temperatures had descended about 5°C from the mid-Cretaceous highs. Evidence for continued seasonality is present in deposits throughout the latter half of the Cretaceous.

The amount of seasonality remains in question. J. A. Wolfe, a paleobotanist with the U.S. Geological Survey, has argued – based upon plant fossils – that in North America at least, climates through the Late Cretaceous were relatively equable. Indeed, at middle paleolatitudes (51°–56° N), Wolfe believes that the mean annual temperature range was just 8°C. This is by contrast with modern conditions in approximately the same region, in which the mean annual temperature range is about 33°C. The latest Cretaceous dinosaur fauna found north of the Arctic Circle (see Chapter 14) is thought to have lived in a setting where the mean annual temperature was between 2° and 8°C. The issue of polar ice during the Late Cretaceous is unresolved to date.

Asteroid Impact

In the late 1970s, Walter Alvarez, a geologist at the University of California at Berkeley (Figure 17.1), was studying K/T marine outcrops now exposed on dry

Figure 17.1 **Scientists studying the K/T boundary in eastern Montana, U.S.A.** Left to right: palynologist R. Tschudy; geochemist C. Orth; geologist W. Alvarez; D. E. Fastovsky. Crouched on scarp: W. A. Clemens, Jr. Just below him: geochemist D. Tripplehorn. Hand in shirt pocket: paleobotanist K. R. Johnson. To his left (studying ground): planetary geologist E. Shoemaker.

land near the town of Gubbio, Italy. He was interested in learning how long it takes to deposit certain kinds of rocks. He knew that **cosmic dust** – that is, particulate matter from outer space – slowly and continuously rains down on the earth. If one knew the rate at which the stuff falls, one could determine – by how much dust is present – how much time had elapsed during the deposition of a particular body of rock.

The outcrop at Gubbio is an interesting one (Figure 17.2). The lower half of the exposure is composed of a rock made up of thin beds of the shells of microorgansims (primarily microscopic creatures called dinoflagellates, coccolithophorids, and foraminifera) from the Cretaceous Period. Abruptly, these beds come to an end and next there follows a thin (2–3 cm) layer of clay, after which there abruptly begins the upper half of the exposure, made up almost exclusively of thin beds of the shells of microorganisms from the Tertiary Period. Obviously, here was a K/T boundary, and right at the thin clay layer!

When Alvarez began his analysis for cosmic dust, he discovered something very strange: Present in the clay layer were large concentrations of the element **iridium**, a rare, platinum-group metal.[3] Iridium is normally found at the earth's surface in very low concentrations; it is found in higher concentrations in the core of the earth and from **extraterrestrial** sources (that is, from outer space). Instead of the expected amount at the earth's surface, about 0.3 parts per billion (ppb), the *iridium anomaly*, as it came to be called, was a wopping 10 ppb at Gubbio! So there was about 30 times as much iridium in that clay layer as Alvarez had expected to find (Figure 17.2).

[3]It is a common misconception that iridium metal is a deadly, toxic metal. In fact, like its chemical relatives gold and platinum, it is quite unreactive. For example, there is a costly fountain pen that can be purchased with an iridium-tipped nib!

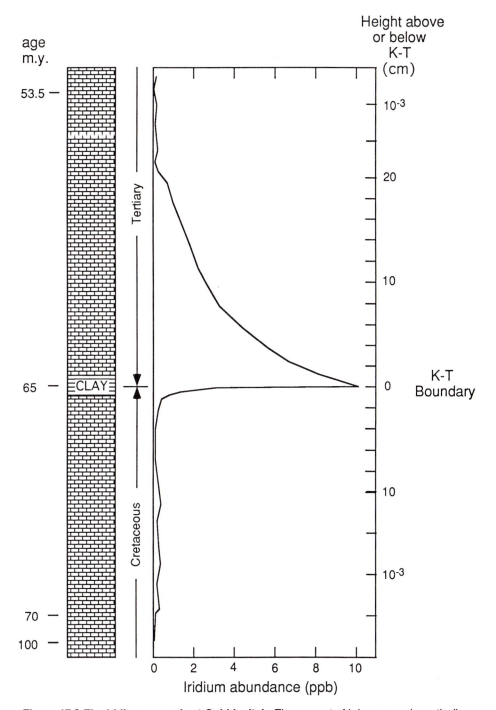

Figure 17.2 **The iridium anomaly at Gubbio, Italy.** The amount of Ir increases dramatically at the clay layer to 9 parts per billion (ppb) and then decreases gradually above it, returning to a background count of about 1 ppb. On the right are numbers representing the thickness of the rock outcrop; on the left, the time intervals and rock types are identified. Note that the vertical scale is linear close to the K/T boundary, but logarithmic away from the boundary, to show results well above and well below the boundary. After Alvarez et al., 1980.

Was the source of the anomaly extraterrestrial, or was it from the core of the earth? Discussion with Luis Alvarez, Walter's father and the Nobel Prize–winning physicist we met in the previous chapter, convinced Walter that the iridium had an extraterrestrial source. The case was sealed when the Alvarezes and two geochemist colleagues at the Lawrence Berkeley Laboratories, F. Asaro and H. Michel, found iridium anomalies at two other K/T sites, one in Denmark and the other in New Zealand. The scientists concluded that 65 Ma, an **asteroid**, a large extraterrestrial body, smacked into the earth, causing the K/T mass extinction. Luis Alvarez described the relationship between an asteroid impact and the iridium layer in this way:

> When the asteroid hit, it threw up a great cloud of dust that quickly encircled the globe. It is now seen worldwide, typically as a clay layer a few centimeters thick in which we see a relatively high concentration of the element iridium – this element is very abundant in meteorites, and presumably in asteroids, but is very rare on Earth. The evidence that we have is largely from chemical analyses of the material in this clay layer. In fact, meteoric iridium content is more than that of crustal material by nearly a factor of 10^4. So, if something does hit the Earth from the outside, you can detect it because of this great enhancement. Iridium is depleted in the Earth's crust relative to normal solar system material because when the Earth heated up [during its formation] and the molten iron sank to form the core, it "scrubbed out" [i.e., removed] the platinum group elements in an alloying process and took them "downstairs" [to the core]. (L. W. Alvarez, 1983, "Experimental evidence that an asteroid impact led to the extinction of many species 65 Myr ago," *Proceedings of the National Academy of Sciences* 80:627)

Based upon what was inferred to be a worldwide distribution (from the three sites sampled), the asteroid was estimated at about 10 km (about 6 miles) in diameter.

Rarely has a scientific idea provoked more controversy. Resistance came in two forms: People challenged the conclusion that an asteroid hit at all, and others questioned what the work of the Alvarezes and their colleagues had to do with extinctions in general and dinosaurs in particular.

In the years since 1980, a tremendous amount of work has been done. Most important, the number of K/T sites with anomalous concentrations of iridium at the boundary has continued to climb, until today well over one hundred K/T boundary sites around the world with anomalous concentrations of iridium have been found (Figure 17.3). In fact, as the Alvarezes continued to update a well-known diagram showing iridium-rich K/T localities on a world map, W. A. Clemens, a distinguished University of California at Berkeley paleontologist, wryly likened the diagram to the ubiquitous signs outside McDonald's restaurants: "Over 100 iridium anomalies served!"

With the increase in anomalies came an increase in understanding of background levels of iridium throughout the world. A particularly important demonstration of the ubiquity of the iridium anomaly came from U.S. Geological Survey mineralogist B. Bohor and Los Alamos National Laboratories geochemist C. Orth and their colleagues, who discovered that the iridium anomaly occurred on land as well as in oceanic sediments, when in 1983 they discovered the anomaly in New

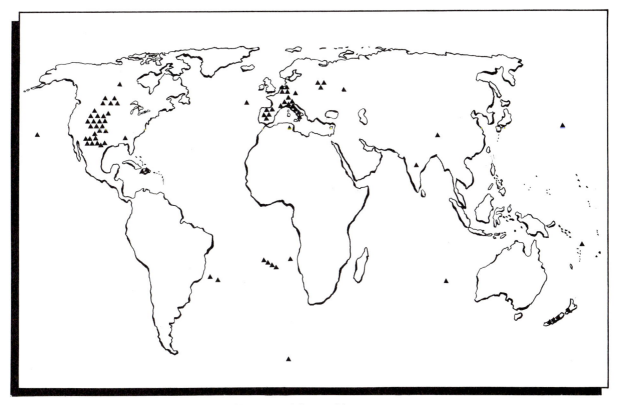

Figure 17.3 The 103 known iridium anomalies around the world.

Mexico and in eastern Montana (and later, in North Dakota, Wyoming, and southern Canada; Figure 17.4). Not only was the Montana site right in the middle of dinosaur-bearing sedimentary rocks, but both terrestrial sites had two new and significant clues about the origin of the iridium: **shocked quartz** and **microtektites**.

Shocked quartz is the name given to quartz that has been placed under such pressure that the crystal lattice (a repeating sequence of atoms), which is normally organized in planes at 90° to each other, becomes compressed and deformed (Figure 17.5). Earlier, U.S. Geological Survey geologist E. Shoemaker had conclusively demonstrated on the much younger Meteor Crater in Arizona that compressed features of this type (properly termed "shock metamorphism") are diagnostic of the tremendous pressures that occur when a meteor or asteroid smashes into the earth.

Microtektites are small, droplet-shaped blobs of silica-rich glass. They are believed to represent material from meteors that is thrown up into the atmosphere in a molten state due to the tremendous energy released when a meteor strikes the earth. They become solid when cooling while still airborne and then fall back to earth. They, too, provided support that an impact had occurred at the K/T boundary.

The global presence of anomalous concentrations of iridium at the K/T boundary in sediments from both the land and the sea, and the presence of

Figure 17.4 **The iridium-bearing clay layer in Montana**, one of the first localities on land where anomalous concentrations of iridium were discovered.

Figure 17.5 **Shocked quartz.** The angled lines across the face of a grain of quartz represent a failure of the crystal lattice along known crystallographic directions within the mineral. Grain is 70 microns (7 x 10^{-5} m) across.
Photograph courtesy of B. F. Bohor, U.S. Geological Survey.

shocked quartz and microtektites in K/T boundary deposits, seemed a clear indication that an asteroid or some other impacting event had occurred 65 Ma. And most geologists were impressed by the rapidly accumulating body of evidence.

Volcanism

But was any other explanation possible for the coincidence of these apparently unusual features? A number of scientists, notably geochemist C. B. Officer of Dartmouth College, were unconvinced by the asteroid scenario. They developed serious reservations about the Alvarez interpretation of the data and instead postulated that it was actually volcanism that caused the features observed worldwide at the K/T boundary.

In 1983, volcanologist W. H. Zoller and colleagues observed that material being ejected from the vent of the Hawaiian volcano Kilauea was rich in iridium (which had evidently come from the core by way of the mantle). Moreover, N. Carter of Texas A&M University and colleagues demonstrated that shock metamorphism can occur through volcanic processes. To add fuel to the fire, it was claimed by a research group in New Zealand that the microtektites in the New Zealand K/T section first studied by the Alvarez group were actually very prosaic earthbound weathering features, and not snazzy impact-derived glass droplets. Recall that the latest Cretaceous was a time of extensive volcanic activity. Officer wondered if the K/T iridium could have originated from volcanism, like that observed in Kilauea.

Then, there was the issue of the lack of an impact crater. Many geologists were not terribly bothered by the fact that no impact crater had been found. After all, they reasoned, had the asteroid hit the ocean (and thus struck oceanic crust), there was a good chance that the evidence would have been subducted and destroyed. That was the simplest answer, but the presence of shocked quartz didn't jibe well with an impact with oceanic crust; such material is very quartz poor. If a continental impact was indicated, then where might it have been? A variety of candidates were served up: a buried enigmatic crater in Iowa ominously named the Manson crater, the entire island of Iceland, the Deccan flood basalts in India, and a mysterious putative impact crater in Siberia. All of these options ran into trouble. The Manson crater (if it is indeed an impact crater at all) is too small (about 35 km in diameter) and too old. Iceland is too young, there is no evidence that Deccan trap volcanism was provoked by an asteroid impact, and the Siberian crater remains largely unexplored.

Smoking Gun

Recent discoveries suggest that volcanism is not a terribly satisfactory explanation for the features observed at the K/T boundary. It has been noted that the Deccan traps produced flood basalts, rather than the kind of explosive event necessary to blast material into the stratosphere so that it could be distributed globally. Moreover, a variety of authorities, including Shoemaker and U.S. Geological Survey igneous petrologist G. Izett, concluded that the shock metamorphic features produced by volcanos are different from those produced by impacts.

But the most significant work has come out of the Caribbean Sea. There, a sequence of glass approximately 1 m thick was discovered, suggesting that the source of the glass had to be somewhere relatively nearby in the Caribbean region in Haiti. The chemical composition of the Haitian glass was shown by University of Rhode Island volcanologist H. Sigurdsson and colleagues to be unlike the glass produced by any known volcano, past or present. Instead, it was rich in sulfur, suggesting that it came from a combination of molten continental crust and evaporites; that is, a carbonate salt precipitated from water by evaporation.

The pieces of the puzzle really started falling into place. Evaporite deposits in the Caribbean region were indicated as the probable location of the impact itself. Still, no crater had been found – or had it? In 1981, geophysicist G. Penfield had reported on a bowl-shaped structure, 180 km in diameter, that he believed to be an impact crater. The structure is located in the Yucatan Peninsula of Mexico, beneath the town of Chicxulub (Figure 17.6). Like the Manson crater, the Chicxulub structure (translated approximately as "devil's tail") is buried under many meters of more recent sediments. Still, drill cores taken through the structure and reported in 1991 by geologist A. Hildebrand and colleagues, then at the University of Arizona, revealed shocked quartz. Moreover, the cores indicated that the structure was formed within a sequence of 1–3 km of evaporitic deposits, as well as some continental crust.

The continental crust and evaporitic deposits geochemically matched the thick glass deposits in Haiti. Moreover, the location was right: close enough to Haiti to produce the thick glass sequence, and close enough to Montana, North Dakota, Wyoming, and southern Canada to blast sand-sized particles to the center of the North American continent. Indeed, earlier, University of Washington sedimentologist J. Bourgeois and colleagues had reported on evidence of "tidal

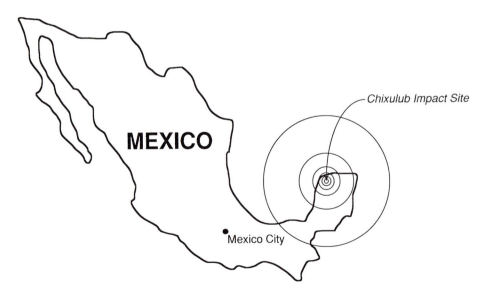

Figure 17.6 Bullseye indicating location of Chicxulub site on the Yucatan Peninsula, Mexico. Evidence is accumulating that the K/T asteroid landed here.

wave" (tsunami) deposits in K/T deposits in the Gulf Coast region of Texas. The Chicxulub site was well situated to produce the tsunami deposits Bourgeois recognized in the sedimentary record. Finally, convincing evidence for the viability of the Chicxulub crater came when the impact that produced it was dated at 65 Ma!

Officer published (and continues to publish) a welter of papers attributing the observations at K/T boundaries around the world to volcanism and not to an asteroid impact. However, as of late 1995, very few geologists would deny that an asteroid impact occurred at the K/T boundary. The evidence appears overwhelming, and the issue is now one of how things were affected after the asteroid arrived. A variety of different scenarios for changes in the physical environment have been proposed, some of which we briefly review here. Most of these scenarios, because of their brevity, would leave little or no trace in the geologic record and thus are speculation based upon sophisticated modeling by computer. Some, you may note, are even contradictory!

BLOCKAGE OF SUNLIGHT. Based upon calculations of the size of the asteroid and of the amount and size of material blasted into the atmosphere, it is generally felt that sunlight would have been blocked for about 3 months. Such estimates are based upon the amount of time that fine particulate material has stayed in the air and the distance that such material has dispersed during modern explosive volcanic events such as Krakatoa, Mount Saint Helens, and Pinatubo. For example, in the case of the Krakatoa eruption, it is well documented that sunsets over London, England, were drastically affected for almost two years by suspended particulate matter that was carried halfway around the globe from the Straits of Sunda in the South Pacific where the volcano was located.

SHORT-TERM GLOBAL WARMING. It is believed that initially, tremendous amounts of energy in the form of heat must have been released upon impact. Such heat release might have caused global temperatures to rise as much as 30 °C for as long as 30 days. The initial global heat release at ground zero might have been 50 to 150 times the energy of the sun as it normally strikes the earth. One group of scientists likened this radiation of heat energy to an oven left on broil.

GLOBAL WILDFIRES. With so much instantaneous heat production, fires might have broken out spontaneously around the globe. Geochemist W. Wolbach of the University of Chicago and a variety of colleagues described soot-rich horizons from five K/T sites in Europe and New Zealand. The amount of the element carbon (in the form of soot) was enriched between 100 and 10,000 times over background, a fact which they attributed to wildfires that overtook the globe as a result of the intensive heat release associated with the impact. Because the soot coincides with the impact layer, they believe that the fires began while the impact ejecta were still airborne.

ACID RAIN. In 1987, R. G. Prinn and B. Fegley of the Massachusetts Institute of Technology proposed that the asteroid impact would have produced acid rain. The idea went approximately as follows: The atmosphere would be rapidly heated by the plume of hot water vapor and rock material thrown up by the impact; a complex series of reactions would ensue, possibly producing nitrous oxide in the atmosphere, which would in turn produce a nitric acid rain. The phenomenon is known: Thermonuclear explosions, lightning, and modern, smaller meteors have all been shown to produce nitrous and nitric acids in the atmosphere. Moreover, because of the evaporites at Chicxulub, some scientists have speculated that large amounts of sulfuric acid could have been produced by the asteroid impact. The $64,000 question is whether or not any of this occurred at the K/T boundary. The University of Rhode Island's S. L. D'Hondt and colleagues, reviewing published estimates of sulfuric and nitric acids potentially produced by the K/T impact, concluded that not enough acid rain would have been produced to cause the extinctions.

LONG-TERM GLOBAL COOLING.   It is hypothesized that after the initial atmospheric heating and the various horrors that it may have visited upon the earth, the vast amount of particulate matter presumed to have been in the atmosphere must have blocked sunlight and caused global cooling. Indeed, even dust from the comparatively minuscule eruption of Krakatoa caused a global decrease in temperature of about 1°C. Estimates vary (of course!) as to how much cooling the earth experienced at the K/T boundary. An extreme estimate suggests that light levels at which photosynthesis could not take place would have occurred for several months and, for a shorter time, the light would drop below the level necessary for animals to see. The heat transfer properties of water (see Chapter 2) would mitigate the extreme effects of the atmospheric cooling in the oceans, but far more severe effects could be experienced on land. Deep-freezing scenarios engendered by this type of reasoning were instrumental in focusing thought on a more timely problem: nuclear winter.

LONG-TERM GLOBAL WARMING.   Equally plausible is the possibility of an enhanced "greenhouse effect," causing global warming. In this scenario, the extermination of much of the world's biomass decreases the amount of organic carbon taken up by the biota. This, in turn, would produce more CO_2 in the atmosphere, which would absorb heat energy radiated from the earth, and *voila*! – "greenhouse" conditions are induced.

The problem with all of these hypotheses of climate modification is that the geological record doesn't really bear them out. For example, based upon fossil plants, Wolfe has estimated that North American terrestrial deposits underwent an increase in the mean annual temperature of approximately 10 °C in the first 500,000 to 1,000,000 years after the K/T boundary. The ocean record is more equivocal, however. Some records, based upon stable isotopes, indicate a slight warming after the boundary; others do not. The entire problem can be explained away by arguing that it is again a question of resolution: 3 months or even 10 years of an unusually cool or unusually hot climate simply can't be recognized 65 million years later. Though this may be so, we must keep in mind that the very recognition of the clay layer as being impact-generated requires the recognition of a very brief instant in time 65 million years ago, an instant far briefer than 10 years or even 3 months.

BIOLOGICAL RECORD OF THE LATEST CRETACEOUS

No amount of comets, volcanos, asteroids, meteors, cooling, warming, ice ages, or natural catastrophes can be used to explain any extinction until we understand the extinction itself. Obviously, some organisms must have gotten through the K/T boundary, or we wouldn't be here today. So, the pattern of **selectivity** – that is, who survived the extinction and who did not – becomes an important issue in understanding an extinction and determining its possible cause. We divide the biota into three crude ecological categories: oceanic, aerial, and terrestrial. Finally, because this is a book about dinosaurs, we devote a special section to their disappearance (although in doing so we imply an ecological importance for them that they never had).

A STRATIGRAPHIC NOTE OF CAUTION. It is the *absence* of the Cretaceous fossils and the appearance of the Tertiary fossils that identify the K/T boundary. Suppose that the extinctions were actually **isochronous** (*iso* – same; *chronos* – time); that is, they occurred at the same time. Then, everywhere we observed an extinction, we would know that we were looking at the same moment in time. Now, suppose that the extinctions were actually **diachronous** (*dia* – separate; *chronos* – time); that is, they occurred at different moments in time. Now the K/T boundary would actually occur in one place at one time and in another place at another time. Using only fossils, we'd never know.

So what is needed here is an independent means of assessing time; that is, a means apart from the fossils, against which the stratigraphic distributions of the fossils can be compared. Most people immediately think of radiometric dates (see Chapter 2). These, when applicable, show that the K/T boundary is isochronous worldwide to within about 500,000 years. This means that the record of extinctions on land can be compared with that in the oceans, as long as it is understood that there is a half-million-year margin for error. Five hundred thousand years is a long time ecologically and a lot can happen during that time that doesn't involve earth-shattering, heart-palpitating global catastrophes.

Is there a more refined way, *independent* of fossils, of identifying the K/T boundary globally? The answer is, yes, and here is where the iridium anomaly assumes a scientific importance beyond its catastrophic allure. *If the iridium anomaly was caused in the manner discussed, then it represents a global isochron and, where it is preserved, identifies – precisely and exactly, and independently of fossils – that moment in time 65 Ma that we call the K/T boundary.* If our ideas about its origin are correct (and the evidence strongly inclines us to believe that they are), then anywhere one finds the iridium layer, she is looking at approximately the same 3 months 65 Ma ago. Considered in this way, the iridium anomaly is the most significant stratigraphic marker in all earth history. Of course, because the resolution and global distribution of no other stratigraphic marker comes close to that of the iridium anomaly, its isochroneity can never be independently tested.

Oceans

Much of what is known of the marine fossil record is based upon epeiric sea deposits, upon shelf sediments that have been preserved on land so that they can

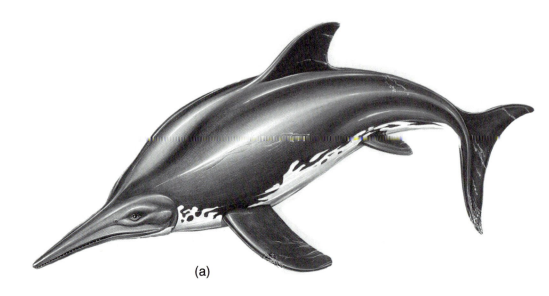

(a)

(b)

(c)

Figure 17.7 **Some Cretaceous marine creatures**.
(a) An ichthyosaur; **(b)** a mosasaur; and **(c)** the shell of
an ammonite (*Acanthoscaphites*). *Mosasaur and Loon*
© *1989 William Stout.*

be studied, and upon drill cores, which sample oceanic sediment throughout the world's oceans. Each of these types of deposits can provide important data on marine extinctions, but none provides a comprehensive global record.

EPEIRIC SEAS AND CONTINENTAL SHELVES. As the seas receded during the last 2–3 million years of the Cretaceous, the large expanses of the continents that were inundated by ocean waters became exposed. Because many groups lived and died in epeiric seas, the absence of seas during this critical time interval means that we don't have data for many important marine organisms.

Among vertebrates, the patterns of extinction and of survivorship of marine teleost fishes, for example, are poorly understood. How well or how badly these organisms fared remains largely conjectural, although it is apparent that as a group, they did not suffer the kind of wholesale decimation seen in other groups. Not so in the case of plesiosaurs, the long-necked, Loch Ness–type fish-eaters of the Jurassic and Cretaceous that appear to have been a group waning in abundance and diversity by the end of the Cretaceous. It is generally thought that the dolphinlike, marine ichthyosaurs (Figure 17.7a) disappeared well before the K/T boundary.

The Late Cretaceous was a time of the appearance and radiation of an important group of marine-adapted lizards called mosasaurs (Figure 17.7b). What is known suggests that as a group, mosasaurs thrived until the end of the Cretaceous.

Among fossil invertebrates, perhaps the most famous group is the ammonites (Figure 17.7c). These were free-swimming carnivorous mollusks, distantly related to the living chambered *Nautilus*, and they were common creatures during much of the Paleozoic and all of the Mesozoic. It was originally thought that ammonites as a group died out before the end of the Cretaceous; however, recent study by University of Washington paleontologist P. Ward has demonstrated that a few ammonites lived right up to the K/T boundary before going extinct.

Another important group of invertebrates is the bivalves. Obviously these did not all die out; there are many familiar bivalves today, including mussels, pectens (best known on the Shell Oil logo), and oysters (the gullible victims of the "Walrus and the Carpenter"). During the Cretaceous, bivalves occupied the same important position in the ecosystem that they occupy today, although there was a somewhat different cast of characters. Studies by University of Chicago paleontologists D. M. Raup and D. Jablonski have shown that 63% of all bivalves went extinct sometime within the last 10 million years of the Cretaceous. The record is unfortunately not more precise than this, but it does show that the extinction took place without regard for latitude: Bivalves in temperate regions were just as likely to go extinct as those in the tropics.

Recently, bivalves and ammonites have been found in a K/T sequence preserved on Seymour Island, Antarctica. Here, however, there have been problems developing a reliable stratigraphy, and the degree of reworking is unclear. Scientists studying these deposits, however, think that a gradual extinction pattern may be indicated for these high-latitude southern faunas.

FORAMINIFERA. The record of foraminifera – single-celled, shell-bearing marine organisms that are either **planktonic** (living in the water column) or **benthic** (living within sediments) – has dominated discussions of K/T boundary events (see Figure 2.16). Foraminifera are numerous, have high environmental specificity (that is, different types live in distinctly different environments), and they evolved relatively rapidly – all of which make them important marine biostratigraphic indicators. The pattern of their extinction has been the subject of contentious debate. Many paleontologists studying foraminifera, including J. Smit of the Netherlands' *Geologisch Instituut*, B. Huber of the National Museum of Natural History (Smithsonian Institution), R. Olsson of Princeton University, and D'Hondt of the University of Rhode Island, have argued persuasively, since as early as the late 1970s (and in many studies thereafter), that the planktonic foraminifera extinction was abrupt, with only a few species of foraminifera crossing the boundary into the Paleocene. G. Keller and N. McLeod, of Princeton University, do not agree. At sites that they studied, they recognize a series of extinctions before the boundary and Cretaceous survivors after the boundary. Huber, D'Hondt, Olsson, and their associates believe that the "survivors" are actually reworked (and therefore Cretaceous in age); they see a major extinction of planktonic foraminifera at the K/T boundary, with only a few lineages surviving into the Tertiary.

One of the most exciting results came from a series of studies done by paleoceanographers J. C. Zachos, M. A. Arthur, and W. E. Dean. Zachos and his colleagues studied stable isotopes of foraminifera with an eye toward monitoring **primary productivity** across the K/T boundary. Primary productivity is the amount of organic matter synthesized by organisms from inorganic materials and sunlight. Obviously, it is the base of both the oceanic and the terrestrial food chains, and with oceans covering 75% of the earth's surface (or even more during the many high sea levels experienced during earth history), it would not be an exaggeration to state that the earth's ecosystems are largely dependent upon oceanic primary productivity.

Today, oceanic productivity can be measured and characterized by the $^{13}C/^{12}C$ ratio between surface and deep waters. Such a ratio can be obtained in ancient sediments by comparing the isotopic signatures of planktonic foraminifera with those of benthic foraminifera. What Zachos, Arthur, and Dean observed was truly astounding: At the K/T boundary itself, a "rapid and complete breakdown" in primary productivity occurred. They inferred that at the K/T boundary, levels of primary productivity precipitously dropped to less than 10% of what they had been and, for the succeeding 1.5 million years, remained at levels well below those preceding the original drop in primary productivity. They called this nearly dead ocean a "Strangelove ocean" after Peter Sellers' black comedy character Dr. Strangelove, a brittle, grotesque, parodic character clearly unconcerned about a scorched, postnuclear world.[4]

Terrestrial Record

Virtually all of what we know of the K/T boundary on land also comes from the Western Interior of North America (Figure 17.8). Many of the conclusions that we have reached on the basis of the North American land record could be overturned if complete, or nearly complete, terrestrial K/T sequences are found and studied elsewhere.

---

[4]The name was actually coined by oceanographers W. S. Broecker and T. H. Peng, who in 1982 (years before the work of Zachos and colleagues) speculated that the oceans after an asteroid impact would be a bit like the earth after Dr. Strangelove.

Figure 17.8 The K/T boundary in eastern Montana. The boundary is midway up the butte, at the dark band. The rocks below the band are Cretaceous sediments; above, the layered rocks are Tertiary sediments.

PLANTS. The plant record in the Western Interior has two major components, a **palynoflora** (spores and pollen) and a **megaflora** (the visible remains of plants, especially leaves; see Figure 17.9). Both records agree nicely with each other, indicating that a major extinction occurred geologically instantaneously at the K/T boundary.

The pollen record, as studied by D. J. Nichols, F. Fleming, and R. J. Tschudy of the U.S. Geological Survey, and by A. Sweet of the Canadian Geological Survey, shows a distinctive pattern. Precisely coincident with the iridium anomaly in Montana, North Dakota, and Wyoming, Cretaceous pollen species go extinct. Pollen that is typical of early Paleocene time does not immediately follow, however. Instead, as documented by Wolfe and G. R. Upchurch (then both at the U.S. Geological Survey), there is a high concentration of fern spores just after the iridium anomaly, suggesting that immediately after the extinction of the Cretaceous plants (represented by the pollen), there was a "bloom" of fern growth, which in time gave way to a more diverse and traditional flora of flowering plants in the early Paleocene.

The situation reported by Sweet and his colleagues in southern and central Canada is somewhat different. These researchers have been able to track a series of five major changes in pollen composition across the boundary at a variety of latitudes. They correlated these changes to latitudinally related climate changes, as well as to the extinctions. Sweet and his colleagues concluded that no single extinction occurred at the K/T boundary, but rather that a series of extinction events occurred before, during, and after the K/T boundary. These results were

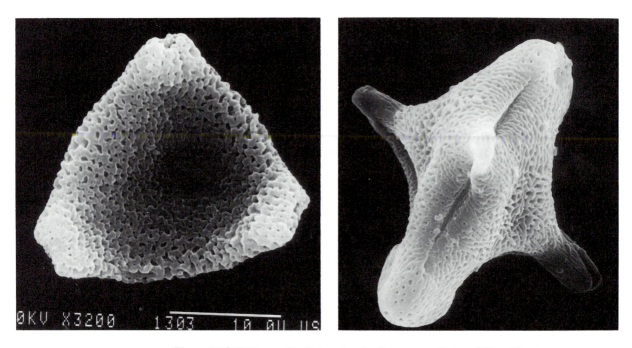

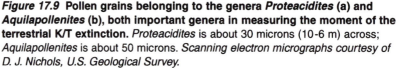

Figure 17.9 **Pollen grains belonging to the genera** *Proteacidites* **(a) and** *Aquilapollenites* **(b), both important genera in measuring the moment of the terrestrial K/T extinction.** *Proteacidites* is about 30 microns (10-6 m) across; *Aquilapollenites* is about 50 microns. *Scanning electron micrographs courtesy of D. J. Nichols, U.S. Geological Survey.*

echoed by those of the Canadian Geological Survey's E. McIver. Her comprehensive study of plant remains across the K/T boundary in Canada suggests that the plants were thriving in frost-free environments, that droughts were local phenomena causing local extinctions, and that the K/T extinctions, when they occurred, involved a "sequential," rather than "instantaneous," decrease in diversity.

The megafloral record has most recently been carefully documented by K. R. Johnson and L. J. Hickey, at that time both at Yale University. Basing their study upon 25,000 plant specimens, they found that although some environmental changes caused extinctions earlier than the K/T boundary, a major and significant extinction took place precisely at the boundary. This correlated exactly with the pollen extinction and iridium anomaly. They observed an extinction of 79% of the dicot flowering plant taxa, suggesting that (as suspected) Wolfe and Upchurch's fern "bloom" may have been a response to the absence of flowering plants that would normally have occupied the ecosystem.

On the other hand, an interesting but very preliminary record from a high-latitude flora in New Zealand indicates little evidence of the major megafloral extinctions that have characterized the Western Interior of North America. These data may show some of the same patterns obtained from the bivalves at high southern latitudes.

VERTEBRATES. Thanks to the efforts of W. A. Clemens of the University of California at Berkeley and his field crews in eastern Montana, who have been collecting vertebrates for almost 30 years, the fates of vertebrates in eastern Montana are as well known as any in the world. Of particular interest is the work of two researchers associated with Clemens: J. D. Archibald and L. J. Bryant. Archibald and Bryant made a thorough study of 150,000 vertebrate specimens in the region, asking the simple question: Of the vertebrates living in eastern Montana up to 3 million years before the K/T boundary, who survived the boundary and who did not?

Their initial study revealed differential survivorship across the K/T boundary, and subsequent review of the data by P. M. Sheehan (Milwaukee Public Museum) and D. E. Fastovsky (University of Rhode Island), as well as by Archibald himself, reinforced a striking survivorship pattern: Those organisms that lived in aquatic environments (i.e., rivers and lakes) showed up to 90% survival, whereas those organisms living on land showed as little as a 10% survivorship (Figure 17.10). Thus, the extinction seems not to have drastically affected aquatic organisms such as fishes, turtles, crocodiles, and amphibians, but apparently wreaked havoc among terrestrial organisms such as mammals and, of course, dinosaurs.

DINOSAURS. One of the first modern studies on patterns of dinosaur extinction was performed in 1986 by University of Minnesota paleontologist R. E. Sloan and colleagues. They attempted to integrate data collected by them from a place called Bug Creek, Montana, with other North American data on dinosaur diversity. They concluded:

> Dinosaur extinction in Montana, Alberta, and Wyoming was a gradual process that began 7 million years before the end of the Cretaceous and accelerated rapidly in the final 0.3 million years of the Cretaceous, during the interval of apparent competition from rapidly evolving immigrant ungulates [hoofed mammals]. This interval involves rapid reduction in both diversity and population density of dinosaurs. (R. E. Sloan et al., 1986, "Gradual dinosaur extinction and simultaneous ungulate radiation in the Hell Creek Formation," *Science* 232:629)

The work provoked interest and, of course, criticism. Some scientists claimed that in a statistical sense, the data presented did not justify the conclusions. Others claimed that geological considerations were not taken into account. Still others claimed that the sampling was not performed properly. The study was important, however, because it represented a first attempt to document with precision events just preceding the K/T boundary. Future studies would have to incorporate a sophistication never before achieved with dinosaur populations.

What are some of the problems with reconstructing changes in dinosaur populations over time? For one thing, dinosaurs are, by comparison with foraminifera for example, large beasts and, more important, are not particularly common. For this reason, the possibility of developing a 150,000-specimen database is impractical, and statistically rigorous studies of dinosaur populations are very hard to carry out. Just counting dinosaurs can be difficult. Mostly, one doesn't find complete specimens, and adjustments have to be made. For example, dinosaurs have many pieces: If you happen to find three vertebrae at a particular site, they might be from one, or two, or three individuals. The only way to be sure is to find them

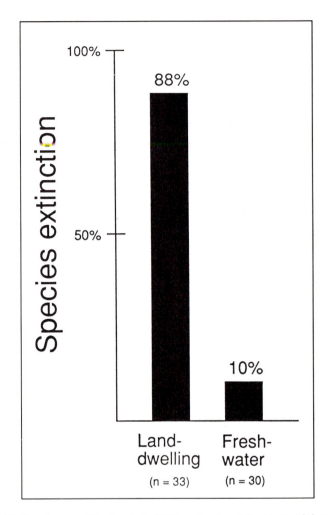

Figure 17.10 **Sheehan and Fastovsky's 1992 estimate of the terrestrial extinction at the K/T boundary in eastern Montana.** According to their view, only 10% of aquatic vertebrates went extinct and 88% of land-dwelling vertebrates went extinct.

articulated. Suppose they are not; then one must speak of "minimum numbers of individuals," in which case the three vertebrae would be said to represent one individual: That would be the minimum number of individual dinosaurs that could have produced the three vertebrae. On the other hand, if one found two left femora, then the minimum number of individuals represented would be two!

It would be nice to use in a study all the specimens that have been collected in the last 170 years of dinosaur studies. Unfortunately, we have already encountered the limitations of using museum collections for ancient ecological reconstructions (Chapter 14). So any study that really is designed to get an accurate census of dinosaur abundance or diversity at the end of the Cretaceous must begin by counting specimens in the field, which is a labor-, time-, and cost-intensive proposition.

Then, of course, the taxonomic level at which one chooses to count dinosaurs can create problems. Recall from Chapter 16 the problem of incongruent taxonomic levels. Suppose that two specimens are found; one is clearly a hadrosaurid, and the other is an indeterminate ornithischian. The indeterminate ornithischian might be a hadrosaurid, in which case we should count two hadrosaurids. But then again it might not (because its identity is indeterminate), in which case calling it a hadrosaurid would give us more hadrosaurids in our survey than actually existed. Calling both specimens "ornithischians" is quite correct, but not very informative, if we hope to track the survivorship patterns of *different types* of dinosaurs.

Finally, within the sediments themselves, problems of correlation exist. Suppose that we record the last (highest) dinosaur in the Jordan, Montana, area and then record the last (highest) dinosaur in the Glendive, Montana, area, about 150 km away from Jordan. Can these two dinosaurs be said to have died at the same time? How could one possibly know? Suppose that in fact these dinosaurs died 200 years apart. Recognizing a particular 200-year interval from a vantage point of 65 million years is virtually impossible. Yet, 200 years is a long time when one is considering an instantaneous global catastrophe.

In 1986, Sheehan began assembling a team of specialists to study the end-of-Cretaceous diversity of dinosaurs in eastern Montana and western North Dakota. The three-year study consisted of a quantitative census of dinosaur diversity during the last 2.2 million years of the Cretaceous. By highly controlled methods of data acquisition, Sheehan and his colleagues were able to show that within 250,000 years of the K/T boundary, the relative proportions of the total dinosaur population represented by the eight families of dinosaurs had not changed. This meant that the percentage of the overall dinosaur fauna occupied by each of the eight families was stable through the last 2.2 million years of the Cretaceous. Any change, if it occurred at all, had to have occurred sometime during the last 250,000 years of the Cretaceous. The study, unfortunately, could not distinguish whether the extinction, when it came, took every day of that last 250,000 years of the Cretaceous, whether it took only the last 125,000 years of that last 250,000 years of the Cretaceous, whether it took 1,000 years of that last 250,000 years, or whether it took only the last minute of that last 250,000 years. What was significant about this study, however, was that here, for the first time, were statistically rigorous data strongly suggesting that the end of the dinosaurs was *geologically* instantaneous.

The study by Sheehan and colleagues drew no less criticism than did that of Sloan and his colleagues. In the case of the Sheehan study, the major criticism was directed at the use of family-level data. Several paleontologists suggested that although family-level diversity apparently did not change, *generic*-level diversity might have changed. The idea, however, that generic-level diversity does show a decrease in the closing years of the Cretaceous in locales other than the Western Interior of North America is presently being studied by P. Dodson, who has hypothesized that a worldwide decrease in generic-level diversity may not be reflected in North America – that North America was a kind of last holdout for the dinosaurs, which, as a group, were waning globally.

Our view is that the very limited data from the Western Interior of the United States strongly imply an abrupt end for the non-avian dinosaurs. Only time and much further study will enable us to integrate dinosaur-bearing localities from around the world into what is already known of North America.

THE EXTINCTION HYPOTHESES

Much – indeed, most – of what has been proposed to explain the extinction of the dinosaurs does not even possess the basic prerequisites for a viable scientific theory. Recall from Chapter 3 that scientific hypotheses must measure up to two minimum criteria: They must be testable and they must explain as many events as possible.

   1. The hypothesis must be testable. As we have seen (Chapter 3), for a hypothesis to be considered in a scientific context, it must have predictable, or observable, consequences. Without testability, there is no way to falsify a hypothesis, and in the absence of falsifiability, we are then considering belief systems rather than scientific hypotheses. If an event occurred and left no traces that could be observed (by whatever means available), science is simply not an appropriate means by which to investigate the event. And even though science cannot be used to identify the event, the event may still have taken place.

   2. The hypothesis must explain as many events as possible. This criterion is rooted in the principle of parsimony (again, see Chapter 3). We would like to explain an event or series of events. If each step of the event (or events) requires an additional ad hoc explanation, our hypotheses lose strength. The explanation that explains the most observations is to be preferred. For this reason, if we can explain all that we observe at the K/T boundary with a single hypothesis, we have produced the most parsimonious hypothesis. As scientists, we prefer the most parsimonious hypothesis, but it should be remembered that nature is not always most parsimonious. However true, that fact is not one that can be addressed by scientific means, and thus scientists always opt for the most parsimonious hypothesis: that hypothesis that explains the most observations.

Extinction Hypotheses

In Table 17.1 we present about 80 years of published proposals designed to explain the extinction of the dinosaurs. The majority were published within the last 30 years. This exhaustive list is abstracted from that published in 1990 by English paleontologist M. J. Benton. All have been published as serious contributions (but see Box 17.1), and many have been given far greater credence than was their due. *Remember, however, any valid hypothesis (or valid hypotheses) must adhere to the two criteria listed previously in order to qualify as a scientific explanation of K/T events!*

   With that stipulation, it is clear that most of what has been proposed for the dinosaur extinction is really untestable. Many of the ideas – parasites, blindness, epidemics – fail for leaving no visible trace: They are not within the realm of science to address. Beyond this, however, virtually all of these hypotheses do not explain all, or even most, of what is known of K/T events. Most (such as mammals' causing the extinction by eating dinosaur eggs) are dinosaur specific, as if dinosaurs – and dinosaurs alone – were the key components in the K/T boundary extinctions. Any theory that purports to explain K/T events in a meaningful way must also explain these other events.

With that in mind, the hypothesis that an asteroid impact caused the events at the K/T boundary becomes a much more interesting and plausible hypothesis. In the following, we consider the asteroid impact hypothesis in the context of the criteria outlined previously.

DOES THE IDEA THAT AN ASTEROID IMPACT CAUSED THE K/T EXTINCTIONS HAVE PREDICTABLE CONSEQUENCES? Clearly, it does. First, if the asteroid was a global event, evidence for it should be visible globally. After 15 years of research, the evidence for global influence of the K/T boundary asteroid impact is overwhelming (Figure 17.3). That being the case, what kind of predictable consequences are there in terms of the extinction?

In the case of the bivalves, the fact that the extinctions took place regardless of latitude is strong evidence that those extinctions were due to a global effect that was apparently unrelated to climate. Had climate been involved as a causal agent, one might expect to see latitudinal changes in the patterns of extinction, but such is not the case.

Another type of evidence comes in the form of the rate at which the extinctions took place. If the asteroid really caused the extinction, the event should have been what Clemens called a "short, sharp shock."[5] There should be no evidence of biotic abundance and diversity dwindling in the years preceding the boundary. Rather, the fossil record should indicate that it was "business as usual" for most lineages right up until the moment of the impact, at which point the extinctions should be global and isochronous. Therefore, patterns of *gradual* extinction should falsify the asteroid impact as a causal agent; patterns of *abrupt* or *catastrophic* extinction should corroborate the hypothesis.

At present, the evidence is mounting that the extinction was abrupt. The complexity of reconstructing events is such that no conclusion is foregone, but in general, detailed sophisticated investigations now underway seem to be indicating abrupt extinctions. This pattern is best seen in ammonites, in foraminifera, in plants (both in pollen and in the megaflora), and in dinosaurs. Other groups where the pattern is more equivocal include bivalves and mammals. The bivalve data, however, are not sufficiently refined at present to distinguish between a gradual and an abrupt extinction. In the case of the mammals, a number of workers have published accounts of "stepwise" extinctions for the group.

DOES THE ASTEROID IMPACT HYPOTHESIS EXPLAIN ALL THE DATA? Initially, researchers struggling to evaluate the asteroid impact hypothesis as a causal agent for the K/T extinctions had some trouble explaining the selectivity of the extinctions. Why did some things go extinct and some not? Was it just dumb luck, just a random process, the way a few creatures seem to survive natural catastrophes? Or, is there some pattern in the selectivity of the extinctions that can be explained in terms of an asteroid?

Sheehan and T. A. Hansen noted that those marine creatures that suffered the most extinctions were those that depended directly upon primary productivity for their food source. Such creatures included not only the planktonic foraminifera and other planktonic marine microorganisms, but also ammonites, other cephalopods, and a variety of mollusks. On the other hand, Sheehan and Hansen

---

[5] Quoting Pooh-Bah in Gilbert and Sullivan's *The Mikado*.

TABLE 17.1 PROPOSED CAUSES FOR THE EXTINCTION OF THE DINOSAURS

(after Benton, 1990)

I. PROPOSED BIOTIC CAUSES

A. Medical Problems

1. Slipped disks in the vertebral column causing dinosaur debilitation.
2. Hormone problems.
   a) Overactive pituitary glands leading to bizarre and nonadaptive growths.
   b) Hormonal problems leading to eggshells that were too thin, causing them to collapse in on themselves in a gooey mess.
3. Decrease in sexual activity.
4. Blindness due to cataracts.
5. A variety of diseases, including arthritis, infections, and bone fractures.
6. Epidemics leaving no trace but wholesale destruction.
7. Parasites leaving no trace but wholesale destruction.
8. Change in ratio of DNA to cell nucleus causing scrambled genetics.
9. General stupidity.

B. Racial Senescence

B. Racial Senescence – This is the idea, no longer given much credence, that entire lineages grow old and become "senile," much as individuals do. Thus, in this way of thinking, late-appearing species would not be as robust and viable as species that appeared during the "early" and "middle" stages of a lineage. The idea behind this was that the dinosaurs as a lineage simply got old and the last-living members of the group were not competitive for this reason.

C. Biotic Interactions

1. Competition with other animals, especially mammals, which may have out-competed dinosaurs for niches, or which perhaps ate their eggs.
2. Overpredation by carnosaurs (who presumably ate themselves out of existence).
3. Floral changes.
   a) Loss of marsh vegetation (presumably the single important source of food).
   b) Increase in forestation (leading to loss of dinosaur habitats).
   c) General decrease in the availability of plants for food with subsequent dinosaur starvation.
   d) The evolution in plants of substances poisonous to dinosaurs.
   e) The loss from plants of minerals essential to dinosaur growth.

II. PROPOSED PHYSICAL CAUSES

A. Atmospheric Causes

1. Climate becoming too hot, so they fried.
2. Climate becoming too cold, so they froze.
3. Climate becoming too wet, so they got waterlogged.
4. Climate becoming too dry, so they desiccated.
5. Excessive amounts of oxygen in the atmosphere causing
   a) Changes in atmospheric pressure and/or atmospheric composition that proved fatal; or
   b) Global wildfires that burned up the dinosaurs.
6. Decrease in the amount of oxygen in the atmosphere affecting the breathing capabilities of dinosaurs.
7. Low levels of CO_2 removing the "breathing stimulus" of endothermic dinosaurs.
8. High levels of CO_2 asphyxiating dinosaur embryos.
9. Volcanic emissions (dust, CO_2, rare-earth elements) poisoning dinosaurs one way or another.

B. Oceanic and Orographic Causes
1. Marine regression producing loss of habitats.
2. Draining of swamp and lake habitats.
3. Stagnant oceans producing untenable conditions on land.
4. Spillover into the world's oceans of Arctic waters that had formerly been restricted to polar regions, and subsequent climatic cooling.
5. The opening of Antarctica and South America, causing cool waters to enter the world's oceans from the south, modifying world climates.
6. Reduced topographic relief and loss of habitats.

C. Other
1. Fluctuations in gravitational constants leading to indeterminate ills for the dinosaurs.
2. Shift in the earth's rotational poles leading to indeterminate ills for the dinosaurs.
3. Extraction of the moon from the Pacific Basin perturbing dinosaur life as it had been known for 160 million years(!).
4. Poisoning by uranium from the earth's soils.

D. Extraterrestrial Causes
1. Increasing entropy leading to loss of large life forms.
2. Sunspots modifying climates in some destructive way.
3. Cosmic radiation and high levels of ultraviolet radiation causing mutations.
4. Destruction of the ozone layer, causing (3).
5. Ionizing radiation as in (3).
6. Electromagnetic radiation and cosmic rays from the explosion of a supernova.
7. Interstellar dust cloud.
8. Oscillations about the galactic plane leading to indeterminate ills for the dinosaurs.
9. Impact of an asteroid.

noted that organisms that depended not *only* on primary productivity, but that could also survive on **detritus** (that is, the scavenged remains of other organisms), fared consistently better. In the marine rocks that formed the basis of their report, Sheehan and Hansen observed that detritus feeders were apparently unaffected by the extinction and, in some cases, even experienced an evolutionary diversification across the K/T boundary.

With subsequent work demonstrating the drop in global primary productivity, the question then became: Could the primary-productivity–detritus-feeding extinction selectivity be traced in the terrestrial realm as well? In short, apparently so. Sheehan and Fastovsky noted that the strong selectivity in the survival of land-dwelling and aquatic vertebrates (Figure 17.10) is very much paralleled by their feeding strategies: Aquatic vertebrates, living in river environments and in flood-plain lakes and wetlands, tend to utilize detritus as a major source of nutrients, whereas land-dwelling vertebrates are far more dependent upon primary productivity. This is because river and lake systems can serve as a "sink" or repository for detrital material; organisms living in such environments that can utilize this resource are therefore "buffered" or protected against short-term drops in primary productivity.

So what happened at the K/T boundary? Scientists who view the asteroid impact as the cause of K/T events usually envision some kind of dramatic and

BOX 17.1

THE REAL REASON THE DINOSAURS BECAME EXTINCT

NOT EVERY PUBLISHED HYPOTHESIS has been serious. In 1964, for example, E. Baldwin suggested that the dinosaurs died of constipation. His reasoning went as follows: Toward the end of the Cretaceous, there was a restriction in the distribution of certain plants containing natural laxative oils necessary for dinosaur regularity. As the plants became geographically restricted, those unfortunate dinosaurs living in places where the necessary plants no longer existed acquired stopped-up plumbing and died hard deaths. The same year, humorist W. Cuppy noted that "the Age of Reptiles ended because it had gone on long enough and it was all a mistake in the first place," a view with which the goat in *Jurassic Park* would have probably agreed.

The November 1981 issue of the *National Lampoon* offered its explanation, entitled "Sin in the Sediment." Moral degeneracy seemed to be the issue:

> It's pretty obvious if you just examine the remains of the dinosaurs. . . . Dig down into older sediments and you'll see that the dinosaurs were pretty well off until the end of the Mesozoic. They were decent, moral creatures, just going about their daily business. But look at the end of the Mesozoic and you begin to see evidence of stunning moral decline. Bones of wives and children all alone, with the philandering husband's bones nowhere in sight. Heaps of fossilized, unhatched, aborted dinosaur eggs. Males and females of different species living together in unnatural defiance of biblical law. Researchers have even excavated entire orgies – hundreds of animals with their bones intertwined in lewd positions. Immorality was rampant.

In 1983, the distinguished sedimentary geologist R. H. Dott, Jr., published a short note in which he vented his frustrations with the pollen season, suggesting that it was pollen in the atmosphere that killed the dinosaurs. He called his contribution "Itching Eyes and Dinosaur Demise."

The issues raised by the *National Lampoon* were compelling enough to again be raised in 1988 by the *Journal of Irreproducible Results*. There, in an offering called "Antediluvian Buggery: The Role of Deviant Mating Behavior in the Extinction of Mesozoic Fauna," L. J. Blincoe informed us:

> A thorough but cursory review of fossil specimens . . . has revealed a unique fossil found in the Cretaceous "beds" of Mongolia in 1971 [see Figure 15.1]. The fossil featured two different species of dinosaur . . . in close association at the moment of their deaths. Prejudiced by their preconceived notions of dinosaur behavior, paleontologists have almost unanimously interpreted this find as evidence of a life and death struggle. . . . However, an alternative theory has now been developed which not only explains this unusual fossil, but also answers the riddle of the dinosaurs disappearance. Quite simply, when their lives were ended by sudden catastrophe, these two creatures were locked together . . . in a passionate embrace. They were, in fact, prehistoric lovers.
>
> The implications of this startling interpretation are clear: dinosaurs engaged in trans-species sexual activity. In doing so they wasted their procreative energy on evolutionarily pointless copulation that resulted in either no offspring or, perhaps on rare occasions, in bizarre, sterile mutations (the fossil record is replete with candidates for this latter category). (*Journal of Irreproducible Results* 33[3]:24–26)

The cartoonists got in on the act. In 1982, Gary Larson produced one of his most famous and popular Far Side cartoons, "The real reason the dinosaurs became extinct." Ten years later, O'Donnell in the New Yorker gave us "Extinction of the Dinosaurs Fully Explained." Even Hallmark cards contributed a proposal.

The real reason dinosaurs are extinct.

© Used by Permission Hallmark Cards, Inc., 1991.

The real reason dinosaurs became extinct

The Far Side cartoon by Gary Larson is
reprinted by permission of Chronicle Features,
San Francisco, CA. All rights reserved.

EXTINCTION of the DINOSAURS
FULLY EXPLAINED

**Cartoonists' views of the
extinction of the dinosaurs.**

Drawing by O'Donnell ©1992
The New Yorker Magazine, Inc.

short-term disturbance to the ecosystem. Such a disturbance could be, as we have seen, a dust cloud blocking sunlight for a few months, global wildfires, an acid rain, or some combination of all these ills. This disturbance to the ecosystem seems to have had, at the very least, a deadly effect on global primary productivity, which in turn seems to have decimated organisms solely dependent upon primary productivity. In this view, extraterrestrial processes become an integral part of the very earthly process of organic evolution. Considered this way, the abrupt and externally controlled removal of large numbers of organisms enabled other groups to evolve under conditions of ecological opportunities that formerly did not exist. In the case of the mammals, it seems clear that the absence of dinosaurs after 165 million years of terrestrial importance was the event that allowed the mammals to evolve and occupy the place in the global ecosystem that they presently hold.

DISSENSION.   It cannot be said that the idea of the asteroid as the cause of the extinctions is accepted in all quarters. Some scientists, as we have seen, conclude that volcanism was responsible for K/T events. They would, however, not quibble with the interpretation that many organisms died out suddenly; they would simply ascribe the ultimate cause to volcanos, and not to an asteroid.

Moreover, as we have seen, faunas from southern high latitudes do not show the evidence of catastrophic demise reflected in northern lower-latitude faunas and floras. Some scientists have claimed that these data falsify the asteroid hypothesis, because the patterns – however rudimentarily they are understood – do not seem to imply a catastrophic pattern of extinction. Other scientists have pointed out, however, that if the asteroid impact occurred equatorially (as it is now believed), the region of the earth most likely to be buffered from the full effects of the impact would be extreme southern latitudes.

Interestingly, much of the resistance to the idea has come from vertebrate paleontologists who do not feel that the data on vertebrate extinctions bear out the conclusions that have been reached. To some, it is a question of stratigraphic resolution; there are too few fossils and too many gaps in sedimentation to accurately assess rates of extinction and evolution. To others, studies such as Sheehan and colleagues' study of dinosaur evolution are too coarse. For these scientists, 250,000 years is so much time that an extinction could take place gradually in that time and yet not be observable in the geological record as anything other than instantaneous. Likewise, that study was carried out at the level of dinosaur families, which is considered by some scientists to be too coarse to adequately estimate rates of extinction of the dinosaurs. Moreover, Clemens and Archibald have both argued strongly that the mammalian extinction began well in advance of the asteroid impact. Their point was that regardless of what did cause the K/T extinctions, such a pattern is not in agreement with an abrupt extinction by an asteroid.

Archibald has suggested that the marine regression at the end of the Cretaceous was responsible for the demise of the dinosaurs. He ascribes the extinction of dinosaurs to habitat partitioning, by which he means that as the seas receded, the habitats in which the dinosaurs had been living were partitioned, or divided, and thus rendered ecologically inadequate for the creatures.

OUR VIEWPOINT . . . AND YOURS?   For us, the problem with such ideas is that, in large part, they do not constitute a unified hypothesis or suite of hypotheses that

Figure 17.11 **The last day of the Cretaceous Period, 65 Ma.** Planetary geologist P. H. Schultz of Brown University concluded that the asteroid struck the earth at an angle of about 30°.

explain what is known of K/T boundary events. To our minds, there has not yet been formulated a theory alternative to the asteroid impact hypothesis that explains all – or even most – of the data as fully as does the asteroid impact hypothesis.

This, however, gets right to the core of the issue. We know that many different organisms went extinct at, or very close to, the K/T boundary (which, incidentally, is why the K/T boundary – the end of the Mesozoic – was originally recognized). That being the case, the *simplest* explanation for this event is preferred. For many earth historians, the coincidence of particular events suggests (perforce) that they are causally related.

There is another view, however. If one is convinced that the patterns of extinction are *not* consistent with the effects of an asteroid, then much of the attraction of the asteroid hypothesis is lost. Moreover, it can rightly be argued that a theory is not necessarily preferable because it is simple;[6] nothing about nature *requires* simplicity, and the very complexity of K/T events suggests complex causes.

Ultimately, acceptance of any hypothesis resides in one's view of the data. *Our* view is that the patterns of extinction are concordant with the asteroid hypothesis. For us, therefore, the extinction of the dinosaurs remains best explained by the impact of an asteroid with the earth (Figure 17.11). We opt for the theory that we think explains all of the data most satisfactorily.

We have attempted to present the data as clearly as possible. It now falls to you to draw your own conclusions about what happened to the dinosaurs 65 Ma.

---

[6]Think of it this way: One might posit that the cause of World War I was the shooting of Archduke Franz Ferdinand in Sarajevo. And, although this is a simple explanation (the shooting did coincide with the outbreak of World War I, after all), historians would be loath to assert that this was the cause of the war. Its causes were far more complicated and were rooted in burgeoning nationalism, in the rise of the proletariat class, and in an interlocking series of multinational treaties.

Important Readings

Alvarez, L. W. 1983. Experimental evidence that an asteroid impact led to the extinction of many species 65 Myr ago. Proceedings of the National Academy of Sciences 80:627–642.

Alvarez, L. W., W. Alvarez, F. Asaro, and H. V. Michel. 1980. Extraterrestrial cause for the Cretaceous–Tertiary extinction. Science 208:1095–1108.

Archibald, J. D. 1984. Bug Creek Anthills (BCA), Montana: faunal evidence for Cretaceous age and non-catastrophic extinctions. Geological Society of America Abstracts with Programs 16:432.

Archibald, J. D., and L. J. Bryant. 1990. Differential Cretaceous/Tertiary extinctions of nonmarine vertebrates: evidence from northeastern Montana; pp. 549–562 *in* V. L Sharpton and P. D. Ward (eds.), Global catastrophes in Earth history: an interdisciplinary conference on impacts, volcanism, and mass mortality. Geological Society of America Special Paper 247.

Archibald, J. D., and W. A. Clemens, Jr. 1980. Evolution of terrestrial faunas during the Cretaceous–Tertiary transition. Memoires de la Société Géologique de la France 139:67–74.

Benton, M. J. 1990. Scientific methodologies in collision: the history of the study of the extinction of the dinosaurs. Evolutionary Biology 24:371–400.

Clemens, W. A., and L. G. Nelms. 1994. Paleoecological implications of Alaskan terrestrial vertebrate fauna in latest Cretaceous time at high paleolatitudes. Geology 21:503–506.

Clemens, W. A., Jr., J. D. Archibald, and L. J. Hickey. 1981. Out with a whimper not a bang. Paleobiology 7:293–298.

D'Hondt, S. L., M. E. Q. Pilson, H. Sigurdsson, A. K. Hanson, Jr., and S. Carey. 1994. Surface-water acidification and extinction at the Cretaceous–Tertiary boundary. Geology 22:983–986.

Dodson, P. 1991. Maastrichtian dinosaurs. Geological Society of America Abstracts with Programs 23, no. 5:A184–185.

Hildebrand, A. R., G. T. Penfield, D. A. Kring, M. Pilkington, A. Camargo Z., S. B. Jacobson, and W. B. Boynton. 1991. Chicxulub Crater: a possible Cretaceous/Tertiary impact crater on the Yucatan Peninsula, Mexico. Geology 19:867–871.

Jablonski, D. 1986. Background and mass extinctions: the alternation of macroevolutionary regimes. Science 231:129–133.

Jablonski, D., and D. J. Bottjer. 1991. Environmental patterns in the origins of higher taxa: the post-Paleozoic fossil record. Science 252:1831–1833.

Johnson, K. R., and L. J. Hickey. 1990. Megafloral change across the Cretaceous/Tertiary boundary in the northern Great Plains and Rocky Mountains, U.S.A.; pp. 433–444 *in* V. L. Sharpton and P. D. Ward (eds.), Global catastrophes in Earth history; an interdisciplinary conference on impacts, volcanism, and mass mortality. Geological Society of America Special Paper 247.

Officer, C. B., A. Hallam, C. L. Drake, and J. D. Devine. 1987. Late Cretaceous and paroxysmal Cretaceous/Tertiary extinction. Nature 236:143–149.

Russell, D. A. 1982. The mass extinctions of the late Mesozoic. Scientific American 246:58–65.

Sharpton, V. L., and P. D. Ward P.D., eds. 1990. Global catastrophes in Earth history; an interdisciplinary conference on impacts, volcanism, and mass mortality. Geological Society of America Special Paper 247, 631 pp.

Sheehan, P. M., D. E. Fastovsky, R. G. Hoffmann, C. B. Berghaus, and D. L. Gabriel. 1991. Sudden extinction of the dinosaurs: latest Cretaceous, upper Great Plains, U.S.A. Science 254:835–839.

Sheehan, P. M., D. E. Fastovsky, R. G. Hoffmann, C. B. Berghaus, and D. L. Gabriel. 1992. Reply. Science 256:160–161.

Sigurdsson, H., S. L. D'Hondt, and S. Carey. 1992. The impact of the Cretaceous/Tertiary bolide on evaporite terrane and generation of major sulfuric acid aerosol. Earth and Planetary Science Letters. 109:543–559.

Silver, L. T., and P. H. Schultz, eds. 1982. Geological implications of large asteroid and comets on Earth. Geological Society of America Special Paper 190, 528 pp.

Sloan, R. E., J. K. Rigby, L. Van Valen, and D. L. Gabriel. 1986. Gradual dinosaur extinction and simultaneous ungulate radiation in the Hell Creek Formation. Science. 232:629–633

GLOSSARY

Absolute age. The age of a rock or fossil measured in years before present.

Accrete. To grow by external addition; in the case of plate tectonics, the addition of crustal material to preexisting crust.

Acromion process. A broad and plate-like flange on the forward surface of the shoulder blade.

Actinopterygii. The clade of ray-finned fish.

Advanced. In an evolutionary context, shared or derived (or specific), with reference to characters.

Aestivate. In zoology, to spend summers in a state of torpor.

Altricial. Pertaining to organisms that are born relatively under-developed, requiring significant parental attention for survival.

Alveolus (pl. alveoli). A sac-like anatomical structure.

Amnion. A membrane in some vertebrate eggs that contributes to the retention of fluids within the egg.

Anaerobic. Without oxygen.

Analogous. Adjective form of analogue.

Analogue. In anatomy, structures that perform in a similar fashion but have evolved independently.

Anamniotic. An egg without an amnion.

Ancestral. In an evolutionary sense, relating to forebears.

Angiosperms. Flowering plants.

Ankylopollexia. A clade of ornithopods including *Camptosaurus, Iguanodon, Ouranosaurus,* and hadrosaurids.

Antagonistic muscle masses. Groups of muscles whose movements oppose each other; for example, muscles whose movements open a jaw are said to be antagonistic to muscles that close the jaw.

Anterior. Pertaining to the head-bearing end of an organism.

Antitrochanter. A downward-directed process on the upper edge of the ilium in hadrosaurids and ceratopsians.

Antorbital fenestra. An opening on the side of the skull, just ahead of the eye. This is a character that unites the clade Archosauria.

Arboreal hypothesis. The idea that bird flight originated by gliding down from trees.

Archipelago. A group of islands.

Archosauria. A clade within Archosauromorpha. The living archosaurs include birds and crocodiles.

Archosauromorpha. The large clade of diapsids that includes the common ancestor of rhynchosaurs and archosaurs, and all its descendants.

Articulated. In paleontology, bones are said to be articulated when they are found positioned relative to each other as they were in life.

Artifact. In an experiment, an incorrect result caused by something in the nature of the data.

Ascending process of the astragalus. A wedge-shaped splint of bone on the astragalus that lies flat against the shin (between the tibia and fibula) and points upwards. Diagnostic character of Theropoda.

Assemblage. In paleontology, a group of organisms. The term is used to refer to a collection of fossils in which it is not clear how accurately the collection reflects the complete, ancient formerly living community.

Asteroid. A large extraterrestrial body.

Asthenosphere. Another name for the upper mantle.

Astragalus. Along with the calcaneum, one of two upper bones in the vertebrate ankle.

Atom. The smallest particle of any element that still retains the properties of that element.

Atomic number. The number of protons (which

equals the number of electrons) in an element.

Autotroph. An organism that uses energy from the sun as well as from inorganic nutrients to produce complex molecules for nutrition.

Background extinctions. Continually occurring, isolated extinctions of individual species. As distinct from mass extinctions.

Barb. Feather material radiating from the shaft of the feather.

Barbule. A small hook that links barbs together along the shaft of the feather.

Beak. Sheaths of keratinized material covering the ends of the jaws (synonym: rhamphotheca).

Benthic. With reference to the marine realm (oceans), living within sediments.

Bilateral symmetry. A kind of symmetry in which the right and left halves of the body are mirror images of each other.

Biogeographic. Pertaining to the distribution of organisms in space.

Biomass. The sum total of the weights of organisms in the assemblage or community being studied.

Biostratigraphy. The study of the relationships in time among groups of organisms.

Biota. The sum total of all organisms that have populated the earth.

Body fossil. The type of fossil in which a part of an organism becomes buried and fossilized – as opposed to trace fossil.

Bone. Calcified skeletal tissue.

Bone histology. The study of bone tissues.

Bonebeds. Relatively dense accumulation of bones of many individuals, usually of a very few *kinds* of organisms.

Boss. A large mass or knob of bone, commonly used with reference to structures on the skull.

Brain. A centralized cluster of nerve cells.

Brain endocasts. Internal casts of braincases.

Braincase. Hollow bony box that houses the brain; located toward the upper, back part of the skull.

Cadence. In locomotion, the rate at which the feet hit the ground.

Caliche. Calcium carbonate nodules that form in soils.

Caniniform. Canine-like; something that is caniniform has the shape of the elongate, pointed (canine) teeth in dogs.

Carbohydrates. A family of 5- and 6-carbon organic molecules whose chemical bonds, when broken, release energy.

Carnian. The older of two subdivisions of the Late Triassic; the Carnian runs from 235 Ma to 223 Ma.

Carnosaur. A large-bodied theropod, with a tendency toward small forelimbs and a large head.

Carpal. Wrist bone.

Carpometacarpus. Unique structure in all living, and in most ancient, birds, in which bones in the wrist and hand are fused.

Cast. Material filling up a mold.

Cellular respiration. The breakdown of carbohydrates through a regulated series of oxidizing reactions.

Cenozoic Era. That interval of time from 65 Ma to present.

Centrum. The spool-shaped, lower portion of a vertebra, upon which the spinal cord and neural arch rest.

Cerapoda. The ornithischian clade of Ceratopsia + Pachycephalosauria + Ornithopoda

Character. An isolated or abstracted feature or characteristic of an organism.

Choana (pl. choanae). In the skull, a passageway leading from the nasal cavity to the interior of the mouth.

Chondrichthyes. The gnathostome clade that includes sharks, skates, and rays.

Chronostratigraphy. The study of geological time.

Circumpolar currents. Cold water masses that circulate around the earth's poles.

Clade. Group of organisms in which all members are more closely related to each other than they are to anything else. All members of a clade share a most recent common ancestor that is itself the most basal member of that clade. Synonymous with "monophyletic group" and "natural group."

Cladistic analysis. Analysis of the ancestor–descendant (evolutionary) relationships among organisms using hierarchies of shared, derived characters.

Cladogram. A hierarchical, branching diagram that shows the distribution of shared, derived characters among selected organisms.

Clavicle. Collarbone.

Climate. The sum of all weather conditions. Usually one refers to particular climatic variables, such as precipitation or temperature.

Cnemial crest. A bony flange on the upper end of the front surface of the tibia.

Co-evolution. The idea that two organisms or groups of organisms may have evolved in response to one another.

Collect. To obtain fossils from the earth.

Competitive edge. Some aspect of an organism or group of organisms that enhances the ability of the possessor to compete successfully against those that do not.

Continental crust. Quartz-rich material of which continents are formed.

Continental effects. The effect on climate exerted by continental masses.

Convection. Movement associated with heat flow in fluids as cooler parts of the fluid absorb the energy from warmer parts.

Convergent. In anatomy, pertaining to the independent invention (and thus, duplication) of a structure or feature in two lineages. The streamlined shape of whales and ichthyosaurs is a famous example of convergent evolution.

Coprolite. Fossilized feces.

Coracoid. The lower (and more central) of two elements of the shoulder girdle (the upper being the scapula).

Cosmic dust. Particulate matter from outer space.

Cranial. Referring to the skull (cranium).

Craton. A large body of continental crust that has been stable for millions of years.

Cretaceous–Tertiary boundary. That moment in time, 65 Ma, between the Cretaceous Period and the Tertiary Period.

Crurotarsi. A clade of archosaurs including crocodilians and their close relatives.

Crust. A thin, chemically distinct rind on the earth's lithosphere.

Curate. To incorporate, preserve, and catalogue specimens into museum collections.

Cursorial hypothesis. The hypothesis that flight in birds originated by an ancestor's running along the ground.

Cursorial locomotion. Running locomotion on land.

Deccan Traps. Interbedded volcanic and sedimentary rocks in western and central India of Cretaceous–Tertiary age.

Deltoid crest. A large process at the head of the humerus.

Dense Haversian bone. A type of Haversian bone in which the canals and their rims are very closely packed.

Dental battery. A cluster of closely packed cheek teeth in the upper and lower jaws, whose shearing or grinding motion is used to masticate plant matter.

Deposition. Net addition of sediment to a land surface.

Derived. In an evolutionary context, pertaining to characters that uniquely apply to a particular group and thus are regarded as having been "invented" by that group during the course of its evolutionary history.

Detritus. Loose particulate rock, mineral, or organic matter; debris.

Diachronous. Occurring at different moments in time; not synchronous.

Diastem(a). A gap.

Digitigrade. In anatomy, a position assumed by the foot when the animal is standing, in which the ball of the foot is held high off the ground and the weight rests on the ends of the toes.

Dinosauria. A clade of ornithodiran archosaurs.

Diphyletic. In evolution, having two separate origins.

Disarticulated. Dismembered.

Diversity. The variety of organisms; the number of *kinds* of organisms.

Dominant. In an ecological sense, being the most abundant or having the greatest effect on a particular aspect of the ecosystem. A rather general term without well-constrained meaning.

Dorsal sacral shield. The upper portion of the sacrum, composed of the ilia, vertebrae, and sacral ribs.

Down. A bushy, fluffy, type of feather in which barbules and vanes are not well developed, used for insulation.

Dryomorpha. A clade of iguanodontians including *Dryosaurus, Camptosaurus, Iguanodon, Ouranosaurus,* and hadrosaurids.

Dynamic similarity. A conversion factor that "equalizes" the stride rates of vertebrates of different sizes and proportions, so that speed of locomotion can be calculated.

Ecological diversity. The proportion of an ecosystem that is occupied by a particular lifestyle, such as feeding type or mode of locomotion. For a simple example, one might study an ecosystem by dividing it into herbivores, carnivores, and omnivores.

Ectothermic. Regulating temperature (and thus, metabolic rate) using an external source of energy (heat).

Elbow. The joint between the upper arm and the lower arm.

Electron. A negatively charged subatomic particle. Electrons reside in clouds around the nucleus of an atom.

Element. Discrete part of the skeleton; i.e., an individual bone.

Encephalization. That condition in which an organism bears a head structure that is distinct from the rest of the body and that contains a brain .

Endemic. An organism or fauna is said to be endemic to a region when it is restricted to that region.

Endemism. The property of being endemic.

Edentulous. Without teeth.

Endosymbionts. Organisms that live within another organism in a mutually beneficial relationship.

Endosymbiosis, theory of. The idea that eukaryotic cells evolved as a result of the ingestion, by prokaryotes, of other prokaryotes. The ingested prokaryotes eventually adopted specialized functions as organelles.

Endothermic. Regulating temperature (and thus, metabolic rate) using an internal source of energy.

Epeiric sea. Relatively shallow (at most, a few hundred meters) marine water covering a craton (synonym: epicontinental sea).

Epoccipital. Bone ornamenting the rim of the frill in ceratopsians.

Epoch. Subdivisions of a period, several million years in duration.

Era. A very large block of geologic time (hundreds of millions of years long), composed of periods.

Erect posture. In anatomy, the condition in which the legs lie parasagittal to (along side of) the body and do not extend laterally from it.

Eukaryote. Complex cell that has a nucleus and a variety of internal chambers called organelles.

Eurypoda. The ornithischian clade Stegosauria + Ankylosauria.

Evaporite. Rock composed of minerals precipitated through desiccation.

Evolution. In biology, descent with modification.

External mandibular fenestra. An outward-facing opening toward the rear of the mandible (commonly between the dentary, surangular and angular bones) found in many archosaurs.

Extinction. When the birth rate fails to keep up with the death rate.

Extinction boundary. The moment in time when organisms or groups of organisms became extinct.

Extraterrestrial. From outer space.

Fauna. A group of animals presumed to live together within a region.

Femur. The upper bone in the hindlimb (thighbone).

Fibula. The smaller of the two lower leg bones in the hindlimb; the bone that lies alongside the shin bone (tibia).

Flight feather. Elongate feather with well-developed, asymmetrical vanes; usually associated with flight.

Flood basalt. Episodic lava flow from fissures in the earth's crust.

Flux. A measure of change; rate of discharge times volume.

Footprint. Trace fossil left by the feet of vertebrates.

Foramen magnum. The opening at the base of the braincase through which the spinal cord travels to connect to the brain.

Foraminifera (sing. foraminifer). Single-celled, shell-bearing protists that live in the oceans.

Fossil. Anything buried.

Fourth trochanter. A ridge of bone along the shaft of the femur (thighbone) for muscle attachment.

Fractionation. As discussed here, the separation of isotopes that occurs during physical or chemical processes.

Frill. In ceratopsians, a sheet of bone extending dorsally and rearward from the back of the skull, made up of the parietal and squamosal bones.

Furcula. Fused clavicles (collarbones); the "wishbone" in birds and certain non-avian theropods.

Gastralia. Belly ribs.

Gastrolith. Smoothly polished stone in the stomach, used for grinding plant matter.

Genasauria. The ornithischian clade of Thyreophora + Cerapoda.

General. In phylogenetic reconstruction, referring to a character that is nondiagnostic of a group; in this context, synonymous with **primitive**.

Geographic distribution. In biology, the spatial placement of organisms.

Ghost lineage. Lineage of organisms for which there is no physical record (but whose existence can be inferred).

Gigantothermy. Modified mass homeothermy, which mixes large size with low metabolic rates and control of circulation to peripheral tissues.

Gizzard. A muscular chamber just in front of the glandular part of the stomach.

Gondwana. A southern supercontinent composed of Australia, Africa, South America, and Antarctica.

Goyocephalia. A clade of highly derived pachycephalosaurs.

Habitat partitioning. In biology, the division by organisms of available ecospace into nonoverlapping domains.

Half-life. The amount of time that it takes for 50% of a volume of unstable isotope to decay.

Hard part. In paleontology, all hard tissues, including bones, teeth, beaks, and claws.

Hard parts tend to be preserved more readily than soft tissues.

Haversian canal. In bone histology, a canal composed of secondary bone.

Heterotroph. In biology, an organism that must ingest all nutrients necessary for survival.

Hierarchy. As applied here, the ordering of objects, organisms, and categories by rank. The military and the clergy are both excellent examples of hierarchies; in these, rank is a reflection of power and, one hopes, accomplishment. Another hierarchical system is money, which is ordered by value.

High pressure zone. In meteorology, a zone in the atmosphere where large, moist, and dense air masses accumulate.

Hip socket. A depressed area in the pelvis where the femur (thighbone) articulates.

Homeotherm. Organism whose core temperatures remain constant.

Homologous. Two features are homologous when they can be traced back to a single structure in a common ancestor.

Homologue. A homologous feature.

Humerus. The upper arm bone.

Hyposphene–hypantrum articulation. Extra articular surface on the neural arches connecting successive elements in the backbone.

Hypothesis of relationship. A hypothesis about how closely or distantly organisms are related.

Ichnofossil. Impression, burrow, track, or other modification of the substrate by organisms.

Ichnotaxa. Taxa established on the basis of trace fossils .

Igneous. Rocks or minerals derived from molten material.

Ilium. The uppermost of three bones that make up the pelvis.

Impact ejecta. The material thrown up when an asteroid strikes the earth.

Incongruent. Not equivalent; conflicting.

Indigenous. Restricted to a particular geographic region.

Induration. The process by which rock is hardened.

Inner core. In the structure of the earth, the densest, iron-rich central portion.

In-place. Not reworked.

Interspecific. Among different species.

Intraspecific. Within the same species.

Iridium. A nontoxic, platinum-group metal, rare at the earth's surface.

Ischium. The most posterior of three bones that make up the pelvis.

Isochronous. Occurring at the same time (synonym: synchronous).

Isotopes. In chemistry, elements that have the same atomic number but different mass numbers.

Jacket. In paleontology, a rigid, protective covering placed around a fossil, so that it can be moved safely out of the field. Commonly made up of strips of burlap soaked in plaster.

Jugal. One of the bones in the cheek region of the skull.

K-strategy. The evolutionary strategy of having few offspring that are cared for by the parents. The symbol *K* stands for the carrying capacity of the environment.

K/T (boundary). Common abbreviation for that moment in time, 65 Ma, which marks the boundary between the Cretaceous and Tertiary periods.

Keel. A flange or sheet of bone, as in the keeled sternum of birds; named for its resemblance to the keel on a sailboat.

Knee. The joint between the upper hind leg (thigh) and lower leg (shin).

Lambeosaurines. The hollow-crested hadrosaurid dinosaurs.

Laurasia. A northern supercontinent made up of the Siberian craton and the Old Red Sandstone Continent.

Lepidosauromorpha. One of the two major clades of diapsid reptiles; the other clade is Archosauromorpha.

Lesser trochanter. A crestlike ridge on the femur.

Lines of arrested growth (LAGs). Lines that are inferred to represent times of nongrowth, visible in the cross section of bones.

Lissamphibia. Modern amphibians: frogs, salamanders, and caecilians.

Lithosphere. The rigid, outermost layer of the earth, 100 km thick.

Lithostratigraphy. The general study of all rock relationships.

Low pressure zones. Regions of greater vertical accumulation of dry air.

Lower mantle. The layer of the earth immediately outside the inner core, about 2,700 km thick.

Lower temporal fenestra. The lower opening of the skull just behind the eye.

Mandible. The lower jaw.

Marginocephalia. The clade of dinosaurs that includes the most recent common ancestor of pachycephalosaurs and ceratopsians and all of its descendants.

Mass extinctions. Global and geologically rapid extinctions of many kinds of, and large numbers of, species.

Mass homeotherm. An organism that has a relatively constant body temperature because of its large size.

Mass number. In chemistry, the total number of neutrons *plus* the total number of protons for a given element.

Mass spectrometer. An instrument able to separate and measure tiny amounts of isotopes.

Matrix. In paleontology, the rock that surrounds fossil bone.

Maxilla. The upper jawbone that contains the cheek teeth.

Megaflora. The visible remains of plants, especially leaves.

Mesosphere. Another name for lower mantle.

Mesotarsal. A linear type of ankle joint in which hinge motion in a fore–aft direction occurs between the upper ankle bones (the astragalus and calcaneum) and the rest of the foot.

Mesozoic. That interval of time from 245 Ma to 65 Ma.

Metabolism. The sum of the physical and chemical processes in an organism.

Metacarpal. Bone in the palm of the hand.

Metamorphic rock. Rock that is formed by the intense folding and recrystallization of other kinds of rocks.

Metapodial. A general name for metacarpals and metatarsals.

Metatarsal. Bone in the sole of the foot.

Microtektite. A small, droplet-shaped blob of silica-rich glass thought to have crystallized from impact ejecta.

Minimal divergence time (MDT). The minimal amount of time missing between the two descendent species and their common ancestor; calculated by comparing phylogeny and age of fossils.

Mold. Ichnofossil that consists of the impression of an original fossil.

Monophyletic group. A group of organisms that has a single ancestor and contains all of the descendants of this unique ancestor (synonymous with **clade** and "natural group").

Morphology. The study of shape.

Motile. Moving.

Nares. Openings in the skull for the nostrils.

Natural group. *See* **monophyletic group**.

Neutron. Electrically neutral subatomic particle that resides in the nucleus of the atom.

Node. A bifurcation or two-way split point in a phylogenetic diagram (cladogram).

Non-avian dinosaurs. All dinosaurs *except* birds.

Notochord. An internal rod of cellular material that, primitively at least, ran longitudinally down the backs of all chordates. May be though of as a precursor to the vertebral column.

Nucleus. Central core of the atom.

Obligate biped. Tetrapod that must walk or run on its hind legs.

Obturator foramen. A large hole in the pubis near the hip socket.

Obturator process. A flange down the shaft of the ischium.

Occipital condyle. A knob of bone at the back of the skull with which the vertebral column articulates.

Occlusion. Contact between upper and lower teeth; necessary for chewing.

Oceanic circulation. The direction and patterns of oceanic currents.

Oceanic crust. Crust that underlies the ocean basins and is relatively thin (~10 km thick).

Ontogenetic. Pertaining to ontogeny.

Ontogeny. Biological development of the individual; the growth trajectory from embryo to adult.

Opisthopubic. The condition in which at least part of the pubis has rotated backward to lie close to, and parallel with, the ischium.

Orbit. Eye socket.

Organelle. Special structure within a eukaryotic cell.

Ornithischia. One of the two monophyletic groups composing Dinosauria.

Ornithodira. The common ancestor of pterosaurs and dinosaurs, and all its descendants.

Orographic. Pertaining to modification of climate by mountain-ranges.

Ossify. To turn into bone.

Osteichthyes. Bony fishes that include ray-finned and lobe-finned gnathostomes.

Osteoderm. Bone within the skin; may be small nodule, plate, or a pavement of bony dermal armor.

Outer core. Liquid iron layer around inner core of the earth, about 2,000 km thick.

Oxidation. Bonding of oxygen.

Palate. The part of the skull that separates the nasal cavity (for breathing) from the oral cavity (for eating); usually strengthened by a paired series of bones.

Palatine. One of the bones of the palate.

Paleoclimate. Ancient climate.

Paleoecology. The study of ancient interactions among organisms.

Paleosol. Ancient soil profile.

Paleozoic. That interval of time from 570 Ma to 245 Ma.

Palpebral. A rodlike bone that crosses the upper part of the eye socket.

Palynoflora. Spores and pollen.

Pangaea. The mother of all supercontinents, formed from the union of all present-day continents.

Parasagittal posture. Posture in which the legs are held under the body.

Parasagittal process. A flange of bone, lateral to the midline of the skull, that helps to subdivide the nasal cavity.

Parascapular spine. An enlarged spine over the shoulder.

Parsimony. A principle that states that the simplest explanation that explains the greatest number of observations is preferred to more complex explanations.

Pectoral girdle. The bones of the shoulder; the attachment site of the forelimbs.

Pedestal. In paleontology, a pillar of matrix underneath the fossil.

Pelvic girdle. The bones of the hips; the attachment site of the hindlimbs.

Perforate acetabulum. A hole in the hip socket.

Period. Subdivision of an era, consisting of tens of millions of years.

Permineralization. The geological process in which the spaces in fossil bones become filled with a mineral.

Phalanx (pl. phalanges). Small bone of the fingers and toes that allows flexibility.

Phanerozoic. That interval of time from 570 Ma to the present; it also refers to the time in earth history during which shelled organisms have existed.

Photosynthesis. The process by which organisms use energy from the sun to produce complex molecules for nutrition.

Phylogenetic. Pertaining to phylogeny.

Phylogeny. The study of the fundamental genealogical connections among organisms.

Phylum. A grouping of organisms whose make-up is supposed to connote a very significant level of organization shared by all of its members.

Phytosaur. Long-snouted, aquatic, fish-eating member of Crurotarsi.

Pineal. The so-called "third eye," a light sensitive window to the braincase that has been lost in mammals and birds.

Planktonic. Living in the water column.

Plate tectonics. The concept that the earth's surface is organized into large, mobile blocks of crustal material.

Plate. Large, mobile block of the crust of the earth.

Pleurocoel. A well-marked excavation on the sides of a vertebra.

Pleurokinesis. Mobility of the upper jaw.

Pneumatic. Having air sacs or sinuses.

Pneumatic foramina. Openings for air sacs to enter the internal bone cavities.

Poikilotherm. Organism whose core temperature fluctuates.

Postacetabular process. The part of the ilium behind the hip socket.

Post-temporal opening. Opening along the back of the skull that transmits the dorsal head vein out of the brain cavity.

Preacetabular process. The part of the ilium in front of the hip socket.

Precipitation. Rain or snow.

Precocial. The condition in which the young are rather adultlike in their behavior.

Predentary. The bone that caps the front of the lower jaws in all ornithischians.

Prepare. To clean a fossil; to get it ready for viewing by freeing it from its surrounding matrix.

Prepubic process. A flange of the pubis that points toward the head of the animal.

Primary bone. Bone tissue that was deposited or laid down first.

Primary productivity. The sum total of organic matter synthesized by organisms from inorganic materials and sunlight.

Primitive. *See* **ancestral.**

Process. Part of bone that is commonly ridge-, knob-, or blade- shaped and sticks out from the main body of the bone.

Productivity. The amount of biological activity in an ecosystem.

Prokaryotic. Small cells with no nucleus or other internal cell partitions.

Propalinal jaw movement. Fore and aft movement of the jaws.

Prospect. To hunt for fossils.

Proton. Electronically charged (+1) subatomic particle that resides in the nucleus of the atom.

Pubis. One of the three bones that make up the pelvic girdle.

"Pull of the Recent." The inescapable fact that as we get closer and closer to the Recent, fossil biotas become better and better known.

Pygostyle. A small, compact, pointed structure made of fused tail bones in birds.

r-strategy. The evolutionary strategy where organisms have lots of offspring and no parental care. The symbol *r* stands for unrestricted.

Radiometric. The dating method to determine unstable isotopic age estimations.

Radius. One of the two lower arm bones; the other is the ulna.

Recombination. The production of new combinations of DNA in the offspring with each reproduction event.

Recrystallization. The process whereby the original mineral is dissolved and reprecipitated, commonly retaining the exact original form of the original mineral.

Red bed. Rock of orange-red color due to an abundance of iron oxides.

Regression. Retreating of seas due to lowering of sea level.

Relative dating. The type of geological dating that, although not proving ages in years before present, provides ages relative to other strata or assemblages.

Remodel. In bone histology, to resorb or dissolve primary bone and deposit secondary bone.

Replace. To exchange original mineral with another mineral.

Reptilia. The old name for turtles, lizards, snakes, and crocodiles – a nonmonophyletic group; now properly used for the monophyletic group of these four groups *and* birds.

Resolution. The degree to which one can distinguish detail.

Respiratory turbinate. A thin, convoluted or complexly folded sheet of bone located in the nasal cavities of living endothermic vertebrates.

Retroarticular process. A very short projection of the lower jaw beyond the jaw joint.

Rework. To actively erode sediment and fossils from wherever they were originally deposited and redeposit them somewhere else.

Rhamphotheca. The cornified covering of the beak.

Robust. (1) In the context of hypothesis testing, a hypothesis is said to be robust when it has survived repeated tests; that is, despite meaningful attempts, it has failed to be falsified. (2) In anatomy, strong and stout.

Rock. A heterogeneous aggregate of minerals.

Rostral. Referring to the rostrum, or snout region of the skull.

Rostral bone. A special bone that fits on the front of the snout in ceratopsians, giving the upper jaws of these dinosaurs a parrotlike profile.

Sacrum. The part of the backbone where the hip bones attach.

Sarcopterygii. The lobe-finned fish.

Saurischia. One of the two monophyletic groups within Dinosauria; the other is Ornithischia.

Scapula (pl. scapulae). The shoulder blade.

Scenario. A story; in evolutionary terms, the combination of phylogenetic patterns and evolutionary explanations.

Sclerotic ring. A ring of bony plates that support the eyeball within the skull.

Seasonality. Highly marked seasons.

Secondarily evolve. To reevolve a feature.

Secondary bone. Bone laid down in the form of Haversian canals.

Sedimentary rock. A rock which represents the lithification of sediment.

Segmentation. The division of the body into repeating units.

Selectivity. With reference to extinctions, those who survived and those who did not.

Semi-erect posture. The posture in which the upper part of both the arms and the legs are directed at about 45° away from the body.

Semilunate carpal. A distinctive, half-moon-shaped bone in the wrist.

Sessile. Stationary.

Sexual dimorphism. Size, shape, and behavioral differences between the sexes.

Sexual selection. Selection not between all of the individuals within a species, but between males alone.

Shaft. The hollow main vane of a feather.

Shocked quartz. Quartz that has been placed under such pressure that the crystal lattice becomes compressed and deformed.

Sigmoidal. Having an S shape.

Sinus. A cavity.

Skeleton. The supporting part of the body; in vertebrates, the skeleton consists of tissue hardened by mineral deposits that we call bone.

Skull. That part of the vertebrate skeleton that houses the brain, special sense organs, nasal cavity, and oral cavity.

Skull roof. The bones that cover the top of the braincase.

Soft tissue. In vertebrates, all of the body parts except bones, teeth, beaks, and claws.

Specific. Diagnostic of a monophyletic group; uniquely evolved.

Sprawling posture. Posture in which the upper part of the arms and legs splay out approximately horizontally from the body.

Spreading ridge. An elongate, raised linear feature at which new oceanic crust is produced as new material emerges from the asthenosphere.

Stable isotope. An isotope that does not spontaneously decay.

Stapes. The middle-ear bone that transmits sound from the tympanic membrane to the side of the braincase.

Sternum. The breastbone.

Stratigraphy. The study of the relationships of strata and the fossils they contain.

Subatomic. "Smaller-than-atomic."

Subducted. In tectonics, referring to crustal material that is gong down into the upper mantle.

Subnarial foramen. An opening in the skull beneath the nostril area.

Superposition. The geological principle in which the oldest rocks are found at the bottom of a stack of strata and the youngest rocks are found at the top.

Survivorship. The pattern of survival measured against extinction.

Sympatric. Living in the same place at the same time.

Synsacrum. A single, locked unit consisting of sacral vertebrae.

Taphonomy. The study of all of what happens to organisms after death.

Tarsal. Ankle bone.

Tarsometatarsus. The name for the three metatarsals fused together with some of the ankle bones.

Taxon (pl. taxa). A group of organisms, designated by a name, of any rank within the biotic hierarchy.

Tectonic. Pertaining to the behavior and positioning of the continental and oceanic plates in geologic time.

Temporal distribution. The distribution of organisms in time.

Testable hypothesis. A hypothesis that makes predictions that can be compared and assessed by observations in the natural world.

Tetrapoda. A monophyletic group of vertebrates primitively bearing four limbs.

Thecodontia. An unnatural group that at one time was thought to contain the separate ancestors of crocodilians, pterosaurs, dinosaurs, and birds.

Therapsida. The clade of synapsids that includes mammals, some of their close relatives, and all their most recent common ancestors.

Thorax. In vertebrates, the part of the body between the neck and abdomen.

Thyreophora. The armor-bearing ornithischians; stegosaurs, ankylosaurs, and their close relatives.

Tibia. One of the two lower bones in the hindlimb; the other is the fibula.

Tibiotarsus. The fused unit of the tibia and the upper ankle bones.

Torpor. A trancelike state in which the temperature of an animal is considerably lowered.

Trace fossil. Impressions in the sediment left by an organism.

Trachea. The windpipe.

Trackway. Group of footprints left as a tetrapod walked across a substrate.

Transgression. Advancing sea margin due to a rise in sea level.

Turn (a fossil). To separate the fossil from the surrounding rock at the base of the pedestal and to turn it over.

Tympanic membrane. The eardrum.

Type section. The outcrop of rock where a stratigraphic unit was originally described. For example, the type section of the Hell Creek Formation is an outcrop of rock found in the Hell Creek Recreation Area, Garfield County, Montana.

Ulna. One of the two lower arm bones; the other is the radius.

Unaltered. When original mineralogy is unchanged.

Ungual phalange. An outermost bone of the fingers and toes.

Unstable isotope. An isotope that spontaneously decays from an energy configuration that is not stable to one that is more stable.

Upper mantle. The layer of the earth external to the lower mantle, 250 km thick.

Upper temporal fenestra. The opening in the

skull roof above the lower temporal fenestra.

Vane. The sheet of feather material that extends away from the shaft.

Vertebrae. The repeated structures that compose the backbone and that, along with the limb skeleton, support the rest of the body.

Wedge. The evolutionary pattern of waxing and waning dominance among groups of organisms.

Wind. Moving air.

Zygopophysis. A fore-and-aft projection from the neural arches of vertebrae.

SUBJECT INDEX

TAXONOMIC INDEX OF GENERA

Author Index

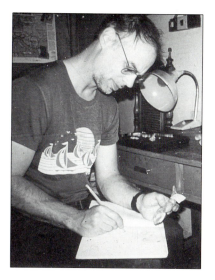

David Fastovsky is on the geology faculty of the University of Rhode Island. He received his Ph.D from the University of Wisconsin–Madison after studying at Reed College and the University of California at Berkeley. He specializes in vertebrate paleontology and terrestrial paleoenvironments, and is the author of numerous scientific publications mainly concerned with the extinction and evolution of dinosaurs and the ancient environments in which they lived. He has undertaken extensive field work in Montana, North Dakota, Arizona, Mexico, and Mongolia.

David Weishampel received his undergraduate degree in geology at the Ohio State University and was awarded his Ph.D from the University of Pennsylvania. He is currently in the Department of Cell Biology and Anatomy at the Johns Hopkins University School of Medicine. He was senior editor of *The Dinosauria* (University of California Press), has written a book on dinosaurs for children (*Plant-Eating Dinosaurs*), and has authored many scientific papers. His areas of particular interest are the evolution and paleobiology of Ornithopoda and the evolutionary interaction between plants and vertebrate herbivores. He has done field work in Montana, Utah, Kansas, Alberta, Maryland, West Virginia, Romania, and Mongolia.